W9-DGJ-292

NUMBER	EQUATION	PAGE
9.1	$r = \dfrac{\Sigma(X - \overline{X})(Y - \overline{Y})}{\sqrt{[\Sigma(X - \overline{X})^2][\Sigma(Y - \overline{Y})^2]}}$	193
9.2	$r = \dfrac{N\Sigma XY - (\Sigma X)(\Sigma Y)}{\sqrt{[N\Sigma X^2 - (\Sigma X)^2][N\Sigma Y^2 - (\Sigma Y)^2]}}$	194
9.3	$r_s = 1 - \dfrac{6\Sigma D^2}{N(N^2 - 1)}$	203
10.1	$Y' = a + b_y X$	216
10.3	$b_y = \dfrac{\Sigma(X - \overline{X})(Y - \overline{Y})}{\Sigma(X - \overline{X})^2}$	218
10.5	$b_y = \dfrac{N(\Sigma XY) - (\Sigma X)(\Sigma Y)}{N(\Sigma X^2) - (\Sigma X)^2}$	218
10.9	$S_{\text{est } y} = \sqrt{\dfrac{\Sigma(Y - Y')^2}{N}}$	227
10.12	$\underset{\substack{\text{Total} \\ \text{variation}}}{\Sigma(Y - \overline{Y})^2} = \underset{\substack{\text{Unexplained} \\ \text{variation}}}{\Sigma(Y - Y')^2} + \underset{\substack{\text{Explained} \\ \text{variation}}}{\Sigma(Y' - \overline{Y})^2}$	230
10.13	$r^2 = \dfrac{\text{explained variation}}{\text{total variation}} = \dfrac{\Sigma(Y' - \overline{Y})^2}{\Sigma(Y - \overline{Y})^2}$	231
11.1	$r_{XY.Z} = \dfrac{r_{XY} - r_{XZ}r_{YZ}}{\sqrt{(1 - r_{XZ}^2)(1 - r_{YZ}^2)}}$	252
11.2	$Y' = a + b_1 X_1 + b_2 X_2.$	254
12.4	$p(A \text{ or } B) = p(A) + p(B)$ when A and B are *mutually exclusive*	278
12.6	$P + Q = 1.00,$ when the events are *mutually exclusive* and *exhaustive*	280
12.9	$p(A \text{ and } B) = p(A)p(B)$ when the events are *independent*	281
12.10	$p(A \text{ and } B) = p(A)p(B|A) = p(B)p(A|B)$	283
14.8	estimated $\sigma_{\overline{X}} = s_{\overline{X}} = \dfrac{s}{\sqrt{N - 1}} = \sqrt{\dfrac{\Sigma(X - \overline{X})^2}{N(N - 1)}}$	344

(continued on inside of back cover)

Fundamentals of Social Statistics

Kirk W. Elifson
Georgia State University

Richard P. Runyon and Audrey Haber

 Addison-Wesley Publishing Company
Reading, Massachusetts · Menlo Park, California
London · Amsterdam · Don Mills, Ontario · Sydney

Library of Congress Cataloging in Publication Data
Elifson, Kirk W., 1943-
  Fundamentals of social statistics.
  Bibliography: p.
Includes index.
  1. Social sciences—Statistical methods.
2. Statistics. I. Runyon, Richard P. II. Haber,
Audrey. III. Title.
HA29.E482   519.5   81-3640
ISBN 0-201-10430-X   AACR2

ISBN 0-201-10430-X
ABCDEFGHIJ-AL-898765432

Preface

Seventeen years ago I enrolled in the first of many statistics courses. While I knew at the time that it frequently left me and my classmates confused and floundering to the extent that we learned little about statistics, I was unable to express the faults I found with the assigned text. The book was devoid of interesting and relevant examples, omitted computational steps, introduced but did not define new concepts, and seldom presented the relevance or logic of the techniques it purported to teach. I frequently found myself juggling numbers that were not placed into a meaningful context rather then engaging in any real problem solving.

My first experience with statistics would have been far more pleasant had the instructor assigned a book that was less vulnerable to these criticisms. One of the primary reasons for my writing this book was to create a text that is not deficient in these areas. More often than not, I have found that my experience as a statistics professor has been shaped by my first statistics course. When course material is presented in clear, logical, and complete discussion—not in abstract terms, but through concrete examples—students understand better and enjoy the course more. Students *can* learn statistics and enjoy the experience if the material is presented appropriately. I hope that you find this text more useful than the one I first used.

This book represents a major revision of the best-selling behavioral science statistics textbook, *Fundamentals of Behavioral Statistics,* by Richard P. Runyon and Audrey Haber. The book has been used by tens of thousands of students and is now in its fourth edition. Approximately four years ago George M. Abbott, formerly of Addison-Wesley, offered me the ultimate challenge of my career as a sociologist. He asked me to rewrite one of the most widely acclaimed behavioral science statistics textbooks for use by social scientists. While he wanted me to preserve the strength of the book that had made it such a success with students, professors and practitioners, he envisioned a textbook that would meet the specific needs of those in sociology and the related areas of criminology, family and urban studies, political science, and social welfare.

Although the principles of statistics underlying the behavioral and social sciences are the same, social scientists use a number of statistical techniques that behavioral scientists seldom use. Therefore, it was necessary to add two new chapters to Runyon and Haber's original text and delete one chapter. An entirely new chapter on contingency tables and their related statistics was added as was a chapter on multivariate data analysis (including such topics as multivariate contingency tables, partial correlational analysis, and multiple regression and correlation analysis). New sections have been added throughout the book including material in the correlation and regression chapters plus a complete discussion of chi square and its related measures of association, such as Cramer's V, the contingency coefficient, and phi.

But, equally as important, the text, including the chapters on topics common to both disciplines, was refocused for the social sciences. Social science applications and examples are used throughout the book and in the exercises at the end of each chapter.

Features of the Book

A number of important features have been included in this text. Key terms appear in boldface. Terms that need emphasis have been italicized. Visual devices, including charts, figures, and graphs are also incorporated to ensure maximum understanding by the student. New equations are discussed fully rather than mechanically applied, and ample examples are initially provided to ensure comprehension.

Each chapter begins with a listing of contents to provide the student with an overview of the included material. A glossary of key terms and chapter summary are included at the end of each chapter. Abundant realistic exercises allow the students to generalize concepts and lay the groundwork for later chapters. Answers are provided at the end of the book for approximately 80 percent of the exercises. In addition, a student workbook, developed by John C. Touhey of Florida Atlantic University, is designed to provide review practice and feedback for the student. The workbook incorporates a programmed review of terms, symbols, and concepts. Selected computational exercises for application purposes and test questions are also included.

The Appendix includes a review of basic mathematics, a glossary of symbols keyed to the first page on which the symbols appear, and a complete set of tables accompanied by explicit directions for their use. Other features of the book include a master glossary of key terms, a chronological listing (on the front and back covers of the text) of the most frequently used equations, and a comprehensive index.

Acknowledgments

I owe a debt of gratitude to the many colleagues and professors who have taught me statistics. My former students have also taught me much about statistics and about how it should be taught.

Several persons having past or present affiliation with Addison-Wesley Publishing Company were particularly helpful. George M. Abbott conceived the idea of a social science version of *Fundamentals of Behavioral Statistics*. Ronald R. Hill proved to be a genius of motivational strategies and without him this book might never have been completed. Debra Osnowitz and Laura Lane coordinated the book's production and both necessarily turned less than adequate prose into lucid English.

Good reviewers are invaluable to an author. The following professors read all or part of the manuscript and offered valuable suggestions: Candace Clark, Seton Hall University; Allan Johnson; Peter Marsden, University of North Carolina; Jerry Miller, University of Arizona; G. Melton Mobley, Emory University; Duane Monette, Northern Michigan University; John H. Neel, Georgia State University; John Reed, University of North Carolina; Judith Tanur, SUNY at Stony Brook; and Carol Seyfrit, University of Maryland. I am particularly indebted to Rosalind Dworkin, University of Houston; Charles L. Jaret, Georgia State University; and Julie Wolfe-Petrusky, University of Utah, whose assistance was particularly helpful in later revisions. Sherry P. Hood cheerfully and professionally typed the manuscript more times than she and I wish to recall.

I am grateful to the Literary Executor of the late Sir Ronald A. Fisher, F.R.S., to Dr. Frank Yates, F.R.S., and to Longman Group Ltd., London, for permission to reprint Table III from their book *Statistical Tables for Biological, Agricultural and Medical Research* (Sixth edition, 1974).

Finally, I wish to thank the three women with whom I live: To my wife, Joan, whose contributions were more than substantial. She is a competent statistician in her own right and was always willing to listen, to encourage, and when I was particularly fortunate, to write a paragraph or two on a topic about which she knew more than I. To my young twin daughters, Kristin and Shelley, who seemed to know when to leave me alone and when to bother me. More than once they walked into my home office asking me if I had completed another chapter— long before they were able to write a word themselves.

November, 1981   K.W.E
Atlanta, Georgia

Contents

I Introduction

1 The Definition of Statistical Analysis 4

1.1 Introduction 5
1.2 Definitions of Terms Commonly Used in Statistics 7
1.3 Descriptive Statistics 10
1.4 Inferential Statistics 11
1.5 A Word to the Student 12
Summary 13
Terms to Remember 14
Exercises 15

2 Basic Mathematical Concepts 18

2.1 Introduction 19
2.2 The Grammar of Mathematical Notation 20
2.3 Summation Rules 22
2.4 Types of Numbers and Scales 24
2.5 Continuous and Discrete Scales 30
2.6 Rounding 33
2.7 Ratios 35
2.8 Proportions, Percentages, and Rates 36
Summary 39

Terms to Remember   40
Exercises   41

II   Descriptive Statistics

3   Frequency Distributions and Graphing Techniques   50

3.1   Grouping of Data   51
3.2   Cumulative Frequency and Cumulative Percentage Distributions   56
3.3   Graphing Techniques   57
3.4   Misuse of Graphing Techniques   57
3.5   Nominally Scaled Variables   59
3.6   Ordinally Scaled Variables   60
3.7   Interval- and Ratio-Scaled Variables   60
3.8   Forms of Frequency Curves   64
3.9   Other Graphic Representations   67
    Summary   71
    Terms to Remember   71
    Exercises   73

4   Percentiles   80

4.1   Introduction   81
4.2   Cumulative Percentiles and Percentile Rank   82
4.3   Percentile Rank and Reference Group   87
4.4   Centiles, Deciles, and Quartiles   87
    Summary   89
    Terms to Remember   89
    Exercises   90

5   Measures of Central Tendency   92

5.1   Introduction   93
5.2   The Arithmetic Mean $(\overline{X})$ 95
5.3   The Median (Md_n) 100
5.4   The Mode (M_0) 103
5.5   Comparison of Mean, Median, and Mode   104
5.6   The Mean, Median, Mode, and Skewness   105
    Summary   106

| | Terms to Remember | 107 |
| | Exercises | 107 |

6 Measures of Dispersion 112

6.1	Introduction	113
6.2	The Range	114
6.3	The Interquartile Range	115
6.4	The Mean Deviation	116
6.5	The Variance (s^2) and Standard Deviation (s)	118
6.6	Interpretation of the Standard Deviation	123
	Summary	124
	Terms to Remember	126
	Exercises	126

7 The Standard Deviation and the Standard Normal Distribution 130

7.1	Introduction	131
7.2	The Concept of Standard Scores	131
7.3	The Standard Normal Distribution	134
7.4	Characteristics of the Standard Normal Distribution	134
7.5	Transforming Raw Scores to z Scores	136
7.6	Illustrative Problems	139
7.7	Interpreting the Standard Deviation	144
	Summary	145
	Terms to Remember	145
	Exercises	146

8 An Introduction to Contingency Tables 150

8.1	Introduction	151
8.2	Dependent and Independent Variables	151
8.3	The Bivariate Contingency Table	153
8.4	Percentaging Contingency Tables	155
8.5	Existence, Direction, and Strength of a Relationship	157
8.6	Introduction to Measures of Association for Contingency Tables	160
8.7	Nominal Measures of Association: Lambda	161
8.8	Nominal Measures of Association: Goodman and Kruskal's Tau	166

8.9	Ordinal Measures of Association	172
8.10	Goodman and Kruskal's Gamma	174
8.11	Somer's *d*	179
	Summary	181
	Terms to Remember	181
	Exercises	183

9   Correlation   188

9.1	The Concept of Correlation	189
9.2	Calculation of Pearson's *r*	192
9.3	Pearson's *r* and *z* Scores	194
9.4	The Correlation Matrix	197
9.5	Interpreting Correlation Coefficients	198
9.6	Spearman's Rho (r_s)	203
	Summary	205
	Terms to Remember	205
	Exercises	206

10   Regression and Prediction   212

10.1	Introduction to Prediction	213
10.2	Linear Regression	214
10.3	Residual Variance and Standard Error of Estimate	224
10.4	Explained and Unexplained Variation	229
10.5	Correlation and Causation	232
	Summary	234
	Terms to Remember	235
	Exercises	236

11   Multivariate Data Analysis   242

11.1	Introduction	243
11.2	The Multivariate Contingency Table	244
11.3	Partial Correlation	248
11.4	Multiple Regression Analysis	254
	Summary	261
	Terms to Remember	262
	Exercises	263

III Inferential Statistics: Parametric Tests of Significance

12 Probability   270

12.1	An Introduction to Probability	271
12.2	The Concept of Randomness	272
12.3	Approaches to Probability	273
12.4	Formal Properties of Probability	275
12.5	Probability and Continuous Variables	285
12.6	Probability and the Normal-Curve Model	285
12.7	One- and Two-Tailed p Values	288
	Summary	290
	Terms to Remember	292
	Exercises	293

13 Introduction to Statistical Inference   300

13.1	Why Sample?	301
13.2	The Concept of Sampling Distributions	302
13.3	Binomial Distribution	305
13.4	Testing Statistical Hypotheses: Level of Significance	312
13.5	Testing Statistical Hypotheses: Null Hypothesis and Alternative Hypothesis	315
13.6	Testing Statistical Hypotheses: The Two Types of Error	317
13.7	A Final Word of Caution	321
	Summary	322
	Terms to Remember	322
	Exercises	324

14 Statistical Inference and Continuous Variables   328

14.1	Introduction	329
14.2	Sampling Distribution of the Mean	333
14.3	Testing Statistical Hypotheses: Parameters Known	336
14.4	Estimation of Parameters: Point Estimation	340
14.5	Testing Statistical Hypotheses with Unknown Parameters: Student's t	344
14.6	Estimation of Parameters: Interval Estimation	350

14.7   Confidence Intervals and Confidence Limits   351
14.8   Test of Significance for Pearson's *r:* One-Sample Case   358
14.9   Test of Significance for Goodman's and
           Kruskal's Gamma   361
           Summary   362
           Terms to Remember   363
           Exercises   364

15   Statistical Inference with Two Independent Samples and Correlated Samples   370

15.1   Sampling Distribution of the Difference between
           Means   371
15.2   Estimation of $\sigma_{\overline{X}_1 - \overline{X}_2}$ from Sample Data   373
15.3   Testing Statistical Hypotheses: Student's *t*   373
15.4   The *t*-Ratio and Homogeneity of Variance   376
15.5   Statistical Inference with Correlated Samples   378
           Summary   382
           Terms to Remember   383
           Exercises   383

16   An Introduction to the Analysis of Variance   390

16.1   Multigroup Comparisons   391
16.2   The Concept of Sums of Squares   392
16.3   Obtaining Variance Estimates   396
16.4   Fundamental Concepts of Analysis of Variance   398
16.5   Assumptions Underlying Analysis of Variance   399
16.6   An Example Involving Three Groups   400
16.7   The Interpretation of *F*   402
           Summary   404
           Terms to Remember   405
           Exercises   405

IV   Inferential Statistics: Nonparametric Tests of Significance

17   Power and Efficiency of a Statistical Test   412

17.1   The Concept of Power   413

17.2	Calculation of Power: One-Sample Case	419
17.3	The Effect of Sample Size on Power	420
17.4	The Effect of α-Level on Power	421
17.5	The Effect of the Nature of H_1 on Power	422
17.6	Parametric versus Nonparametric Tests: Power	423
17.7	Calculation of Power: Two-Sample Case	424
17.8	The Effect of Correlated Measures on Power	425
17.9	Power, Type I, and Type II Errors	426
17.10	Power Efficiency of a Statistical Test	428
	Summary	428
	Terms to Remember	429
	Exercises	429

18 Statistical Inference with Categorical Variables: Chi Square and Related Measures 432

18.1	Introduction	433
18.2	The χ^2 One-Variable Case	434
18.3	The χ^2 Test of the Independence of Variables	438
18.4	Limitations in the Use of χ^2	441
18.5	Nominal Measures of Association Based on χ^2	444
	Summary	447
	Terms to Remember	448
	Exercises	448

V Appendixes

A	Review of Basic Mathematics	454
B	Glossary of Symbols	464
C	Tables	472
D	Glossary of Terms	510
E	References	526

VI Answers to Selected Exercises 532

Index	557

Introduction

1   The Definition of Statistical Analysis

2   Basic Mathematical Concepts

The Definition of Statistical Analysis

1

1.1   Introduction

1.2   Definitions of Terms Commonly Used in Statistics

1.3   Descriptive Statistics

1.4   Inferential Statistics

1.5   A Word to the Student

1.1 Introduction

Think for a moment of the thousands of incredibly complex things you do during the course of a day. You are absolutely unique. No one else possesses your physical features, your intellectual makeup, your personality characteristics, and your value system. Yet, like billions of others of your species, you are among the most finely tuned and enormously sophisticated statistical instruments ever devised by natural forces. Every moment of your life provides testimony to your ability to receive and process a variety of information and then to use this information instantly to determine possible courses of action.

To illustrate, imagine you are driving in heavy traffic. You are continuously observing the road conditions, noting the speed of cars in front of you compared to your own speed, the position and rate of approach of vehicles to your rear, and the presence of automobiles in the oncoming lane. If you are an alert driver, you are constantly summarizing this information—usually without words or even awareness.

Imagine next that, without warning, the driver of the car in front of you suddenly jams on the brakes. In an instant, you must act upon this prior information. You must brake the car, turn left, turn right, or pray. Your brain instantly considers alternative courses of action: If you jam on the brakes, what is the possibility that you will stop in time? Is the car behind you far enough away to avoid a rear-end collision?

Can you avoid an accident by turning into the left lane or onto the right shoulder? Most of the time, your decision is correct. Consequently, most of us live to a ripe old age.

In this situation, as in many others during the course of a lifetime, you have accurately assessed the possibilities and taken the right course of action. And you make such decisions thousands of times each and every day of your life. For this reason you should regard yourself as a mechanism for making statistical decisions. In this sense, you are already a statistician.

In daily living, our statistical functioning is usually informal and loosely structured. Consider the times you have contemplated the *likelihood* of someone you are attracted to, but do not know well, rejecting your invitation to have lunch. We *behave* statistically, although we may be totally unaware of the formal laws of probability which will be presented in Chapter 12.

In this course, we will attempt to provide you with some of the procedures for collecting and analyzing data, and making decisions or inferences based upon these analyses. Since we will frequently be building upon your prior experiences, you will often feel that you have made a similar analysis before: "Why, I have been calculating averages almost all my life—whenever I determine my test average in a course or the mileage my car gets," and "I compute range whenever I figure how much my time varies on my favorite two-mile jog." If you constantly draw upon your previous knowledge and relate course materials to what is familiar in daily life, statistics need not, and should not, be the bugaboo it is often painted to be.

What, then, is statistics all about? Although it would be virtually impossible to obtain a general consensus on the definition of statistics, it is possible to make a distinction between two definitions of statistics.

1. Statistics are commonly regarded as a *collection* of numerical facts that are expressed in terms of summarizing statements and that have been collected either from several observations or from other numerical data. From this perspective, statistics constitute a collection of statements such as, "The average age at marriage of females in 1981 was 21.3," or "Seven out of ten people prefer American-made cars," or "The murder rate in Atlanta over the past year was. . . ."

2. Statistics may also be regarded as a *method* of dealing with data. This definition stresses the view that statistics is a tool concerned with the collection, organization, and analysis of numerical facts or observations.

The second definition constitutes the subject matter of this text.

A distinction may also be made between the two functions of the statistical method: **descriptive** statistical techniques and **inferential** or **inductive** statistical techniques.

Descriptive statistics presents information in convenient, summary form. Inferential statistics, on the other hand, is concerned with generalizing this information or, more specifically, with making inferences about populations that are based upon samples taken from the populations.

In describing the functions of statistics, certain terms such as samples and populations have already appeared with which you may or may not be familiar. Before elaborating on the differences between descriptive and inferential statistics, it is important to learn the meaning of certain terms that will be used repeatedly throughout the text.

1.2 Definitions of Terms Commonly Used in Statistics

Variable   Any characteristic of a person, group, or environment that can vary *or* denote a difference. Thus weight, occupational prestige, sex, group cohesion, and race are all variables since they can vary or denote a difference. A variable is contrasted with a constant, the value of which never changes (for example, pi which is equal to 3.14).

Data   Numbers or measurements that are collected as a result of observations, interviews, etc. They may be head counts such as the number of individuals stating a preference for a Republican presidential candidate; or they may be scores, as on a job satisfaction scale. Note that *data* is the plural form of *datum*. So you would say, "The data *are* available to anyone who wishes to see them."

Population   A complete set of individuals, objects, or measurements having some common observable characteristic. We can distinguish between a **finite population** and an **infinite population**. All babies born in a particular year would constitute a finite population for it would be possible to compile a list of their names. An infinite population cannot be listed and is necessarily theoretical in nature. It would be impossible, for example, to compile a complete list of *all* babies regardless of when they were born or will be born.

Element   A single member of a population. Thus if a population consists of all the presently divorced males in a particular county, each divorced male constitutes an element.

Parameter   Any characteristic of a *population* that is measurable, for example, the proportion of registered Democrats among Americans of voting age. A parameter is only measurable in a finite population; however, later in the text you will learn procedures that allow for an estimation of a parameter in either a finite or infinite population. In this text we shall follow the practice of employing Greek letters (for example, μ, σ) to represent population parameters. You will learn, for example, that if μ (pronounced "mu") equals 42, we are referring to an average for an entire population, and if σ (pronounced "sigma") equals 3.1, we are referring to a measure of variability for an entire population.

Sample   A subset or part of a population. For example, the newly elected members of the United States House of Representatives constitute a sample of the population of all the current members of the House of Representatives.

Random sample*   Sample in which all elements have an equal chance of being selected. This requires that all the elements can be identified or listed. It would be possible, for example, to draw a random sample of prison inmates in the Pontiac State Penitentiary since a complete list of the inmates is available.

Statistic   A number that describes a characteristic of a sample. Commonly, we use a statistic which is calculated from a sample in order to estimate a population parameter; for example, a sample of Americans of voting age is used to estimate the proportion of Democrats in the entire population of voters. We shall use italic letters (for example, $\overline{X}$, s) to represent sample statistics. For example, if an author specifies that $\overline{X}$ (the average of the X scores) equals 4.12, you know immediately that the referent is a sample and not a population.

---

* Simple random sampling is more precisely defined in Section 12.2 as selecting samples in such a way that each sample of a given size has precisely the same probability of being selected.

Example   Imagine that a public opinion polling firm has been contracted to conduct a study for a U.S. senator from Wisconsin concerning the percentage of the state's registered voters who approve of nuclear power as an energy source. As part of the polling process, a number of individuals are randomly selected from the voter registration lists and carefully interviewed. The people selected are *elements,* and all the elements selected constitute the *random sample.*

The *variable* of interest is the registered voters' attitude toward nuclear power (not everyone will feel the same way and the answers will vary). The *data* consist of all the responses to the question asked of each person included in the sample. When the data are analyzed according to certain rules to yield summary statements such as the percentage who favor the use of nuclear power as an energy source, the resulting numerical value (in this case a percentage) is a *statistic.* The *population* we are interested in generalizing about is all the state's registered voters. The "true" percentage favoring nuclear power among all the state's registered voters constitutes a *parameter* (see Fig. 1.1). Note that it is highly unlikely that the parameter will ever be known, since finding it would require interviewing every registered voter in the state. Since this is usually not feasible for economic and other reasons, it is rare that an

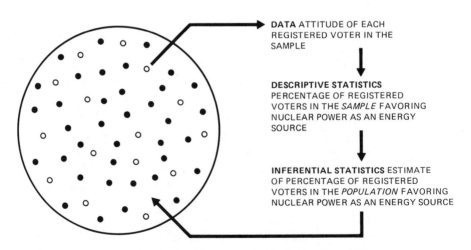

DATA ATTITUDE OF EACH REGISTERED VOTER IN THE SAMPLE

DESCRIPTIVE STATISTICS PERCENTAGE OF REGISTERED VOTERS IN THE *SAMPLE* FAVORING NUCLEAR POWER AS AN ENERGY SOURCE

INFERENTIAL STATISTICS ESTIMATE OF PERCENTAGE OF REGISTERED VOTERS IN THE *POPULATION* FAVORING NUCLEAR POWER AS AN ENERGY SOURCE

1.1 A *random sample* is selected from some population. The hollow and solid dots represent all registered voters in Wisconsin. The hollow dots represent the sample. *Data* are collected and summarized, using descriptive statistics. In *inferential statistics,* we attempt to estimate one or more population *parameters* (for example, the percentage of all registered voters in Wisconsin favoring nuclear power as an energy source).

exhaustive study of populations in public opinion polling is undertaken. Consequently, parameters are rarely known; but, as we shall see, they are commonly estimated from sample statistics.

It is, of course, possible to define a very small population by narrowing the definition of "common observable characteristic" to something like "all students attending their Sociology 201 class at Knox College today." In this event, calculating a parameter (such as the class average on the examination given today) would pose no difficulty. For such small populations, there is no reason to select samples and to calculate sample statistics.

From the point of view of the instructor, observations on the students in Sociology 201 at Knox College may be of supreme importance. However, such limited populations would be of little theoretical interest. In the real workaday world, the social scientist is usually interested in making statements having general validity over a wider set of people (or elements) and thus must *estimate* the (population) parameter. This estimate is based on the statistic that is calculated on the sample. Thus we might use the observations on the students at Knox College as a basis for estimating parameters of the larger population.

Let us return to the two functions of statistical analysis for a closer look.

1.3 Descriptive Statistics

When social scientists conduct a study, they usually collect a great deal of information or data about the problem at hand. The data may take a variety of forms: the number of medical schools in each state, the SAT scores of a group of college students, or the number of completed questionnaires from a research project. In their original form, as collected, these data are usually a confusing hodgepodge of scores, frequency counts, etc. In performing the descriptive function, the statistician employs rules and procedures for presenting data in a more usable and meaningful form.

Rules are also followed when calculating various *statistics* from masses of raw data. Imagine that a team of social scientists administered a questionnaire concerning drug usage and attitudes to a group of high school students. What are some of the things they might do with the resulting information?

1. They might rearrange the scores and group them in various ways in order to be able to see at a glance an overall picture of the data (Chapter

3, "Frequency Distributions and Graphing Techniques"). For example, how many of the students use drugs regularly? How many have never used drugs?

2. They might construct tables, graphs, and figures to permit visualization of the results (Section 3.3, "Graphing Techniques," in Chapter 3).

3. They might convert raw scores to other types of scores that are more useful for specific purposes. Thus these scores could be used to rank the students, compare a student with the class average, etc. Other types of conversion will also be described in the text (Chapter 4, "Percentiles," and Chapter 7, "The Standard Deviation and the Normal Distribution").

4. They might calculate averages, to learn something about the typical attitudes of the students (Chapter 5, "Measures of Central Tendency").

5. Using the average as a reference point, they might describe the spread of scores about this central point. Statistics that measure this spread of scores are known as measures of variability or measures of dispersion (Chapter 6, "Measures of Dispersion").

6. A relationship between two different variables may be obtained. Statistics are available for describing the relationship between two variables. Such statistics are extremely useful to the social scientist. For example, it might be important to determine the relationship between the number of books in the home and classroom grades or that between students' attitudes toward the police and the extent to which they reported having committed delinquent acts. Once these relationships are known, the social scientist may use information from one variable to predict another (Chapter 9, "Correlation," and Chapter 10, "Regression and Prediction").

1.4 Inferential Statistics

The social scientists' task is not nearly over when they have completed the descriptive function. On the contrary, they are often nearer to the beginning than to the end of the task. The reason for this is clear when we consider that the purpose of the research is often to explore hypotheses of a general nature, rather than simply to compare limited samples.

Imagine that you are an educational psychologist who is interested in determining the effectiveness of a videotape lecture approach versus the usual procedure of the professor delivering lectures in person. Consequently, you design a study involving two conditions: *experimental* and *control.* The question you wish to answer is which approach (videotape teaching versus the in-person lecture approach) results in better student performance on a final examination. The students in the experimental group view videotapes of the lectures attended by the control group. The students in the control group serve as a basis for comparison and are taught using the usual lecture approach. After all students have been tested, you might find that "on the average" the experimental group that viewed the videotapes did not perform as well on the final examination as did the control group. In other words, the average of the experimental group on the final examination was lower than that of the control group. You then ask the question, "Can I conclude that the teaching technique produced the difference between the two groups?" To answer this question, it is not sufficient to rely solely upon *descriptive statistics.*

"After all," you reason, "even if both techniques worked, it is highly unlikely that the two groups' final examination averages would have been *identical. Some* difference would have been observed." The operation of chance factors such as the time at which the classes are scheduled is certain to produce some differences in the final examination scores between the two groups. The critical question, from the point of view of inferential statistics, becomes: Is the difference between groups great enough to rule out chance factors in the experiment as a sufficient explanation? Stated another way, if you were to repeat the experiment, would you be able to predict with confidence that the same differences (that is, the control group would consistently perform better than the experimental group) would systematically occur?

As soon as you raise these questions, you move into the area of statistical analysis known as *inferential* (or *inductive*) statistics. As you will see, much of this text is devoted to procedures the researcher uses to arrive at conclusions extending beyond the sample statistics themselves.

1.5 A Word to the Student

The study of statistics need not and should not be an unpleasant experience. If you approach it with the proper frame of mind, you will find that statistics can be a very interesting field of study. It has numerous applications. H. G. Wells, the nineteenth-century social critic, remarked, "Statistical thinking will one day be as necessary for efficient citizenship as the ability to read and write." Keep this thought constantly in mind. The course will be much more interesting and profitable to you if you develop the habit of "thinking statistically."

Attempt to apply statistical concepts to all daily activities, no matter how routine. When you are stopped at an intersection you cross frequently, note the time the traffic light remains red. Obtain some estimate of the length of the green cycle. If it is red 3 minutes and green 2, you would expect that the chances are three in five that it will be red when you reach the intersection. Start collecting data. Do you find that it is red 60% of the time as expected? If not, why not? Perhaps you have unconsciously made some driving adjustments in order to change the statistical probabilities. Watch commercials on television and read newspaper advertisements carefully. When the advertiser claims, "This toothpaste is more effective against cavities," ask, "More effective than what? What is the evidence?" If you make statistical thinking an everyday habit, not only will you find that the study of statistics becomes more interesting, but the world you live in will appear different and perhaps more interesting.

Summary

In this chapter, we have distinguished between two definitions of statistical analysis, one stressing statistics as a collection of numerical facts and the other emphasizing statistics as a method or tool concerned with the collection, organization, and analysis of numerical facts. The second definition constitutes the subject matter of this text.

A distinction is made between two functions of the statistical method: descriptive and inferential statistical analyses. The former is concerned with the organization and presentation of data in a convenient, summary form. The latter addresses the problem of making broader generalizations or inferences from sample data to populations.

A number of terms commonly used in statistical analysis were defined.

Terms to Remember

Data   Numbers or measurements that are collected as a result of observations, interviews, etc.

Descriptive statistics   Procedures used to organize and present data in a convenient, summary form.

Element   A single member of a population.

Finite population   A population whose elements or members can be listed.

Inferential or inductive statistics   Procedures used to arrive at broader generalizations or inferences from sample data to populations.

Infinite population   A population whose elements or members cannot be listed.

Parameter   Any characteristic of a finite population that can be estimated and is measurable or of an infinite population that can be estimated.

Population   A complete set of individuals, objects, or measurements having some common observable characteristic.

Random sample   Sample in which all elements have an equal chance of being selected.

Sample   A subset or part of a population.

Statistic   A number that describes a characteristic of a sample.

Statistics   Collection of numerical facts expressed in summarizing statements; method of dealing with data: a tool for collecting, organizing, and analyzing numerical facts or observations.

Variable   Any characteristic of a person, group, or environment that can vary or denotes a difference.

Exercises

1. In your own words, describe what you understand the study of statistics to be.

2. Indicate whether each of the following represents a variable, a constant, or both depending on the usage:

   a) Number of days in the month of August.

   b) The number of labor strikes per year in the United States since 1975.

   c) The murder rate in Peoria last year.

   d) Scores obtained on a 100-item multiple-choice examination.

   e) Maximum score possible on a 100-item multiple-choice examination.

   f) Mandatory retirement age.

   g) Percentage of people in the United States who have tried marijuana.

   h) Number of individuals living on subsistence incomes.

   i) Countries in the United Nations.

3. In the example cited in Section 1.2, imagine that the survey results indicated that, overall, 57% of the senator's constituents favored nuclear power as an energy source. May we assume that if the respondents were reinterviewed six months later the percentage favoring would remain the same? What factors might have caused a shift in attitudes?

4. Bring in newspaper articles citing recent survey or poll results. In how many articles is the method of sampling mentioned? Do the articles reveal where the financial support for the surveys came from? Why is this information important? Why is it so commonly not revealed?

5. List four populations and indicate why they are finite or infinite.

6. When surveying small populations (for example, school superintendents in Rhode Island), social scientists frequently study all the elements rather than draw a sample. Can you think of any other populations with few elements?

7. We wish to generalize from a sample to the:

   a) data   **b)** population   **c)** statistic   **d)** variable

8. Suppose you are interested in comparing the number and type of welfare recipients in three rural counties. County A had 4 Aid-to-Families-with-Dependent-Children (AFDC) recipients, 3 Disability

recipients, and 3 Aid-to-Aged recipients. County B had 6 AFDC, 5 Disability, and 2 Aged recipients. County C had 10 AFDC, 4 Disability, and 4 Aged recipients. How might you arrange these data to facilitate the communciation of information? Identify the population and data.

9. How could you determine the average welfare payment received by recipients in your state? Would you have to contact each person? Identify the population. Would the average in this example be a statistic or a parameter? Why?

10. Professor Norman Yetman of the University of Kansas has concerned himself with possible economic discrimination against black athletes by the media.* He has analyzed their opportunities to appear on commercials, make guest appearances, and obtain off-season jobs. Two of the findings follow:

    a) In 351 commercials associated with New York sporting events in the autumn of 1966, black athletes appeared in only 2.

    b) An analysis of media advertising opportunities for athletes on a professional football team in 1971 revealed that 8 of 11 whites had an opportunity in contrast to only 2 of 13 black athletes.

    Indicate whether these two examples provide:

    a) parameters

    b) data

    c) descriptive statistics

    d) inferential statistics

11. List five social science variables and briefly discuss how they might be measured.

12. Differentiate between the following pairs of words:

    a) sample–population

    b) statistic–parameter

    c) inferential statistics–descriptive statistics

    d) experimental group–control group

    e) finite population–infinite population

13. If a social scientist were simply to draw up a table giving family background information (number of brothers and sisters, history of

---

* *Source:* Gary Lehman, Associated Press sportswriter, August 22, 1975.

divorces in the immediate family, etc.) about all of the alcoholics in a study, he or she would be using statistics in a

    a) descriptive manner

    b) inferential manner

    c) statistical manner

    d) hypothetical manner

14. If the researcher were to investigate the hypothesis that "every alcoholic has at least one case of alcoholism or heavy drinking in his or her family history," then he or she would be using

    a) descriptive statistics

    b) inferential statistics

    c) statistical statistics

    d) hypothetical statistics

Basic Mathematical Concepts

2

2.1 Introduction

2.2 The Grammar of Mathematical Notation

2.3 Summation Rules

2.4 Types of Numbers and Scales

2.5 Continuous and Discrete Scales

2.6 Rounding

2.7 Ratios

2.8 Proportions, Percentages, and Rates

2.1 Introduction

"I'm not much good at math. How can I possibly pass a statistics course?" We have heard these words from the lips of countless undergraduate students. For many, this is probably a concern stemming from prior discouraging experiences with mathematics. A brief glance through the pages of this text may only serve to increase this anxiety, since many of the formulas appear rather complicated and may seem impossible to master. Therefore, it is most important to set the record straight right at the beginning of the course.

You do not have to be a mathematical genius to master the statistical principles in this text. The degree of mathematical sophistication necessary for a firm grasp of the fundamentals of statistics is often exaggerated. As a matter of actual fact, statistics requires a good deal of arithmetic computation, sound logic, and a willingness to stay with a point until it is mastered. To paraphrase Carlyle, success in statistics requires an infinite capacity for taking pains. Beyond this modest requirement, little is needed but the mastery of several algebraic and arithmetic procedures that most students learned early in their high school careers. In this chapter, we review the grammar of mathematical notations, discuss several types of numerical scales, and adopt certain conventions for rounding numbers.

If you wish to brush up on basic mathematics, Appendix A contains a review of all the math necessary to master this text.

2.2 The Grammar of Mathematical Notation

Throughout the textbook, you will be learning new mathematical symbols. For the most part, we shall define these symbols when they first appear. However, there are four notations that will appear so frequently that their separate treatment at this time is justified. These notations, are X, Y, N, and the Greek letter Σ (pronounced "sigma").

While defining these symbols and demonstrating their use, we will also review the grammar of mathematical notation. It is not surprising that many students become so involved in the numerous mathematical formulas and symbols that they fail to realize that mathematics has its nouns, adjectives, verbs, and adverbs.

Mathematical nouns   In mathematics, we commonly use symbols to stand for quantities. The notation we use most commonly in statistics to represent quantity (or a score) is X, although we shall occasionally use Y. In addition, X and Y are used to identify variables; for example, if age and educational attainment are two variables in a study, X might be used to represent age and Y to represent educational attainment. Another frequently used noun is the symbol N, which represents the number of scores or measurements with which we are dealing. Thus, if we have ten scores,

$$N = 10.$$

Mathematical adjectives   When we want to modify a mathematical noun, we commonly use subscripts that indicate a specific score in a series and identify it more precisely. Thus if we have a series of scores or quantities, we may represent them as X_1 (refers to the first X score and is read as "X sub 1"), X_2, X_3, X_4, etc. We shall frequently encounter X_i, in which the subscript may take on any value that we want.

Mathematical verbs   Notations that direct the reader to do something have the same characteristics as verbs in the spoken language. One of the most important verbs is the symbol already alluded to as Σ. This notation directs us to sum all quantities or scores following the symbol. Thus

$$\sum (X_1,\ X_2,\ X_3,\ X_4,\ X_5) = X_1 + X_2 + X_3 + X_4 + X_5$$

indicates that we should add together all these quantities from X_1 through X_5. Other verbs we shall encounter frequently are $\sqrt{\ }$, directing us to find the square root, and exponents* (X^a), which tell us to raise a quantity to the indicated power. In mathematics, mathematical verbs are commonly referred to as *operators*.

Mathematical adverbs   These are notations that, as in spoken language, modify the verbs. We shall frequently find that the summation signs are modified by adverbial notations. Let us imagine that we want to indicate that the following quantities are to be added:

$$X_1 + X_2 + X_3 + X_4 + X_5 + \cdots + X_N.$$

Symbolically, we would represent these operations as follows:

$$\sum_{i=1}^{N} X_i.$$

The notations above and below the summation sign indicate that i takes on the successive values from 1, 2, 3, 4, 5, up to N. Stated verbally, the notation reads: We should sum all quantities of X starting with $i = 1$ (that is, X_1) and proceeding through to $i = N$ (that is, X_N).

The following shorthand version excludes the notations above and below the summation sign and is often used when all of the quantities from 1 to N are to be added. It is equivalent to the complete version:

$$\sum X = \sum_{i=1}^{N} X_i = X_1 + X_2 + X_3 + X_4 + X_5 + \cdots + X_N.$$

Sometimes this form of notation may direct us to add only selected quantities; thus

$$\sum_{i=2}^{5} X_i = X_2 + X_3 + X_4 + X_5.$$

Stated verbally, the notation reads: We should sum all quantities of X starting with $i = 2$ (that is, X_2) and proceeding through to $i = 5$ (that is, X_5). The shorthand version would not allow us to convey this information.

---

* An exponent indicates how many times a number is to be multiplied by itself.

2.3 Summation Rules

The summation sign is one of the most frequently used operators in statistics. Let us summarize a few of the rules governing the use of the summation sign.

Generalization 1   *The sum of a constant added together N times is equal to N times that constant.* Symbolically,

$$\sum_{i=1}^{N} c = Nc.$$

Let c be a constant. Thus, if $c = 10$ and $N = 5$,

$$\sum_{i=1}^{N} c = (10 + 10 + 10 + 10 + 10) = 5(10) = 50.$$

Similarly, if $\overline{X}$ is the constant and $\overline{X} = 20$, $N = 15$,

$$\sum_{i=1}^{N} \overline{X} = N\overline{X} = (15)(20) = 300.$$

Generalization 2   *The sum of a constant times each value of a variable is equal to the sum of the values of the variable times that constant.* Symbolically,

$$\sum_{i=1}^{N} cX_i = c\sum_{i=1}^{N} X_i.$$

Thus, if $c = 5$, and $X_1 = 2$, $X_2 = 3$, $X_3 = 4$,

$$\sum_{i=1}^{N} cX_i = (c)X_1 + (c)X_2 + (c)X_3$$

$$= (5)(2) + (5)(3) + (5)(4) = 45.$$

Also,

$$c\sum_{i=1}^{N} X_i = c(X_1 + X_2 + X_3)$$

$$= 5(2 + 3 + 4) = 45.$$

Generalization 3 *The sum of the values of a variable plus a constant is equal to the sum of the values of the variable plus N times that constant.* Symbolically,

$$\sum_{i=1}^{N} (X_i + c) = \sum_{i=1}^{N} X_i + Nc.$$

Imagine a sample in which $N = 3$ and $X_1 = 3$, $X_2 = 4$, and $X_3 = 6$. *The sum of the three values of the variable may be shown by*

$$\sum_{i=1}^{N} X_i = X_1 + X_2 + X_3$$

$$= 3 + 4 + 6.$$

To show the sum of the values of a variable when a constant ($c = 2$) has been added to each.

$$\sum_{i=1}^{N} (X_i + c) = (3 + c) + (4 + c) + (6 + c)$$

$$= 3 + 4 + 6 + (c + c + c)$$
$$= 3 + 4 + 6 + (2 + 2 + 2)$$
$$= 13 + 3c = 13 + 3(2)$$
$$= 19.$$

Generalization 4 *The sum of the values of a variable when a constant has been subtracted from each is equal to the sum of the values of the variable minus N times the constant.* Symbolically,

$$\sum_{i=1}^{N} (X_i - c) = \sum_{i=1}^{N} X_i - Nc.$$

To show the sum of the values of a variable when a constant has been subtracted from each, let $X_1 = 3$, $X_2 = 4$, $X_3 = 6$, and $c = 2$.

$$\sum_{i=1}^{N} (X_i - c) = (3 - c) + (4 - c) + (6 - c)$$

$$= 3 + 4 + 6 - (c + c + c)$$
$$= 3 + 4 + 6 - 3c$$
$$= 13 - 3(2) = 7.$$

Example These generalizations are often useful in statistics, particularly when we want to sum numbers with large quantitative values, for example,

$$X_1 = 100,465, \quad X_2 = 100,467, \quad X_3 = 100,469, \quad X_4 = 100,472.$$

To sum these four numbers, you can subtract 100,000 from each quantity, sum the remainder, and then add 4 times 100,000. Thus,

$$\sum_{i=1}^{4} X_i = 465 + 467 + 469 + 472 + 4(100,000)$$

$$= 1873 + 400,000$$

$$= 401,873.$$

Finally, note the difference between the following two summations:

$$\sum_{i=1}^{N} X_i^2 \quad \text{and} \quad \left(\sum_{i=1}^{N} X_i \right)^2.$$

Thus, if $X_1 = 2$, $X_2 = 3$, and $X_3 = 4$,

$$\sum_{i=1}^{N} X_i^2 = X_1^2 + X_2^2 + X_3^2 \quad \text{and} \quad \left(\sum_{i=1}^{N} X_i \right)^2 = (X_1 + X_2 + X_3)^2$$

$$\begin{aligned} &= 2^2 + 3^2 + 4^2 &&= (2 + 3 + 4)^2 \\ &= 4 + 9 + 16 &&= (9)^2 \\ &= 29 &&= 81. \end{aligned}$$

2.4 Types of Numbers and Scales

Cultural anthropologists, psychologists, and sociologists have repeatedly called attention to the common human tendency to explore the world that is remote from our experiences long before we have investigated that which is closest to us. So, while we probe distant stars and describe with great accuracy their apparent movements and relationships, we virtually ignore the very substance that gave us life: air (which we inhale and exhale over four hundred million times a year). In the authors' experience, a similar pattern exists in relation to the student's familiarity with numbers and concepts of them.

In our numerically oriented western civilization, student Cory uses numbers long before he is expected to calculate the batting averages of the latest baseball hero. Nevertheless, ask him to define a number, or to describe the ways in which numbers are employed, and you will likely be met with an expression of bewilderment. "I have never thought about it before," he will frequently reply. After a few minutes of soul searching and deliberation, he will probably reply that numbers are symbols denoting amounts of things that can be added, subtracted, multiplied, and divided. These are all familiar arithmetic concepts, but do they exhaust all possible uses of numbers? At the risk of reducing our student to utter confusion, you may ask: "Is the symbol 7 on a baseball player's uniform such a number? What about your home address? Channel 2 on your television set? Do these numbers indicate amounts of things? Can they reasonably be added, subtracted, multiplied, or divided? Can you multiply the number on any football player's back by any other number and obtain a meaningful value?" A careful analysis of our use of numbers in everyday life reveals a very interesting fact: Most of the numbers we use do not have the arithmetical properties we usually give to them; that is, they cannot be meaningfully added, subtracted, multiplied, and divided. A few examples are the serial number of a home appliance, a zip code number, a telephone number, a home address, an automobile registration number, and the catalog numbers on a book in the library.

The important point is that numbers are used in a variety of ways to achieve many different ends. Much of the time, these ends do not include the representation of an amount or a quantity. In fact, there are three different ways in which numbers are used:

1. To name (**nominal numbers**),
2. To represent position in a series (**ordinal numbers**), and
3. To represent quantity (**cardinal numbers**).

Measurement is the assignment of numbers to objects or events according to sets of predetermined (or arbitrary) rules. The different levels of measurement that we will discuss represent different levels of numerical information contained in a set of observations (data), such as: a series of house numbers, the order of finish in a horse race, a set of I.Q. scores, or the price per share of various stocks. The type of scale obtained depends on the kinds of mathematical operations that can be legitimately performed on the numbers. In the social sciences, we encounter measurements at every level.

It should be noted that there are other schema for classifying numbers and the ways they are used. We use the nominal–ordinal–cardinal classification because it best handles the types of data we obtain in the social sciences. As you will see, we count, we place in relative position, and we obtain numerical scores. It should also be noted that assignment to categories is not always clear-cut or unambiguous. There are times when even experts cannot agree. To illustrate: Is the number in your street address nominal or ordinal? The answer depends on your need. For certain purposes it can be considered nominal, such as when used as a *name* of a dwelling. At other times it can be considered ordinal, since the numbers place your house in a position relative to other houses on the block. Thus 08 may be to the left of 12 and to the right of 04.

The fundamental requirements of observation and measurement are acknowledged by all the physical and social sciences as well as by any modern-day corporation interested in improving its competitive position. The things that we observe are often referred to as **variables**. Any particular observation is called the **value of the variable.** Let us look at two examples. If we are studying the price of stocks on the New York Stock Exchange, our variable is price. Thus, if the selling price of a stock is 8 1/2, the value of the variable is $8.50. If we are interested in determining whether or not an employer discriminates on the basis of racial background, the racial classification of the employees is our variable. This variable may have several values, such as black, Hispanic, and white.

2.4.1 Nominal Scales

Four basic levels of measurement scales are utilized by social and physical scientists. These include the nominal scale, the ordinal scale, the interval scale, and the ratio scale. Each type of scale has unique characteristics, and as we will see later, implications for the type of statistical procedures that can be used with it.

Nominal scales do not involve highly complex measurement, but rather rules for placing individuals or objects into categories. The categories must (1) be homogeneous, (2) be mutually exclusive*, and (3) make no assumption about ordered relationships between categories.

---

* We refer to categories as *mutually exclusive* because it is impossible for a person's score to belong to more than one category.

Consider a study of anti-Semitism. Let's classify the respondents into one of two categories: Jewish and non-Jewish. Such a dichotomy would satisfy each of the three classification rules for a nominal scale. All of the respondents in the Jewish category are *homogeneous* (similar) in terms of the variable religious identification in that only Jewish respondents are included. The respondents in the non-Jewish category are also homogeneous. The categories we have chosen are *mutually exclusive* since no respondent can be placed into both categories; that is, a person is either Jewish or non-Jewish. Finally, we can make no assumptions about the ordered relationships between categories, only that the individuals in the Jewish category *differ* from those in the non-Jewish category in terms of their religious identification. Our data would consist of the number of individuals in each of these two categories. Note that we do not think of the variable religious affiliation as representing an ordered series of values, such as height, prestige, or speed. A person that is female does not have more of the variable sex than one that is male.

Observations of unordered variables constitute the lowest level of measurement because they are the least mathematically versatile of the four levels of measurement scales and are referred to as a nominal scale of measurement. We may assign numerical values to represent the various classes in a nominal scale but these numbers function only as category labels. If we wished to classify persons in terms of their present marital status we might number the categories for convenience, but the numbers would only serve to identify the class. Here we have assigned numbers as we would names and it would be meaningless to add or subtract them. For example, (1) married, (2) divorced, (3) separated, and (4) single. Other examples of nominal level variables include sex, religious affiliation, and race.

The data used with nominal scales consist of frequency counts or tabulations of the number of occurrences in each class of the variable under study. In the aforementioned anti-Semitism study, our frequency counts of Jewish and non-Jewish respondents would comprise our data. Such data are often referred to interchangeably as *frequency data, nominal data,* or *categorical data.*

2.4.2 Ordinal Scales

When we move into the next higher level of measurement, we encounter variables in which the classes *do* represent a rank-ordered series of relationships. Thus, the classes in **ordinal scales** are not only homogeneous and mutually exclusive, but they stand in some kind of *relation* to one another. More specifically, the relationships are expressed in terms of the algebra of inequalities: a is less than b ($a < b$) or a is greater than b ($a > b$). In Fig. 2.1 we see that B and C are higher than A, but we cannot say how much higher because we do not know the distance between A, B, and C since the values are measured on an ordinal scale. We cannot even be sure if the distance AB is greater than, equal to, or less than the distance BC.

2.1 Relationship of three points on an ordinal scale.

Types of relationships encountered in an ordinal scale are: greater, poorer, healthier, more prejudiced, more feminine, more prestigious, etc. The numerals employed in connection with ordinal scales indicate only position in an ordered series and not how much of a difference exists between successive positions on the scale. We can rank order the following educational degrees from higher to lower [for example, (1) doctorate, (2) master's degree, (3) bachelor's degree], but we cannot say that the PhD is twice as high as the bachelor's degree or that a particular difference exists between categories.

Examples of ordinal scaling include rank ordering: the leading causes of death among the elderly, academic departments in a college according to their prestige, baseball teams according to their league standings, officer candidates in terms of their leadership qualities, and potential candidates for political office according to their popularity with the people. Note that the ranks are assigned according to the ordering of individuals within the class. Thus the most popular candidate may receive the rank of 1, the next most popular may receive the rank of 2, and so on, down to the least popular candidate. It does not, in fact, make any difference whether or not we give the most popular candidate the highest numerical rank or the lowest, *so long as we are consistent in placing the individuals accurately with respect to their relative position in the ordered series.*

It is important to realize that many social scientists believe all of the attitude scales employed in the social sciences are ordinal level scales. When you and others are asked, for example, to respond to a question with Likert-type response categories [for example, (1) strongly disagree, (2) disagree, (3) agree, and (4) strongly agree], the responses are ordinal in nature. Consider course evaluations and the following statement: "My professor is well prepared for class." If a friend in the class responds "(1) strongly disagree" and you respond "(3) agree," we can only conclude that you ranked the professor more favorably on this question than did your friend. We cannot claim that your rating of "3" or "agree" indicates that you rated the professor three times as high as your friend did since she rated the professor "1" or "strongly disagree." Again, the numbers associated with the categories only indicate their ordered relationship, not how much of a difference exists between the categories.

2.4.3 Interval and Ratio Scales

Finally, the highest level of measurement in science is achieved with scales employing numbers (**interval** and **ratio scales**). The numerical values associated with these scales permit the use of arithmetic operations such as adding, subtracting, multiplying, and dividing. In interval and ratio scales, equal differences between points on any part of the scale are equal. The only difference between the two scales stems from the fact that the interval scale employs an arbitrary zero point, whereas the ratio scale employs a true zero point. Consequently, only the ratio scale permits us to make statements concerning the ratios of numbers in the scale; for example, 4 feet are to 2 feet as 2 feet are to 1 foot. A good example of the difference between an interval and a ratio scale is a person's height as measured from a table top (interval) versus height as measured from the floor.

With both interval and ratio scales we can state exact differences between categories. This property or characteristic is particularly valuable because many statistical procedures can only be used with interval or ratio measures. Most statisticians consider the often-used variables age, education, and income as ratio variables because each has a true zero point. The first two, age and education, have a true zero point of zero years. We can say that if a wife has completed 14 years of formal education and her husband 7 years, she has had twice as much formal education as he has. We cannot, of course, assume that she has twice as much knowledge because knowledge cannot be measured in standard units.

Some statistical purists enjoy pointing out that common social science variables such as income and education can be conceived of as only ordinal level measures in some instances. They might ask, for example, "Does a $1000 raise mean the same thing to an individual whose annual income exceeds $100,000 as it does to someone whose annual income is $15,000?" Or, "Is the one-unit difference between 8 and 9 years of formal education really equivalent to the one-unit difference between 18 and 19 years of formal education?" The argument being made is that the *meaning* of the $1000 or 1 additional year of formal education differs in these examples. While the implications of these examples are interesting to consider, we will treat variables such as these as ratio measures. Also, apart from the difference in the nature of the zero point, interval and ratio scales have the same properties and will be treated alike throughout the text.

It should be clear that one of the most sought-after goals of the social scientist is to achieve measurements that are at least interval in nature. Indeed, interval scaling is assumed for most of the statistical tests reported in this book. However, although it is debatable that many of our scales achieve interval measurement, most social scientists are willing to make this assumption.

One of the characteristics of higher-order scales is that they can readily be transformed into lower-order scales. Thus the outcome of a 1-mile foot race may be expressed as time scores (ratio scale), for example, 3:56, 3:58, and 4:02. The time scores may then be transformed into an ordinal scale, for example, first-, second-, and third-place finishers. However, the reverse transformation is not possible. If we know only the order of finishing a race, for example, we cannot express the outcome in terms of a ratio scale (time scores). Although it is permissible to transform scores from higher-level to lower-level scales, it is not usually recommended since information is lost in the transformation.

2.5 Continuous and Discrete Scales

Imagine that you are given the problem of trying to determine the number of children per American family. Your scale of measurement would start with zero (no children) and would proceed, by *increments of 1* to perhaps 15 or 20. Note that, in moving from one value on the scale to the next, we proceed by *whole numbers* rather than by fractional amounts. Thus a family has either 0, 1, 2, or more children. In spite of the statistical abstraction that the American family averages 1 3/4 children, we do not know a single couple that has achieved this marvelous state of family planning.

Such scales are referred to as **discrete** or **discontinuous scales,** and they have equality of *counting units* as their basic characteristic. Thus if we are studying the number of children in a family, each child is equal with respect to providing one counting unit. Arithmetic operations such as adding, subtracting, multiplying, and dividing are permissible with discrete scales. We can say that a family with four children has twice as many children as one with two children. Observations of discrete variables are always exact so long as the counting procedures are accurate. Examples of discrete variables are: group size, the number of males in a class, the number of work stoppages in Illinois, and the number of welfare recipients in a county.

You should not assume from this discussion that discrete scales necessarily involve *only* whole numbers. However, most of the discontinuous scales used by social scientists are expressed in terms of whole numbers. For example, a political scientist tabulates the number of people who voted for a particular candidate, or a sociologist tabulates the number of kindergarten children from family units of different sizes. In each of these examples we are clearly dealing with values that proceed by whole numbers.

In contrast, a **continuous scale** is one in which there are an unlimited number of *possible* values between *any two adjacent values of the scale.* Thus if the variable is height measured in inches, then 4 inches and 5 inches would be two adjacent values of the scale. However, there can be an infinite number of intermediate values, such as 4.5 inches or 4.7 inches. If the variable is height measured in tenths of inches, then 4.5 inches and 4.6 inches are two adjacent values of the scale, but there can *still* be an infinite number of intermediate values such as 4.53 inches or 4.59 inches.

A scale in which the variable may take on an unlimited (infinite) number of intermediate values is referred to as a *continuous* scale. It is important to note that, although our measurement of discrete variables may be exact, our measure of continuous variables is always approximate. If we were measuring the attitudes of Philadelphians toward their police, for example, we would find a wide variety of attitudes ranging from "very positive" to "very negative" because no two people can have *exactly* the same attitude (Fig. 2.2). Other social science examples of continuous variables include measures of achievement motivation, political activism, religiosity, and income.

Let's consider one additional point. Continuous variables are often expressed as whole numbers and therefore appear to be discontinuous. Thus you may say that you are 5 feet, 8 inches tall and weigh 150 pounds. However, the decision to express heights to the nearest inch and weights to the nearest pound was yours. You could just as easily have expressed height to the nearest fraction of an inch and weight to the nearest ounce. You do not have such a choice when reporting such things as the number of children in a family. These *must* occur as whole numbers.

Figure 2.2 depicts the differences between continuous and discrete scales.

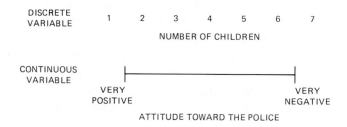

2.2 Discrete and continuous variables.

2.5.1 Continuous Variables, Errors of Measurement, and True Limits of Numbers

In our preceding discussion, we pointed out that continuously distributed variables can take on an unlimited number of intermediate values. Therefore we can never specify the exact value for any particular measurement, since it is possible that a more sensitive measuring instrument can slightly increase the accuracy of our measurements. For this reason, we stated that numerical values of continuously distributed variables are always approximate. However, it is possible to specify the limits within which the true value falls; for example, the **true limits** of a value of a continuous variable are equal to that number plus or minus one-half of the unit of measurement.

Let us look at a few examples. You have a bathroom scale, which is calibrated in terms of pounds. When you step on the scale the pointer will usually be a little above or below a pound marker. However, you report your weight to the nearest pound. Thus if the pointer were approximately three-quarters of the distance between 212 and 213 pounds, you would report your weight as 213 pounds. It would be understood that the true limit of your weight, assuming an accurate scale, falls between 212.5 and 213.5 pounds. If, on the other hand, you are measuring the weight of whales, you would probably have a fairly gross unit of measurement, say 100 pounds. Thus if you reported the weight of a whale at 32,000 pounds, you would mean that the whale weighed between 31,950 pounds and 32,050 pounds. If the scale were calibrated in terms of 1000 pounds, the true limits of the whale's weight would be between 31,500 pounds and 32,500 pounds.

2.6 Rounding

Imagine that we have obtained some data that, in the course of conducting our statistical analysis, require that we divide one number into another. There will be innumerable occasions in this course when you will be required to perform this arithmetic operation. In most cases, the answer will be a value that extends to an endless number of decimal places. For example, if we were to express the fraction 1/3 in decimal form, the result would be 0.33333 + . It is obvious that we cannot extend this series of numbers *ad infinitum*. We must terminate at some point and assign a value to the last number in the series which best reflects the remainder. When we do this, two types of problems will arise:

1. To how many decimal places do we carry the final answer?
2. How do we decide on the last number in the series?

The answer to the first question is usually given in terms of the number of significant figures. For simplicity and convenience, we will adopt the following policy with respect to rounding:

In obtaining the final answer, we should round to two more places than were in the original data. We should not round the intermediate steps.*

Thus if the original data were in whole-numbered units, we would round our answer to the second decimal. If in tenths, we would round to the third decimal, and so forth.

Once we have decided the number of places to carry our final figures, we are still left with the problem of representing the last digit. Fortunately, the rule governing the determination of the last digit is perfectly simple and explicit. If the remainder beyond that digit is greater than 5, increase that digit to the next higher number. If the remainder beyond that digit is less than 5, allow that digit to remain as it is. Let's look at a few illustrations. In each case we shall round to the second decimal place:

6.546   becomes   6.55;

6.543   becomes   6.54;

1.967   becomes   1.97;

1.534   becomes   1.53.

You may ask, "In these illustrations what happens if the digit at the third decimal place is 5?" You should first determine whether or not the digit is exactly 5. If it is 5 plus the slightest remainder, the rule holds and you must add one to the digit at the second decimal place. If it is almost, but not quite 5, the digit at the second decimal place remains the same. If it is *exactly 5, with no remainder,* then an arbitrary convention which is accepted universally by mathematicians applies: Round the digit at the second decimal place to the *nearest even number.* If this digit is already even, then it is not changed. If it is odd, then *add* one to this digit to make it even. Let's look at several illustrations in which we round to the second decimal place:

6.545001   becomes   6.55.   Why?

6.545000   becomes   6.54.   Why?

1.9652   becomes   1.97.   Why?

0.00500   becomes   0.00.   Why?

---

* Since many of you will be using calculators, you should be aware of minor differences that may occur in the final answer. These differences may be attributed to the fact that different calculators will carry the intermediate steps to a different number of decimal places. Thus a calculator that carries the intermediate steps to 4 places will probably produce a slightly different final answer than one that carries to 14 places.

0.01500 becomes 0.02. Why?

16.89501 becomes 16.90. Why?

2.7 Ratios

One of the several ratios commonly used by social scientists is the *sex ratio* or the number of males per 100 females. We find that sex ratios vary between cities, rural-urban areas, and age groups. A city such as Washington, D.C. has a relatively low sex ratio due to the large number of clerical and secretarial positions associated with governmental affairs whereas Anchorage, Alaska, which is a frontier city, is characterized by a very high sex ratio because of the high concentration of males in the work force. A ratio results from dividing one quantity by another and in the case of the sex ratio the quantities are multiplied by 100 to eliminate the decimal. Let's compute the sex ratio for the United States in 1978 when there were approximately 106.0 million males and 112.0 million females:*

$$\text{sex ratio} = \frac{\text{number of males}}{\text{number of females}} \times 100;$$

$$\text{sex ratio} = \frac{106.0 \text{ million males}}{112.0 \text{ million females}} \times 100$$

$$= 94.7.$$

We interpret our computed sex ratio as 94.7 males per 100 females. If we calculated the sex ratio by age group, we would observe the trend evident in the accompanying table. Can you explain why there are so few males per 100 females among the 65 + age group?

Age-specific
sex ratio

	SEX RATIO
Under 14	104.3
14–24	101.5
25–44	96.7
45–64	92.0
65+	68.5

* U.S. Bureau of the Census, Series P–25, No. 800.

Another ratio used by social scientists is the *dependency ratio* which indicates the number of individuals per 100 younger than 15 or older than 64 relative to those between the ages of 15 and 64. High dependency ratios characterize developing nations. Why? In Vietnam, the population totaled 53.3 million persons in 1980, of which 21.9 million or 41% were less than 15 and 2.1 million or 4% were over 64 years old. The dependency ratio in Vietnam is approximately 44.9, whereas this figure for the United States is 38.2:*

$$\frac{\text{dependency}}{\text{ratio}} = \frac{\text{persons younger than 15 or older than 64}}{\text{persons 15 to 64}} \times 100;$$

$$\frac{\text{dependency}}{\text{ratio for Vietnam}} = \frac{21.9 \text{ million} + 2.1 \text{ million}}{53.3 \text{ million}} \times 100$$

$$= 45.0.$$

A more common use of ratios involves comparing one quantity to another. Nonwhites in the United States constituted approximately 29.7 million persons in 1978 and whites 188.9 million, for a total of approximately 218.6 million persons. We can express the ratio of nonwhites to whites as 29.7:188.9 or 1:6.4. We will see that ratios provide the basis for proportions, percentages, and rates in that all are based on the relationship between two categories or quantities.

2.8 Proportions, Percentages, and Rates

A **proportion** is calculated by dividing the quantity in one category by the total of all the categories. In 1976, the most frequent cause of death among persons aged 15 to 24 was accidents. During that year 24,316 persons died following an accident (automobile, household, etc.), and a total of 37,832 in this age group died from the five leading causes of death (see Table 2.1). Given this information, we can calculate that the proportion of accidental deaths out of the total number of deaths in

---

* 1980 World Population Data Sheet of the Population Reference Bureau, Inc. In industrial nations, the numerator is frequently adjusted to persons younger than 21 because few enter the work force at age 15, as is common in developing nations. The ratio for the United States is based on the same equation as the Vietnam figure for comparative purposes.

1976 attributable to the five leading causes for the 15 to 24 age group equalled 0.643 by dividing 24,316 by 37,832.

Table 2.1   Five leading causes of death in 1976 among persons 15 to 24 years of age (both sexes)

	NUMBER	PROPORTION
Accidents	24,316	0.643
Homicide	5,038	0.133
Suicide	4,747	0.126
Malignancies	2,659	0.070
Heart disease	1,072	0.028
Total	37,832	1.000

Source: Facts of Life and Death (1978), Office of the Assistant Secretary for Health, National Center for Health Statistics.

Proportions can range in value from 0 (if there are no cases in a category) to 1.000 (if a category contains all the cases). Note also in Table 2.1 that the proportions associated with the five categories total 1.000 because they are mutually exclusive and exhaustive; hence the categories constitute a nominal level of measurement.

A proportion can be converted to a **percentage** when multiplied by 100. Consider Table 2.1. We can convert the proportion of deaths associated with suicide to a percentage by multiplying the proportion 0.126 by 100, and the proportion associated with accidents can also be converted to a percentage when multiplied by 100. Hence, 12.6% died from suicide, and 64.3% from an accident, or a total of 76.9% of the deaths reported in Table 2.1 resulted from these two causes. If we added the 13.3% who were victims of homicide, the 7.0% who died of malignancies, and the 2.8% who died of heart disease, we could account for 100% of the 37,832 deaths.

We are frequently concerned with *percentage change.* While you probably have been calculating percentage change for many years, a brief review is worthwhile. Two time periods are involved when calculating percentage change. Consider how we might calculate the percentage of population increase in a small village in Wisconsin from the 1970 to 1980 census. If the population totaled 1000 in 1970 and increased to 1710 in 1980, how do we determine the percentage increase? Our first time period is 1970 and the second 1980. Using this information and the following procedure, we can now calculate percentage change:

$$\frac{\text{percentage}}{\text{change}} = \frac{\text{(quantity at time 2)} - \text{(quantity at time 1)}}{\text{(quantity at time 1)}} \, ;$$

$$\frac{\text{percentage}}{\text{change}} = \frac{(1710 - 1000)}{1000}$$

$$= \frac{710}{1000}$$

$$= +71\%$$

The procedure works equally well if the percentage change is negative. If the population of our fictitious village declined from 1000 in 1970 to 900 in 1980, we would anticipate negative growth:

$$\frac{\text{percentage}}{\text{change}} = \frac{(900 - 1000)}{1000} = \frac{-100}{1000} = -10\%.$$

The population decline is 10%.

Note that it is very important that you not disregard the minus sign in the numerator. The denominator will always constitute your base and is the quantity at time 1. Another way of looking at the concept of percentage change is to remember that we are interested in changes from our base (in this case the population in 1970). If the quantity at time 2 is less than the base (time 1), we know that the percentage change is negative. Consider another example. Suppose you earned $8000 in 1980 and the next year received a major promotion which included an annual salary of $16,000. Your salary has doubled and the percentage increase is 100%. Demonstrate this to yourself.

We have seen the value of proportions and percentages; however, one more related value, the **rate**, allows us to examine our data in another way. For example, it allows us to determine the number of deaths by each cause for a given number of persons. We know that 64.3% of the deaths attributable to the five leading causes in the 15 to 24 age group were caused by accidents (Table 2.1). We *do not* know how many deaths per 100,000 persons in the 15 to 24 age group, for example, were caused by accidents. Or, put another way, if you are between 15 and 24 years of age, is it very likely you will die in an accident? A total of approximately 40,600,000 persons were between the ages of 15 and 24 in 1976. We know that 24,316 died of accidents; therefore we can calculate the percentage in this age group that died of an accident. You should find that the proportion of deaths attributable to accidents is approximately 0.0005989, the percentage is 0.05989 (resulted from multiplying the proportion by 100), and the rate per 100,000 is 59.89 (resulted from multiplying the proportion by 100,000).

Rates are always reported for a specific number of cases, normally per 100, per 1000, or per 100,000. The birth rate can be calculated in a variety of ways including rate of live births per 1000 population, rate per 1000 women of child-bearing age (normally 15 to 44), or what is termed an *age-specific* birth rate. So be sure you know how a rate is calculated. The age-specific birth rate is the rate of live births per 1000 women in specific age categories. In 1977 the age-specific birth rate for women 15 to 19 years was 53.7, for women 20 to 24 it was 115.2, for women 25 to 29 it was 114.2, for women 30 to 34 it dropped to 57.5, and for women 35 to 39 it dropped further to 19.2.* The importance of the age-specific birth rate is that we can determine which age groups are having disproportionate numbers of children.

Formally, a **rate** is a ratio of the occurrences in a group category to the total number of elements in the group with which we are concerned:

$$\text{rate} = \frac{\text{number of occurrences in a group category}}{\text{total number of elements in the group}}.$$

This ratio is then multiplied by a given number (normally 100, 1000, or 100,000) to determine the rate per a given number of persons or events. If you were told that the burglary rate per 1000 households in Portland, Oregon is 151, you would know immediately that the rate was calculated by forming a ratio of the number of burglaries in Portland to the total number of households in Portland and multiplying by 1000 (to give the rate per 1000).

Summary

In this chapter, we pointed out that advanced knowledge of mathematics is not a prerequisite for success in this course. A sound background in high school mathematics plus steady application to assignments should be sufficient to permit mastery of the fundamental concepts put forth in this text.

To aid the student who may not have had recent contact with mathematics, we have attempted to review some of the basic concepts of mathematics. Included in this review are: (1) the grammar of

---

* *Source:* U.S. National Center for Health Statistics, *Vital Statistics of the United States,* 1977.

mathematical notations, (2) types of numbers, (3) types of numerical scales, (4) continuous and discrete scales, (5) rounding, and (6) ratios, proportions, percentages, and rates. Students requiring a more thorough review of mathematics may refer to Appendix A.

Terms to Remember

Cardinal numbers   Numbers used to represent quantity.

Continuous scales   Scales in which the variables can assume an unlimited number of intermediate values.

Discrete scales (Discontinuous scales)   Scales in which the variables have equality of counting units.

Interval scale   Scale in which exact distances can be known between categories. The zero point in this scale is arbitrary, and arithmetic operations are permitted.

Measurement   The assignment of numbers to objects or events according to sets of predetermined (or arbitrary) rules.

Nominal numbers   Numbers used to name.

Nominal scale   Scales in which the categories are homogeneous, mutually exclusive, and unordered.

Ordinal numbers   Numbers used to represent position or order in a series.

Ordinal scale   Scale in which the classes can be rank ordered, that is, expressed in terms of the algebra of inequalities (for example, $a < b$ or $a > b$).

Percentage   A proportion that has been multiplied by 100.

Proportion   A value calculated by dividing the quantity in one category by the total of all the components.

Rate A ratio of the occurrences in a group category to the total number of elements in the group with which we are concerned.

Ratio scale Same as interval scale, except that there is a true zero point.

True limits of a number The true limits of a value of a continuous variable are equal to that number plus or minus one-half of the unit of measurement.

Variable Any characteristic of a person, group, or environment that can vary or denote a difference.

Exercises

The following exercises are based on this chapter and Appendix A.

1. It has often been speculated that Californians cope with earthquake hazards by avoidance—simply not thinking about them. Professor Ralph H. Turner, a sociologist and the director of UCLA's Institute for Social Science Research, has collected data that bear on this common belief. Here are the data collected on a number of survey questions concerned with media coverage of earthquake-related information.

   Find the proportion and percentage of individuals responding in each category to each of the questions raised in the survey.

   Do the responses appear to support the common view that Californians avoid thinking about earthquakes?

   a) Do the media provide too little, too much, or sufficient coverage about what to do if an earthquake strikes?

	NUMBER RESPONDING
Too little	386
About right	103
Too much	8
No opinion	3

   b) How about the news media's coverage on preparations for an earthquake?

	NUMBER RESPONDING
Too little	357
About right	121
Too much	14
No opinion	8

c) Do the media provide sufficent information about what the government is doing to prepare for an earthquake?

d) How about the attention the media pay to nonscientific earthquake predictions?

	NUMBER RESPONDING
Too little	413
About right	67
Too much	10
No opinion	10

	NUMBER RESPONDING
Too little	126
About right	142
Too much	215
No opinion	17

2. Find a when $b = 10$, $c = 4$, and $a + b + c = 19$.
3. Find y when $N = 4$ and $20 + N = y + 2$.
4. Find ΣX when $N = 20$, $\overline{X} = 60$, where $\overline{X} = \Sigma X/N$.
5. Find N when $\overline{X} = 90$, $\Sigma X = 360$, where $\overline{X} = \Sigma X/N$.
6. Find N when

$$\Sigma (X - \overline{X})^2 = 640, \qquad s^2 = 16, \qquad \text{where} \qquad s^2 = \frac{\Sigma (X - \overline{X})^2}{N}.$$

7. Find s^2 when

$$\Sigma (X - \overline{X})^2 = 240, \qquad N = 12, \qquad \text{where} \qquad s^2 = \frac{\Sigma (X - \overline{X})^2}{N}.$$

8. Round the following numbers to the second decimal place.

   a) 99.99500 **b)** 46.40501

   c) 2.96500 **d)** 0.00501

   e) 16.46500 **f)** 1.05499

   g) 86.2139 **h)** 10.0050

9. The local chapter of the John Birch Society lost 25 members from the previous year, and membership now totals 38. What is the percentage change?

10. There are 21 females and 18 males in your social welfare class. What is the ratio of males to females?

11. If there were 192 homicides in the Atlanta (Georgia) Standard Metropolitan Statistical Area (SMSA), and the SMSA population totaled 1,902,000, what is the homicide rate per 1000 persons?

12. The accompanying table shows the number of male and female victims of homicide between the years 1968 and 1973.

YEAR	NUMBER OF MALE VICTIMS OF HOMICIDE	NUMBER OF FEMALE VICTIMS OF HOMICIDE
1968	5106	1700
1969	5215	1801
1970	5865	1938
1971	6455	2106
1972	6820	2156
1973	7411	2575

a) Of the total number of male homicide victims during the years 1968 through 1973, find the percentage for each year.

b) Of the total number of female homicide victims during the years 1968 through 1973, find the percentage for each year.

c) Of the total number of homicide victims in 1973, find the percentage that was male.

d) Of the total number of homicide victims in the 1968 to 1973 period, find the percentage that was female.

13. Determine the value of the following expressions in which $X_1 = 4$, $X_2 = 5$, $X_3 = 7$, $X_4 = 9$, $X_5 = 10$, $X_6 = 11$, $X_7 = 14$.

a) $\displaystyle\sum_{i=1}^{4} X_i =$ **b)** $\displaystyle\sum_{i=1}^{7} X_i =$ **c)** $\displaystyle\sum_{i=3}^{6} X_i =$

d) $\displaystyle\sum_{i=2}^{5} X_i =$ **e)** $\displaystyle\sum_{i=1}^{N} X_i =$ **f)** $\displaystyle\sum_{i=4}^{N} X_i =$

14. Express the following in summation notation.

a) $X_1 + X_2 + X_3$ **b)** $X_1 + X_2 + \cdots + X_N$

c) $X_3^2 + X_4^2 + X_5^2 + X_6^2$ **d)** $X_4^2 + X_5^2 + \cdots + X_N^2$

15. Using the figures shown in the accompanying table, answer the following questions.

a) Of all the students majoring in each academic area, what percentage is female?

b) Considering only the males, what percentages are found in each academic area?

c) Considering only the females, what percentages are found in each academic area?

d) Of all students majoring in the five areas, what percentage is male? What percentage is female?

	MALE	FEMALE
Business administration	400	100
Education	50	150
Humanities	150	200
Science	250	100
Social science	200	200

16. Following is a list showing the number of births in the United States (expressed in thousands) between 1950 and 1975. Calculate the percentage of males and females for each year.

YEAR	MALES	FEMALES
1950	1824	1731
1955	2074	1974
1960	2180	2078
1965	1927	1833
1970	1915	1816
1975	1613	1531

Source: U.S. National Center for Health Statistics, *Vital Statistics of the United States,* annual.

17. Determine the square roots of the following numbers to two decimal places.

a) 160   **b)** 16   **c)** 1.60   **c)** 0.16   **e)** 0.016

18. State the true limits of the following numbers.

a) 0   **b)** 0.5   **c)** 1.0   **d)** 0.49   **e)** −5   **f)** −4.5

19. Using the values of X_i given in Exercise 13 above, show that

$$\sum_{i=1}^{N} X_i^2 \neq \left(\sum_{i=1}^{N} X_i \right)^2.$$

20. An interviewer completed 150 interviews this month. How many interviews must be completed next month for a 25% increase?

21. In 1965 a total of 2,857,000 persons were on active military duty, as compared to 2,060,000 in 1977. (*Source:* U.S. Department of Defense, Selected Manpower Statistics, annual.) Assuming these estimated numbers are exact,

a) Calculate the percentage change from 1965 to 1977.

b) The number of military personnel in 1977 is what percentage of the number in 1975?

22. Examine the accompanying table and calculate the percentage change for each of the categories from January 1974 to December 1978.

Supplemental security income for the aged, blind, and disabled: Number of persons receiving federally administered payments and total amount, 1974 and 1978

PERIOD	NUMBER OF PERSONS				AMOUNT OF PAYMENTS
	Total	Aged	Blind	Disabled	Total
January 1974	3,215,632	1,865,109	72,390	1,278,133	$365,149,000
December 1978	4,216,925	1,967,900	77,135	2,171,890	546,567,000

Source: Social Security Administration, *Social Security Bulletin,* May 1979, 42(5).

23. New York State had a total of approximately 8,715,000 males and 9,521,000 females in 1970. What was the sex ratio? (*Source:* U.S. Bureau of the Census, *Census of Population: 1970,* I.)

24. In 1980, Japan had a total population of 116.8 million persons of whom 24% were younger than 15 and 8% were over 64. What was the dependency ratio and how would you interpret your answer? (*Source:* 1980 World Population Data Sheet of the Population Reference Bureau, Inc.)

25. Examine the accompanying table and calculate the women's earnings as a percentage of men's earnings. Determine the percentage change of women's and men's earnings from 1960 to 1975.

Median annual earnings of year-round full-time workers 14 years and over by sex, 1960–1975.

YEAR	ANNUAL EARNINGS (DOLLARS)	
	Women	Men
1960	3,293	5,417
1965	3,823	6,375
1970	5,323	8,966
1975	7,504	12,758

Source: U.S. Department of Labor, Bureau of Labor Statistics, *U.S. Working Women: A Data Book, 1977.*

26. Eugene J. Kanin has studied various aspects of the behavior of sexually aggressive males. When his findings were compared with those for a sample of nonaggressive males, it was hypothesized that nonaggressive males might use other, less aggressive strategies as a means of achieving sexual relations more often than their aggressive counterparts.

The data from a high school sample of 254 nonaggressive and 87 aggressive males reveal the following frequencies with which the respondents admitted to the use of nonaggressive techniques.

	NONAGGRESSIVE MALES*	AGGRESSIVE MALES
Attempted to get girl intoxicated	23	33
Falsely promised marriage	19	7
Falsely professed love	37	39
Threatened to terminate relationship	9	8
Total number of males interviewed	N = 254	N = 87

Source: E. J. Kanin, An examination of sexual aggression as a response to sexual frustration, *Journal of Marriage and the Family,* 1967, 29, 428–433. Copyright 1967 by the American Psychological Association. Reprinted by permission.

* The sum of the frequencies do not equal N since some respondents admitted to using more than one technique and others claimed never to use such techniques.

a) Find the percentage of aggressive and nonaggressive males admitting to the use of each exploitative technique.

b) Does it appear that the hypothesis is supported?

27. The accompanying table shows the methods of suicide used by students at the University of California at Berkeley (1952–1961) and at Yale (1920–1955).

	BERKELEY	YALE
Firearms	8	10
Poisoning	6	3
Asphyxiation	4	5
Hanging	2	6
Jumping from high places	2	1
Cutting instruments	1	0
Total suicides	N = 23	N = 25

Source: R. H. Seiden, Campus tragedy: A story of student suicide, *Journal of Abnormal and Social Psychology,* 1966, 71, 389–399.

a) Calculate the percentage of suicides in which the various means of self-destruction were employed at each institution. Round to the nearest percent.

b) If the total number of students who attended Berkeley during this time period was 77,584, and for Yale it was 63,876, compute each university's suicide rate per 1000 and determine which was higher.

Descriptive Statistics

3   Frequency Distributions and Graphing Techniques

4   Percentiles

5   Measures of Central Tendency

6   Measures of Dispersion

7   The Standard Deviation and the Standard
           Normal Distribution

8   An Introduction to Contingency Tables

9   Correlation

10   Regression and Prediction

11   Multivariate Data Analysis

Frequency Distributions and Graphing Techniques

3

3.1   Grouping of Data

3.2   Cumulative Frequency and Cumulative Percentage Distributions

3.3   Graphing Techniques

3.4   Misuse of Graphing Techniques

3.5   Nominally Scaled Variables

3.6   Ordinally Scaled Variables

3.7   Interval- and Ratio-Scaled Variables

3.8   Forms of Frequency Curves

3.9   Other Graphic Representations

3.1 Grouping of Data

Imagine that you conducted a survey for the State Recreation Division assessing visitor use of a state park. Your job was to distribute the questionnaires to the drivers of cars entering the park during a two-week period and to collect the completed questionnaires as the cars exited the park gate. Prior to a complete analysis of the twenty-five questions that were included on the questionnaires, your boss wants to get a sense of the visitors' ages. Because you collected forms from over 2000 cars it is obvious that you cannot examine all of them in a short period of time, so you pull at **random** (that is, in such a way that each sample of a given size in a population has an equal chance of being selected) 75 completed questionnaires.

Altogether, the questionnaires returned by the drivers of the 75 cars sampled indicate that the cars contained a total of 110 occupants. You write down the ages of the 110 occupants with the results listed in Table 3.1.

As you glance over the data in Table 3.1, it becomes clear that your boss will not be able to make heads or tails out of them unless you organize them in some systematic fashion. It occurs to you to list all the scores from lowest to highest and then place a slash mark alongside of each age every time it occurs (Table 3.2). The number of slash marks, then, represents the frequency of occurrence of each age.

Table 3.1   Ages of 110 park visitors selected at random

40	8	15	29	58	42	49	24	56	5
48	35	33	52	17	35	38	41	42	47
11	22	66	40	34	65	23	49	30	34
16	57	52	53	15	25	31	45	49	51
52	74	47	39	57	48	53	36	48	40
64	13	37	44	44	33	30	26	37	53
21	25	0	60	69	40	52	40	62	44
33	47	24	19	46	52	40	33	73	47
39	27	43	21	23	26	44	39	26	35
69	38	61	37	50	12	25	43	71	42
45	43	35	63	29	65	3	50	45	53

Table 3.2   Frequency distribution of ages of 110 park visitors selected at random

X	f*	X	f	X	f	X	f
0†	/	21	//	42	///	63	/
1	0	22	/	43	///	64	/
2	0	23	//	44	////	65	//
3	/	24	//	45	///	66	/
4	0	25	///	46	/	67	0
5	/	26	///	47	////	68	0
6	0	27	/	48	///	69	//
7	0	28	0	49	///	70	0
8	/	29	//	50	//	71	/
9	0	30	//	51	/	72	0
10	0	31	/	52	/////	73	/
11	/	32	0	53	////	74	/
12	/	33	////	54	0		
13	/	34	//	55	0		
14	0	35	////	56	/		
15	//	36	/	57	//		
16	/	37	///	58	/		
17	/	38	//	59	0		
18	0	39	///	60	/		
19	/	40	//////	61	/		
20	0	41	/	62	/		

*The symbol for frequency in this and succeeding tables is f.
†Less than one year old.

By doing this, you have constructed a **frequency distribution** of
scores. Note that, in the present example, the ages are widely spread
out, a number of ages have a frequency of zero, and there is no visually
clear indication of a pattern. Under these circumstances, it is customary
for most researchers to *group* the scores into what are referred to as

class intervals and then obtain a frequency distribution of grouped scores.

3.1.1 Grouping into Class Intervals

Grouping into class intervals involves "collapsing the scale" and assigning scores to **mutually exclusive** and **exhaustive*** classes where the classes are defined in terms of the grouping intervals used. The reasons for grouping are threefold: (1) Unless automatic calculators are available, it is uneconomical and unwieldy to deal with a large number of cases spread out over many scores. (2) Some of the scores have such low frequency counts associated with them that we are not justified in maintaining these scores as separate and distinct entities. (3) Categories provide a concise and meaningful summary of the data.

On the negative side is, of course, the fact that grouping inevitably results in the loss of information. For example, individual scores lose their identity when we group into class intervals, and some small errors in statistics based upon grouped scores are unavoidable.

The question now becomes, "On what basis do we decide upon the grouping intervals we will use?" Obviously, the interval selected must not be so large that we lose the information provided by our original measurement. For example, if we were to divide the previously collected ages into two classes, those below 36, and those 36 and above, practically all the information about the original ages would be lost. On the other hand, the class intervals should not be so small that the purposes served by grouping are defeated. In answer to our question, there is, unfortunately, no general solution that can be applied to all data. Much of the time the choice of the number of class intervals must represent a judgment based upon a consideration of how the data will be utilized. Often the desired interval size determines the number of class intervals that will be used. Let's assume we will present the data in 5-year age intervals; an examination of the raw scores reveals that 15 class categories will include all the respondents who range in age from less than 1 to 74.

Once we have decided upon an appropriate number of class intervals for a set of data, the procedures for assigning scores to class inter-

---

* We refer to the classes as *mutually exclusive* because it is impossible for a person's score to belong to more than one class. *Exhaustive* means that all scores can be placed within the established categories.

vals are quite straightforward. Although several different techniques may be used, we shall use only one for the sake of consistency. The procedures are as follows.

Step 1 Find the difference between the highest and the lowest score values contained in the original data. Add 1 to obtain the total number of scores or potential scores. In the present example, this result is $(74 - 0) + 1 = 75$.

Step 2 Divide this figure by 15, the number of class intervals, to obtain the number of scores or potential scores in each class interval. If the resulting value is not a whole number (and it usually is not), we prefer to round to the nearest odd number so that a whole number will be at the middle of the class interval. However, this practice is far from universal and you would not be wrong if you rounded to the *nearest number*. In the present example, the number of scores for each class interval is 75/15, or 5. We shall designate the width of the class interval by the symbol i. In the example, $i = 5$.

Step 3 Take the lowest score in the original data as the minimum value in the lowest class interval. Add to this $i - 1$ to obtain the maximum score of the lowest class interval. Thus, the lowest class interval of the data is $0 - 4$.

Step 4 The next higher class interval begins at the integer following the maximum score of the lower class interval. In the present example, the next integer is 5. Follow the same procedures as in step 3 to obtain the maximum score of the second class interval. Follow these procedures for each successive higher class interval until all the scores are included in their appropriate class intervals.

Step 5 Assign each obtained score to the class interval within which it is included. The **grouped frequency distribution** appearing in Table 3.3 was obtained by using these procedures.

You will note that by grouping you have obtained an immediate picture of the distribution of ages of park visitors. For example, you know that the ages of visitors ranged from an infant less than 1 year old to someone 74 years old. Also, the majority of park visitors in your sample are adults, with a clustering of frequencies in the class intervals between the ages of 33 and 53. It is also apparent that the number of scores in the extremes tends to taper off. Thus we have achieved one of

Table 3.3   Grouped frequency distribution of ages based upon data appearing in Table 3.2

CLASS INTERVAL	f	CLASS INTERVAL	f	CLASS INTERVAL	f
0–4	2	25–29	9	50–54	12
5–9	2	30–34	9	55–59	4
10–14	3	35–39	13	60–64	5
15–19	5	40–44	17	65–69	5
20–24	7	45–49	14	70–74	3

$N = 110$

our objectives in grouping: to provide an economical and manageable array of scores.

One word of caution: Most scores with which the social scientist deals are expressed as whole numbers rather than as decimals. This is why our examples involve whole numbers. However, scores are occasionally expressed in decimal form. The simplest procedure is to treat the scores as if the decimals did not exist; in other words, treat each score as a whole number. The decimals can then be reinserted at the final step. If, in the preceding example, the highest score had been 7.4 and the lowest 0.0, the calculations would have been exactly the same. At the last step, however, the highest interval would have been changed to 7.0 to 7.4 and the lowest to 0.0 to 0.4, with corresponding changes in between. The width of the class interval would have been 0.5.

3.1.2 The True Limits of a Class Interval

In our prior discussion of the true limits of a number (Section 2.5.1), we pointed out that the true value of a number is equal to its apparent value plus or minus one-half of the unit of measurement. Of course, the same is true of these values even after they have been grouped into class intervals. Thus although we write the limits of the highest class interval as 70 to 74, the true limits of the interval are 69.5 to 74.5 (that is, the lower real limit of 70 and the upper real limit of 74, respectively). Later, when calculating certain statistics for grouped data, we shall make use of the *true limits* of the class interval.

3.2 Cumulative Frequency and Cumulative Percentage Distributions

It is often desirable to rearrange the data from a frequency distribution into a **cumulative frequency distribution,** which is a distribution that shows the **cumulative frequency** *below* the upper real limit of the corresponding class interval. Besides aiding in the interpretation of the frequency distribution, a cumulative frequency distribution is of great value in obtaining the median and the various percentile ranks of scores, as we shall see in Chapter 4.

The cumulative frequency distribution is obtained in the following manner. Looking at the data in Table 3.4, you will note that the entries in the column labeled f indicate the frequency of park visitors falling within each class interval. Each entry within the cumulative frequency column indicates the number of all cases or frequencies *less than the upper real limit* of that interval. Thus in the third class interval from the top in Table 3.4, the entry 7 in the cumulative frequency column indicates that a total of 7 visitors in your park visitor sample of 110 persons were younger than 14.5, the upper real limit of the interval 10 to 14. The entries in the cumulative frequency distribution are obtained by the simple process of successive addition of the entries in the frequency column. The value of 19 in the cumulative frequency column corresponding to the age interval 20 to 24 was obtained by adding $2 + 2 + 3 + 5 + 7 = 19$. Note that the bottom entry in the cumulative frequency column is always equal to N, or in this instance, 110. If you fail to obtain this result, you know that you have made an error in cumulating frequencies and should check your work.

The **cumulative proportion distribution** column, also shown in Table 3.4, is obtained by dividing each entry in the cumulative f column by N. When each **cumulative proportion** is multiplied by 100, we obtain a **cumulative percentage distribution.** Note that the bottom entry in the cumulative proportion column must equal 1.00 and the **cumulative %** column must equal 100% since all the cases have been included.

We are now able to confirm our earlier observation that the majority of the park visitors were adults. Consider the cumulative % column, for example, where we see that only 17% of the visitors were younger than 24.5 years of age, the upper real limit of 24. The % column clearly shows that the 40 to 44 age category included the largest percentage of visitors.

We should note that psychologists, educators, and other behavioral scientists typically use cumulative frequency and cumulative

Table 3.4   Grouped frequency distribution and cumulative frequency distribution based on data appearing in Table 3.3. $N = 110$

CLASS INTERVAL	f	%	CUMULATIVE f	CUMULATIVE PROPORTION	CUMULATIVE %
0–4	2	2	2	0.02	2
5–9	2	2	4	0.04	4
10–14	3	2	7	0.06	6
15–19	5	5	12	0.11	11
20–24	7	6	19	0.17	17
25–29	9	8	28	0.25	25
30–34	9	9	37	0.34	34
35–39	13	11	50	0.45	45
40–44	17	16	67	0.61	61
45–49	14	13	81	0.74	74
50–54	12	11	93	0.85	85
55–59	4	3	97	0.88	88
60–64	5	5	102	0.93	93
65–69	5	4	107	0.97	97
70–74	3	3	110	1.00	100

percentage tables in which the class intervals are presented from high to low. Sociologists and other social scientists normally present class intervals from low to high as in Table 3.4. Both approaches are equally valid.

3.3   Graphing Techniques

We have just examined some of the procedures involved in making sense out of a mass of unorganized data. As we pointed out, your work is usually just beginning when you have constructed frequency distributions of data. The next step, commonly, is to present the data in pictorial form so that the reader may easily determine the essential features of a frequency distribution and compare one with another if desired. Such pictures, called graphs, should *not* be thought of as substitutes for statistical treatment of data but rather as *visual aids* for thinking about and discussing statistical problems.

3.4   Misuse of Graphing Techniques

As you may be aware, graphs that present data incorrectly can mislead the reader. For example, in Fig. 3.1, the vertical (*Y*-axis or **ordinate**) and horizontal (*X*-axis or **abscissa**) have been lengthened to produce two distinctly different impressions.

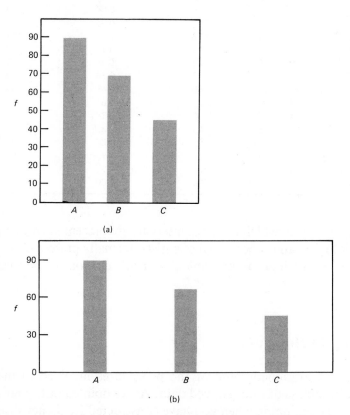

(a)

(b)

3.1 Bar graphs representing the same data but producing different impressions by varying the
relative lengths of the *Y*-axis and the *X*-axis.

It will be noted that Fig. 3.1(a) tends to exaggerate the difference in frequency counts among the three classes, whereas Fig. 3.1(b) tends to minimize these differences. The problem involves selection of scale units to represent the horizontal (*X*) and vertical (*Y*) axes. Clearly, the choice of these units is arbitrary, and the decision to make the *Y*-axis twice the length of the *X*-axis is just as correct as the opposite representation. It is clear that in order to avoid problems it is necessary to present the data in a straightforward manner.

3.5 Nominally Scaled Variables

The **bar graph,** illustrated in Fig. 3.2, is a graphic device used to represent data that are either nominally or ordinally scaled. A vertical bar is drawn* for each category, in which the *height* of the bar represents the number of members of that class. If we arbitrarily set the width of each bar at one unit, the *area* of each bar may be used to represent the frequency for that category. Thus the total area of all the bars is equal to N or 100%.

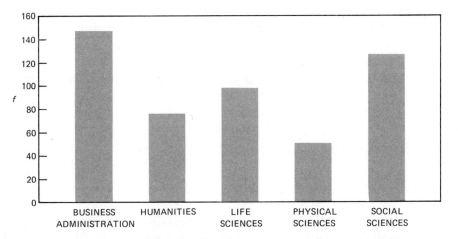

3.2 Number of students enrolled in introductory sociology courses who are majoring in the various academic fields (hypothetical data).

---

* Bar graphs are sometimes drawn horizontally (this has an advantage in cases where the number of cases (or classes) is large and the list may occupy a full page in length). Nevertheless the vertical array (as shown in Fig. 3.2) is more often used (and more easily understood at sight) because of its adaptability to a histogram or a frequency curve (see Figs. 3.4 and 3.6).

In preparing frequency distributions of nominally scaled variables, you must keep two things in mind:

1. No order is assumed to underlie nominally scaled variables. Thus the various categories can be represented along the horizontal axis (*X*-axis) in any order you choose. We prefer to arrange the categories alphabetically to eliminate any possibility of personal factors entering into the decision.

2. The bars should be separated rather than contiguous, so that any implication of continuity among the categories is avoided.

3.6 Ordinally Scaled Variables

It will be recalled that the scale values of ordinal scales carry the implication of an ordering that is expressible in terms of the algebra of inequalities (greater than, less than). In terms of our preceding discussion, ordinally scaled variables should be treated in the same way as nominally scaled variables, except that the categories should be placed in their naturally occurring order along the horizontal (*X*) axis. Figure 3.3 illustrates the use of the bar graph with an ordinally scaled variable.

3.7 Interval- and Ratio-Scaled Variables

3.7.1 Histogram

It will be recalled that interval- and ratio-scaled variables differ from ordinally scaled variables in one important way, that is, equal differences in scale values are equal. This means that we permit the vertical bars to touch one another in graphic representations of interval- and ratio-scaled frequency distributions. Such a graph is referred to as a **histogram,** and it replaces the bar graph used with nominal and ordinal variables. Figure 3.4 illustrates the use of the histogram with a discretely distributed ratio-scaled variable.

We noted previously (Section 3.5) that frequency may be represented either by the area of a bar or by its height. However, there are many graphic applications in which the height of the bar, or the ordinate, may give misleading information concerning frequency. Con-

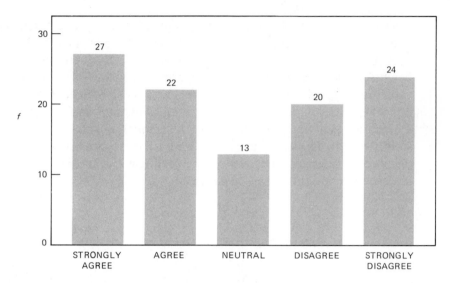

3.3 Distribution of 106 physicians' responses to the question, "Should venereal disease treatment be made available to minors without the knowledge of their parents?" (hypothetical data).

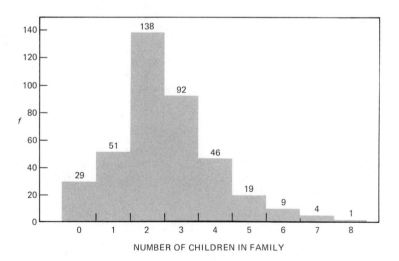

3.4 Frequency distribution of the number of children per family among 389 families surveyed in a small suburban community (hypothetical data).

sider Fig. 3.5 which shows the data grouped into *unequal* class intervals and the resulting histogram. The widths of the bars in Fig. 3.5 have been drawn proportional to the size of the class intervals, and the heights of the bars indicate the relative frequency within each age category. Thus the 30- to 34-year-old category is represented by a bar 2½ times wider than the 14- to 15-year-old category. In this instance the area under each bar does not reflect the frequency of cases, and confusion is sure to result. In general it is advisable to construct histograms using equal intervals whenever we are dealing with variables in which an underlying continuity may be assumed.

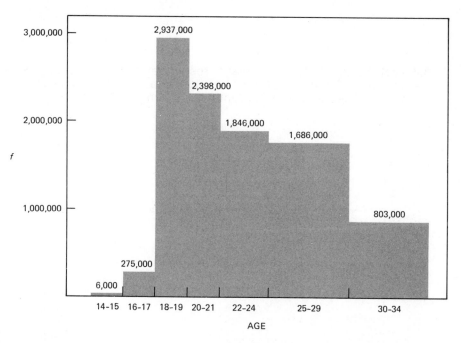

3.5 Frequency distribution of U.S. college students' ages in October, 1976. *Source:* Digest of Educational Statistics, 1977–1978, National Center for Education Statistics, p. 92.

3.7.2 Frequency Curve (Frequency Polygon)

We can readily convert the histogram into another commonly used form of graphic representation, the **frequency curve,** by joining the midpoints of the bars with straight lines. However, it is not necessary to construct a histogram prior to the construction of a frequency curve. All you need to do is place a dot where the tops of the bars would have been, and join these dots. Some people prefer to use the histogram only for discrete distributions and the frequency curve for distributions in which the underlying continuity is explicit or may be assumed. Figure 3.6 shows a frequency curve superimposed on a histogram that is based upon the grouped frequency distribution appearing in Table 3.3. In practice, we prefer to use the histogram for discrete distributions and the frequency curve for distributions in which the underlying continuity is explicit or may be assumed. The heights of the bars are determined by the data in Table 3.3

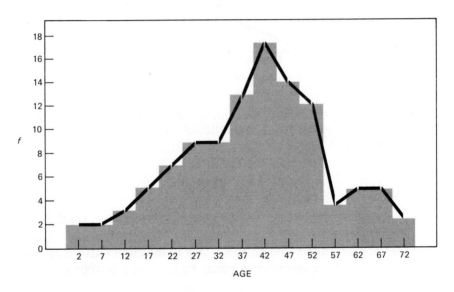

3.6 Frequency curve and histogram based on the data appearing in Table 3.3. The values on the horizontal axis represent the midpoints of the class intervals in Table 3.3.

3.7.3 Cumulative Frequency Curve (Ogive)

In Section 3.2 we demonstrated the procedures for constructing cumulative frequency and cumulative percentage distributions. The corresponding graphic representations are the **cumulative frequency curve** (**ogive**) and the cumulative percentage curve. These are both combined in Fig. 3.7, with the left-hand Y-axis showing cumulative frequencies and the right-hand Y-axis showing cumulative percentages.

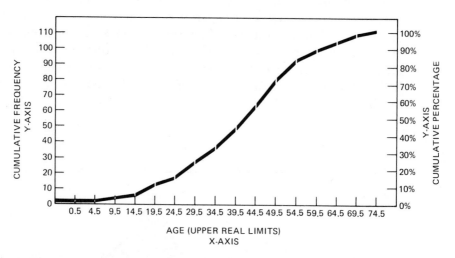

3.7 Cumulative frequency and cumulative percentage curve based upon cumulative frequency and cumulative perecentage distributions appearing in Table 3.4.

There are two important points to remember: (1) the cumulative frequencies are plotted against the *upper real limit* of each interval, and (2) the maximum value on the Y-axis in the cumulative frequency curve is N, and in the cumulative percentage curve it is 100%.

3.8 Forms of Frequency Curves

Frequency curves may take on an unlimited number of different shapes. However, many of the statistical procedures discussed in the text assume a particular form of distribution: namely, the bell-shaped **normal curve.**

In Fig. 3.8, several forms of bell-shaped distributions are shown. Curve (a), which is characterized by a piling up of scores in the center of the distribution, is referred to as a **leptokurtic** distribution. In curve (c), in which the opposite condition prevails, the distribution is referred to as **platykurtic.** And finally, curve (b) takes on the form of a bell-shaped normal curve and is referred to as a **mesokurtic** distribution.

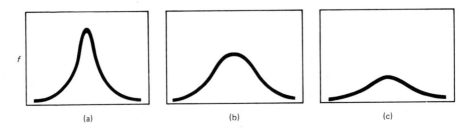

3.8 Three forms of bell-shaped distributions: (a) leptokurtic, (b) mesokurtic, and (c) platykurtic.

The normal curve is referred to as a symmetrical distribution, since, if it is folded in half, the two sides will coincide.* Not all symmetrical curves are bell shaped, however. Three different nonnormal symmetrical curves are shown in Fig. 3.9.

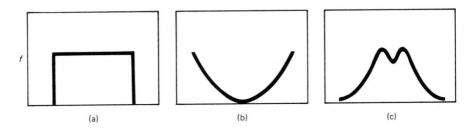

3.9 Illustrations of three nonnormal symmetrical frequency curves.

* A theoretical bell-shaped curve called a standard normal distribution will be discussed in Chapter 7.

Certain distributions have been given names, for example, that in Fig. 3.9(a) is called a *rectangular* distribution and that in Fig. 3.9(b) a U-distribution. Incidentally, the distribution appearing in Fig. 3.9(c) is found when a variable has a high concentration of frequencies around two separate values or if the frequency distributions of two different populations are represented in a single graph.* For example, a frequency distribution of average adult male and female earnings might yield a curve similar to Fig. 3.9(c) because the average annual income of males is considerably higher than that of females. Hence, the two peaks would reflect the income differential between males and females.

When a distribution is not symmetrical and "tails" off at one end, it is said to be **skewed.** If we say that a distribution is **positively skewed** or skewed to the right, we mean the distribution has relatively fewer frequencies at the high end of the horizontal axis. If, on the other hand, we say that a distribution is **negatively skewed** or skewed to the left, we mean there are relatively fewer frequencies at the low end of the horizontal axis. Figure 3.10(a) is positively skewed, and Fig. 3.10(b) is negatively skewed.

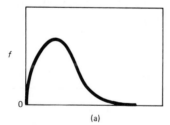 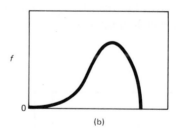

3.10 Illustrations of skewed frequency curves (scores increase from left to right).

It is not always possible to determine by inspection whether or not a distribution is skewed. Consider the frequency curve and

---

* This is sometimes called a bimodal distribution.

histogram we plotted in Fig. 3.6 for the sample of park visitors. Would you say that the curve is skewed to the left or right? Of course in this instance it is rather difficult to be certain. There is a precise mathematical method for determining both direction and magnitude of skew. It is beyond the scope of this book to go into a detailed discussion of this topic.

3.9 Other Graphic Representations

Throughout this chapter we have been discussing graphic representations of frequency distributions. However, other types of data are frequently collected by social scientists. We shall briefly discuss a few graphic representations of such data. A frequently used graphic representation is the *pictograph* as represented by Figs. 3.11 and 3.12. Figure 3.11 clearly presents the attrition of students from the formal educational system. This adaptation of a bar graph allows the reader to see at a glance that considerable attrition occurs between the fifth grade and graduation from college. Indeed, of 10 students who were in the fifth grade in 1968, only 2.4 were likely to earn a college degree in 1980. Figure 3.12 depicts the declining numbers of elevator operators needed in the United States. In 1985, the projection is that only 25,000 persons will be employed as elevator operators due to the increased number of automatic elevators and a decline in the manual models.

The *pie chart* (see Fig. 3.13) involved dividing the "pie" or circle into its component parts to represent the distribution of a nominal level variable. A protractor is used to divide the 360 degrees in a circle into percentages to reflect the actual distribution of the variable. For example, bus travel constitutes 9% of the trips taken by the handicapped, and the corresponding segment of the pie must be 32.4 degrees since 1% equals 3.6 degreees (360/100 = 3.6 and 9 × 3.6 = 32.4).

Finally, an example of a *trend chart* is provided by the U.S. population projections in Fig. 3.14. The four series have been determined by altering the three components of population change—births, deaths, and net immigration. Series II, II-*x,* and III are considered the most realistic projections, with Series I assuming an extremely high birth rate in the future.

FOR EVERY 10 PUPILS IN THE 5th GRADE IN FALL 1968.

9.8 ENTERED THE 9th GRADE IN FALL 1972.

8.7 ENTERED THE 11th GRADE IN FALL 1974.

7.5 GRADUATED FROM HIGH SCHOOL IN 1976.

4.7 ENTERED COLLEGE IN FALL 1976.

2.4 WERE LIKELY TO EARN BACHELOR'S DEGREES IN 1980.

3.11 Estimated retention rates, fifth grade through college graduation: United States, 1968–1980. *Sources:* U.S. Department of Health, Education, and Welfare, National Center for Education Statistics, *Biennial Survey of Education in the United States; Statistics of State School Systems; Fall Statistics of Public Elementary and Secondary Day Schools;* and unpublished data.

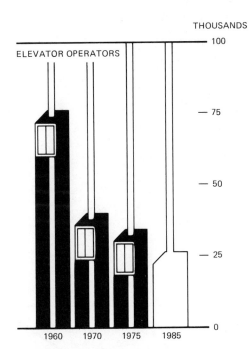

THOUSANDS

3.12 Number of elevator operators,
actual and projected for given years.
Sources: 1960–1970 Decennial
Census, Bureau of the Census;
1970–1985, Bureau of Labor
Statistics. Data for 1970 and 1985
are unpublished; data for 1975 are in
Employment and Earnings, January
1976, p. 11, as cited in Bureau of
Labor Statistics Bulletin 1919.

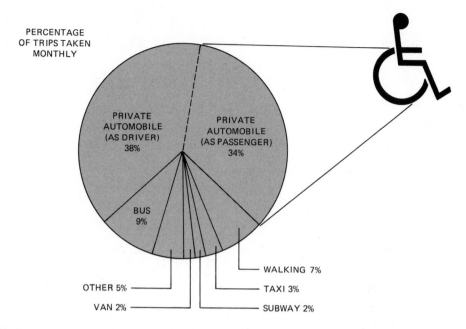

PERCENTAGE
OF TRIPS TAKEN
MONTHLY

PRIVATE
AUTOMOBILE
(AS DRIVER)
38%

PRIVATE
AUTOMOBILE
(AS PASSENGER)
34%

BUS
9%

OTHER 5%

VAN 2%

WALKING 7%

TAXI 3%

SUBWAY 2%

3.13 How the handicapped travel. ©1979 by The New York
Times Company, reprinted by permission.

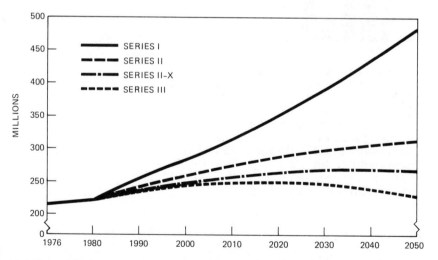

3.14 United States Population—Projections to 2050. *Source:* U.S.
Bureau of the Census, current Population Reports, series p-25.

Summary

This chapter was concerned with the techniques used in making sense out of a mass of data. We demonstrated the construction of frequency distributions of scores and presented various graphing techniques. When the scores are widely spread out, many have a frequency of zero. It is customary to group scores into class intervals. The resulting distribution is referred to as a grouped frequency distribution.

The basis for arriving at a decision concerning which grouping units to use and which procedures to follow to construct a grouped frequency distribution were discussed and demonstrated. It was seen that the true limits of a class interval are obtained in the same way as the true limits of a score. The procedures for converting a frequency distribution into a cumulative frequency distribution, a cumulative proportion distribution, and a cumulative percentage distribution were demonstrated.

We also reviewed the various graphing techniques used in the social sciences. The basic purpose of graphical representation is to provide visual aids for thinking about and discussing statistical problems. The primary objective is to present data in a clear, unambiguous fashion so that the reader may determine at a glance the relationships that we want to portray.

We discussed devices that mislead the unsophisticated reader. The use of the bar graph with nominally and ordinally scaled variables, and the use of the histogram and the frequency curve with continuous and discontinuous ratio- or interval-scaled variables were covered. Various forms of normally distributed data, non-bell-shaped symmetrical distributions, and asymmetrical or skewed distributions were presented. Finally, we discussed and demonstrated several graphic representations of data, other than frequency distributions, commonly used in the social sciences.

Terms to Remember

Abscissa (X-axis)   Horizontal axis of a graph.

Bar graph   A form of graph that uses bars to indicate the frequency of occurrence of observations within each nominal or ordinal category.

Cumulative frequency   The number of cases (frequencies) at and below a given point.

Cumulative frequency curve (ogive)   A curve that shows the number of cases below the upper real limit of an interval.

Cumulative frequency distribution   A distribution that shows the cumulative frequency below the upper real limit of the corresponding class interval.

Cumulative percentage   The percentage of cases (frequencies) at and below a given point.

Cumulative percentage distribution   A distribution that shows the cumulative percentage below the upper real limit of the corresponding class interval.

Cumulative proportion   The proportion of cases (frequencies) at and below a given point.

Cumulative proportion distribution   A distribution that shows the cumulative proportion below the upper real limit of the corresponding class interval.

Exhaustive   A grouped frequency distribution is said to be exhaustive when all possible scores can be placed within the established categories.

Frequency curve (Frequency polygon)   A form of graph, representing a frequency distribution, in which a continuous line is used to indicate the frequency of the corresponding scores.

Frequency distribution   A frequency distribution shows the number of times each score occurs when the values of a variable are arranged in order according to their magnitudes.

Grouped frequency distribution   A frequency distribution in which the values of the variable have been grouped into class intervals.

Histogram   A form of bar graph used with interval- or ratio-scaled frequency distribution.

Leptokurtic distribution   Bell-shaped distribution characterized by a piling up of scores in the center of the distribution.

Mesokurtic distribution   Bell-shaped distribution; "ideal" form of normal curve.

Mutually exclusive   Events A and B are said to be mutually exclusive if both cannot occur simultaneously.

Negatively skewed distribution   Distribution that has relatively fewer frequencies at the low end of the horizontal axis.

Normal curve   A frequency curve with a characteristic bell-shaped form.

Ogive   A cumulative frequency curve that shows the number of cases below the upper real limit of an interval.

Ordinate (Y-axis)   Vertical axis of a graph.

Platykurtic distribution   Frequency distribution characterized by a flattening in the central position.

Positively skewed distribution   Distribution that has relatively fewer frequencies at the high end of the horizontal axis.

Random   A method of selecting samples so that each sample of a given size in a population has an equal chance of being selected.

Skewed distribution   Distribution that departs from symmetry and tails off at one end.

Exercises

1. Give the true limits and the width of interval for each of the following class intervals.

   a) 8–12   **b)** 6–7   **c)** 0–2

   d) 5–14   **e)** $(-2)-(-8)$ **f)** 2.5–3.5

   g) 1.50–1.75 **h)** $(-3)-(+3)$

2. For each of the following sets of measurements, state (a) the best width of class interval (i), (b) the apparent limits of the lowest interval, (c) the true limits of that interval.

   i) 0 to 106   **ii)** 29 to 41   **iii)** 18 to 48

   iv) -30 to $+30$ **v)** 0.30 to 0.47   **vi)** 0.206 to 0.293

3. Given the following list of scores in a statistics examination, use $i = 5$ for the class intervals and (a) set up a frequency distribution; (b) list the true limits of each interval; (c) prepare a cumulative frequency distribution; and (d) prepare a cumulative percentage distribution.

SCORES ON A STATISTICS EXAMINATION									
63	88	79	92	86	87	83	78	40	67
68	76	46	81	92	77	84	76	70	66
77	75	98	81	82	81	87	78	70	60
94	79	52	82	77	81	77	70	74	61

4. Using the data in Exercise 3, set up frequency distributions with the following:

Discuss the advantages and the disadvantages of employing these widths.

    a) $i = 1$ (ungrouped frequency distribution)   **b)** $i = 3$

    c) $i = 10$ **d)** $i = 20$

5. The case loads for social workers vary greatly. The accompanying table lists the number of clients to whom 40 social workers have been assigned: (a) Construct a grouped frequency distribution. (b) List the true limits of each interval. (c) Describe the distribution in words.

68	57	68	79	92	86	87	83	78	67
84	92	63	46	81	92	77	84	86	52
51	46	77	98	81	82	81	87	42	85
90	37	94	52	82	77	81	77	94	49

6. Several entries in a frequency distribution of children's ages are 1 to 5, 6 to 10, 11 to 15. (a) What is the width of the interval? (b) What are the lower and upper real limits of each of these three intervals?

7. Discuss the value of constructing a grouped frequency distribution from a set of ungrouped scores. Can you think of a situation when you would not want to group a set of data? Why?

8. Construct a grouped frequency distribution, using 5 to 9 as the lowest class interval, for the following scores on a job satisfaction scale. List the width and real limits of the highest class interval. If high scores indicate high satisfaction and it is possible to score as low as 0 or as high as 75, how would you characterize this distribution of scores?

67	63	64	57	56	55	53	53	54	54
45	45	46	47	37	23	34	44	27	44
45	34	34	15	23	43	16	44	36	36
35	37	24	24	14	43	37	27	36	26
25	36	26	5	44	13	33	33	17	33

9. Do Exercise 8 again, using 3 to 7 as the lowest class interval. Compare the resulting frequency distribution with those of Exercises 8, 10, and 11.

10. Repeat Exercise 8, using $i = 2$. Compare the results with those of Exercises 8, 9, and 11.

11. Repeat Exercise 8, using $i = 10$. Compare the results with those of Exercises 8, 9, and 10.

12. Give an example of a variable that would have a distribution that would be characterized as

    a) Normal.   **b)** U-shaped

    c) Positively skewed.   **d)** Negatively skewed.

    e) Rectangular.

13. Given the following frequency distribution of the number of previous arrests for 50 inmates at a state prison, draw a histogram, a frequency curve, and a cumulative frequency curve (ogive).

CLASS INTERVAL	f	CLASS INTERVAL	f
0–2	0	15–17	3
3–5	17	18–20	1
6–8	15	21–23	1
9–11	8	24–26	0
12–14	4	27–29	1

14. Given the information in the accompanying table, draw a bar graph and discuss the reasons that might contribute to the variation in the percentages.

Percentage of employed women in each occupation group with year-round full-time jobs in 1975

OCCUPATIONAL GROUP	PERCENTAGE WHO WORKED YEAR ROUND, FULL TIME
Professional-technical	52.0
Managerial-administrative, except farm	64.5
Sales	25.8
Clerical	49.6
Craft	43.1
Operatives, except transport	38.7
Transport equipment operatives	17.4
Nonfarm laborers	32.7
Service, except private household	26.5
Private household	13.1
Farm	25.3

Source: Bureau of Labor Statistics, *U.S. Working Women: A Data Book,* 1977.

15. Given the accompanying table, draw three histograms and compare the results for the three columns of data. Which do you believe more accurately portrays the real picture?

Estimated number of illegitimate live births and illegitimacy rates and ratios: Selected years, 1940–1975

YEAR	NUMBER	RATE PER 1000 UNMARRIED WOMEN AGED 14 TO 44 YEARS	RATIO PER 1000 TOTAL LIVE BIRTHS
1975	447,900	24.8	142.5
1970	398,700	26.4	106.9
1960	224,300	21.6	52.7
1950	141,600	14.1	39.8
1940	89,500	7.1	37.9

Source: Public Health Service, *Facts of Life and Death,* DHEW Publication No. (PHS) 79-1222, November 1978.

16. Describe the types of distributions you would expect if you were to graph each of the following:

    a) Annual incomes of American families.

    b) The SAT or ACT scores of entering freshmen at your school.

    c) The number of minutes spent studying per week by students in your school.

    d) The number of months that criminals are sentenced to prison for burglary in your state. (This assumes your state has not adopted uniform sentencing!)

17. Construct a bar graph to present the data in the accompanying table. (*Hint:* Consider superimposing the black population figures on the total figures.) Also construct a trend chart that summarizes the relationship between the black and total population since 1790.

Total resident population in the United States for selected years: 1790 to 1975

| YEAR | MILLIONS OF PERSONS | |
	Total	Black
1790	3.9	0.8
1860	31.4	4.4
1870	39.8	5.4
1890	62.9	7.5
1900	76.2	8.8
1910	92.2	9.8
1920	106.0	10.5
1930	123.2	11.9
1940	132.2	12.9
1950	151.3	15.0
1960	179.3	18.9
1970	203.2	22.6
1975	212.6	24.4

Source: Bureau of the Census, *The Social and Economic Status of the Black Population in the United States: An Historical View, 1790–1978,* Series P-23, No. 80.

18. Construct a trend chart for the accompanying food stamp data. Has the number of participants increased more rapidly than the total retail value or the cost to participants? Construct a cumulative frequency distribution for the total retail value column.

Federal Food Stamp Program: 1961–1978

YEAR (ENDING JUNE 30)	PARTICIPANTS (1000)	VALUE OF STAMPS ISSUED	
		Total retail value (million $)	Cost to participants (million $)
1961	50	1	
1962	141	35	22
1963	358	50	31
1964	360	73	44
1965	633	85	53
1966	1,218	174	109
1967	1,832	296	190
1968	2,402	452	279
1969	3,222	603	374
1970	6,457	1,090	540
1971	10,549	2,713	1,190
1972	11,594	3,309	1,512
1973	12,107	3,884	1,753
1974	13,524	4,727	2,009
1975	19,197	7,266	2,880
1976	17,982	8,700	3,373
1977	16,097	8,340	3,282
1978 (preliminary)	15,270	8,311	3,146

Source: U.S. Department of Agriculture, Food and Nutrition Service, Agricultural Statistics, as cited in U.S. Bureau of the Census, Statistical Abstract of the United States: 1979, Table 130.

19. Using the following data, construct a pie chart.

Attendance figures for selected spectator sports, 1974

SPORTS	ATTENDANCE (IN THOUSANDS)
Major league baseball	30,026
Professional basketball	9,204
Professional football	10,675
Horse racing	75,800
Greyhound racing	16,274

Source: U.S. Department of Commerce, Social Indicators 1976, 514.

20. Figures on birthrates are usually given in terms of the number of births per thousand in the population. This table shows the birthrate data at the end of each decade since the turn of the century, rounded to the

nearest whole number. (*Source:* Commission on Population Growth and the American Future.)

a) Plot these data on a line chart.

b) Is any general trend in birthrate discernible?

YEAR	NUMBER OF BIRTHS PER 1000 POPULATION
1900	32
1910	30
1920	28
1930	21
1940	19
1950	24
1960	24
1970	18

Percentiles

4

4.1 Introduction

4.2 Cumulative Percentiles and Percentile Rank

4.3 Percentile Rank and Reference Group

4.4 Centiles, Deciles, and Quartiles

4.1 Introduction

A declining birth rate and an increasing life span have resulted in major changes in the age composition of the U.S. population during the past century. Suppose someone told you that in 1985 approximately 24.7 million people in the United States would be 65 years of age or older. What would be your reaction? Would you decide that the elderly are not nearly as numerous as you thought? Or that the United States is fast becoming a nation of elderly? You should conclude that you do not have enough information to offer a knowledgeable comment.

It should be clear that by itself a figure such as 24.7 million is not very meaningful. It takes on much more meaning when it can be compared to the total population and to the population of other age categories. If you were told that 89% of the U.S. population was less than 65 years old, you would have a frame of reference for interpreting the information. You would know the **percentile rank** of the 65-year-old persons. The percentile rank of a score, then, represents the percent of cases in a distribution that had scores (in this case, ages) at or *lower* (younger) *than the one cited*. Thus to say that an age of 65 has a percentile rank of 89 indicates that 89% of the population is less than 65.

Consider one more example before we continue. Suppose you took the Advanced Graduate Record Examination in Sociology and received a score of 510. By itself the score would not mean much, but assume that a score of 510 put you at the 51st percentile. Now you know that 51% of the persons who took the examination on the same day received a lower score than you and 49% received a higher score. Knowing the percentile rank of a score allows you to compare it with other scores.

4.2 Cumulative Percentiles and Percentile Rank

4.2.1 Obtaining the Percentile Rank of Scores from a Cumulative Percentage Graph

In Chapter 3, we learned how to construct cumulative frequency and cumulative percentage distributions. If we were to graph a cumulative percentage distribution, we could read the percentile ranks directly from the graph. Note that the reverse is also true; that is, given a percentile rank, we could read the corresponding scores. Figure 4.1 displays a graphic form (an ogive) of the cumulative percentage distribution presented in Table 4.1.

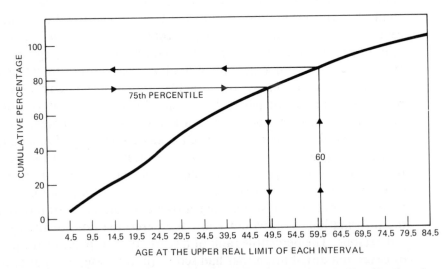

4.1 Graphic representation of a cumulative percentage distribution (projected age composition of the United States population in 1985). Persons over 85 have been excluded. (Note: The cumulative percentage corresponding to a given score is the same as the percentile rank of that score.) *Source:* U.S. Bureau of the Census, Current Population Reports, Series P-25, No. 704, "Projections of the Population of the United States: 1977–2050," U.S. Government Printing Office, Washington, D.C., 1977 (Table 8).

Table 4.1 Grouped Frequency Distribution and Cumulative Frequency Distribution of the Projected Age Composition of the United States in 1985

AGE INTERVAL	f (IN MILLIONS)	%	CUMULATIVE f (IN MILLIONS)	CUMULATIVE %
0–4	18.8	8.2	18.8	8.2
5–9	16.3	7.1	35.1	15.3
10–14	16.6	7.2	51.7	22.5
15–19	18.0	7.8	69.7	30.3
20–24	20.5	8.9	90.2	39.2
25–29	20.6	8.9	110.8	48.1
30–34	19.3	8.4	130.1	56.5
35–39	17.3	7.5	147.4	64.0
40–44	14.1	6.1	161.5	70.1
45–49	11.5	5.0	173.0	75.1
50–54	10.9	4.7	183.9	79.8
55–59	11.1	4.8	195.0	84.6
60–64	10.6	4.6	205.6	89.2
65–69	9.2	4.0	214.8	93.2
70–74	7.3	3.2	222.1	96.4
75–79	5.1	2.2	227.2	98.6
80–84	3.1	1.4	230.3	100.0

Note: Excludes the 2,588,000 persons over 85 years of age because the 85+ age category as reported by the U.S. Bureau of the Census does not have an upper limit and would not be comparable.

To illustrate, imagine that we wanted to determine the percentile rank of 60-year-old persons. We locate 60 along the horizontal axis and construct a perpendicular at that point so that it intercepts the curve. From that point on the curve, we read directly across on the scale to the left and see that the percentile rank is approximately 85. On the other hand, if we wanted to know the score at a given **percentile,** we could reverse the procedure. For example, what is the score at the 75th percentile? We locate the 75th percentile on the vertical axis and read directly to the right until we meet the curve; at this point, we construct a line perpendicular to the horizontal axis, and read the value on the scale of scores. In the present example, it can be seen that the age at the 75th percentile is approximately 49.

4.2.2 Obtaining the Percentile Rank of Scores Directly

We are often called upon to determine the percentile rank of scores without the assistance of a cumulative percentage curve, or with greater precision than is possible with a graphical representation.

Using the grouped frequency distribution found in Table 4.1, let us determine directly the percentile rank of an age of 60 which we previously approximated by the use of the cumulative percentage curve. The first thing we should note is that an age of 60 falls within the interval 60 to 64. The total cumulative frequency below that interval is 195 million. Since a percentile rank of a score is defined symbolically as

$$\text{percentile rank} = \frac{\text{cum } f}{N} \times 100, \qquad (4.1)$$

it is necessary to find the precise cumulative frequency corresponding to an age of 60. It is clear that the cumulative frequency corresponding to an age of 60 lies somewhere between 195 million and 205.6 million, the cumulative frequencies at both extremes of the interval. We must now establish the exact cumulative frequency of an age of 60 within the interval 60 to 64. In doing this, we are actually trying to determine the proportion of distance that we must move into the interval in order to find the number of cases included up to an age of 60.

An age of 60 is 0.5 years above the lower real limit of the interval (that is $60 - 59.5 = 0.5$). Since there are 5 ages within the interval (60, 61, 62, 63, and 64), 60 years of age is 0.5/5 of the distance through the interval. We now make a very important assumption: that is *that the cases or frequencies within a particular interval are evenly distributed throughout that interval.* Since there are 10.6 million cases within the interval, we may now calculate that an age of 60 is $(0.5/5) \times 10.6$, or the 1.06-millionth case within the interval. In other words, the frequency 1.06 million corresponds to 60 years of age. We already know from Table 4.1 that in 1985, 195.0 million persons will be younger than 60 years of age or will be less than the lower real limit of the interval 60 to 64. Adding the two together $(195 + 1.06)$, we find that an age of 60 has a cumulative frequency of exactly 196.06 or, when rounded, 196.1 million. Substituting 196.1 into Eq. (4.1), we obtain

$$\text{percentile rank of 60 years of age} = \frac{196.1}{230.3} \times 100 = 85.15.$$

You will note that our answer, when rounded to the nearest percentile, agrees with the approximation obtained by the use of the graphical representation of a cumulative percentage distribution (Fig. 4.1).

Equation (4.2) presents a generalized equation for calculating the percentile rank of a given score.

$$\text{Percentile rank} = \frac{\text{cum} f_{ll} + \dfrac{X - X_{ll}}{i} (f_i)}{N} \times 100 \qquad (4.2)$$

where

> $\text{cum} f_{ll}$ = cumulative frequency at the lower real limit of the interval containing X,
>
> X = given score,
>
> X_{ll} = score at lower real limit of interval containing X,
>
> i = width of interval,
>
> f_i = number of cases within the interval containing X.

4.2.3 Finding the Score
Corresponding to a Given Percentile Rank

What if you want to know the age above and below which 50% of the population will lie in 1985? We are interested in the age that corresponds to the 50th percentile.

To obtain the answer, we must work in the reverse direction, from the cumulative frequency scale to the scale of ages. The first thing we must determine is the cumulative frequency corresponding to the 50th percentile (which we will later refer to as the median of a distribution). Once we know the cumulative frequency, we will locate the interval that contains the cumulative frequency and then establish the age in question. Here is how we calculate the answer. It follows algebraically from Eq. (4.1) that

$$\text{cum} f = \frac{\text{percentile rank}}{100} \times N. \qquad (4.3)$$

Since we are interested in an age at the 50th percentile and N is 230.3 million, the cumulative frequency of a score at the 50th percentile is

$$\text{cum} f = \frac{50 \times 230.3}{100} = 115.15 \text{ or } 115.2 \text{ million.}$$

Referring to Table 4.1, we see that the frequency 115.2 million is in the age interval with the real limits of 29.5 to 34.5. It is 4.4 million frequencies into the interval since the cum f at the lower real limit of the interval is 110.8 million, which is 4.4 million less than 115.2. There are

19.3 million persons in the interval. Thus the frequency 110.8 million is 4.4/19.3 of the way through an interval with a lower real limit of 29.5 and an upper real limit of 34.5. In other words, it is 4.4/19.3 of the way through the 5 ages in the interval. Expressed in age units, (4.4/19.3) × 5 or 0.228 × 5 = 1.14. By adding 1.14 to 29.5, we obtain the age at the 50th percentile, which is 30.64.

For students desiring a generalized method for determining scores (ages) corresponding to a given percentile, Eq. (4.4) should be helpful.

$$\text{Score at a given percentile} = X_{\text{ll}} + \frac{i\,(\text{cum } f - \text{cum } f_{\text{ll}})}{f_i} \quad \textbf{(4.4)}$$

where

X_{ll} = score at lower real limit of the interval containing cum f,
   i = width of the interval,
cum f = cumulative frequency of the score,
cum f_{ll} = cumulative frequency at the lower real limit of the interval containing cum f,
   f_i = number of cases within the interval containing cum f.

To illustrate the use of the equation, consider an example with which we are already familiar. What age is at the 85.15th percentile? First, by using Eq. (4.3) we obtain

$$\text{cum } f = \frac{85.15 \times 230.3}{100} = 196.1 \text{ million.}$$

If you refer to Table 4.1, you will see that our calculated cum f of 196.1 million is reasonable, given the 85.15th percentile. Try to get in the habit of ensuring that your answers are correct.

The age at the lower real limit of the interval containing the frequency 196.1 million is 59.5; i is 5; cum f to the lower real limit of the interval is 195.0; and the number of cases within the interval is 10.6 million. Substituting these values into equation (4.4), we obtain

$$\text{age at 85.15th percentile} = 59.5 + \frac{5(196.1 - 195.0)}{10.6}$$

$$= 59.5 + 0.5 = 60 \text{ years of age.}$$

Note that this is the age from which we previously obtained the percentile rank 85.15 and that this equation illustrates, incidentally, a good procedure for checking the accuracy of your calculations. In other words, whenever you find the percentile rank of a score (age, etc.), you may take that answer and determine the score corresponding to that percentile value. You should obtain the original score. Similarly, whenever you obtain a score corresponding to a given percentile rank you may take that answer and determine the percentile rank of that score. You should always come back to the original percentile rank. Failure to do so indicates that you have made an error. It is preferable to repeat the solution without reference to your prior answer rather than attempt to find the mistake in your prior solution. Such errors are frequently of the proofreader type which defy detection, are time consuming to locate, and highly frustrating.

4.3 Percentile Rank and Reference Group

Just as a score is meaningless in the abstract, so is a percentile rank. A percentile rank must always be expressed in relation to some reference group. Thus if a friend claims that he obtained a percentile rank of 93 in a test of mathematical aptitude, you might not be terribly impressed if the reference group were made up of individuals who completed only the eighth grade. On the other hand, if the reference group consisted of individuals holding doctorate degrees in mathematics, your reaction would unquestionably be quite different.

Separate norms are frequently published for many standardized tests. Table 4.2 shows the raw score equivalents for selected percentile points on several advanced tests of the Graduate Record Examination. Note that a person scoring 500 would obtain a percentile rank of 63 on the Advanced Test in Sociology, but only have a percentile rank of 42 on the Advanced Test in History.

4.4 Centiles, Deciles, and Quartiles

Occasionally you will encounter the terms centile, decile, and quartile. All refer to a specific division of the scale of percentile ranks. A scale of percentile ranks is comprised of 100 units. A **centile** is equivalent to a percentile and is a percentage rank that divides a distribution into 100 equal parts. The 13th centile is specified as C_{13} and is equivalent to the 13th percentile. A **decile** is a percentage rank that divides a distribution into 10 equal parts. The 9th decile (specified as D_9) is equivalent to the 90th centile (percentile). The 1st decile is the 10th centile, etc. You

Table 4.2   Advanced tests (total scores) interpretive data. Percentile ranks used on score reports. Percentage of examinees scoring lower than selected scaled scores.

SCALED SCORE	EDUCATION	FRENCH	HISTORY	POLITICAL SCIENCE	PSY- CHOLOGY	SOCIOLOGY
800						
780						99
760		99				99
740		99			99	99
720		98	99		98	98
700		96	98		96	97
680		94	97	99	94	95
660	99	92	95	98	90	94
640	98	89	92	97	86	92
620	97	84	88	95	81	89
600	94	79	84	91	74	86
580	91	74	78	87	67	83
560	86	66	70	80	60	79
540	80	59	61	74	52	74
520	73	51	52	66	43	69
500	65	42	42	58	36	63
480	56	33	33	50	28	58
460	48	25	25	41	22	51
440	41	18	17	33	17	45
420	33	13	11	26	12	39
400	26	9	7	20	9	33
380	20	5	4	15	6	28
360	16	3	2	11	3	22
340	12	1	1	7	2	17
320	8			4	1	13
300	6			3		9
280	3			1		6
260	2					3
240	1					
Number of Examinees	40,925	3633	14,208	11,027	54,655	9975
Percent Women	67	78	34	26	51	57
Percent Men	33	21	66	74	49	43
Mean	454	519	514	473	530	457
Standard Deviation	93	92	83	90	95	120

Source: Copyright by the Educational Testing Service, Princeton, N.J. Reproduced by permission. All rights reserved.

Note: Based on the performance of examinees tested between October 1, 1974, and September 30, 1977.

should convince yourself that there are 9 deciles in any scale of percentile ranks. Why are there not 10?

A **quartile** is a percentile rank that divides a distribution into 4 equal parts. There are 3 quartiles in any scale of percentile ranks and they include the 1st quartile (Q_1) which is equivalent to the 25th centile, the 2nd quartile (Q_2) which is equivalent to the 50th centile or 5th decile, and the 3rd quartile (Q_3), which is equivalent to the 75th centile.

Summary

In this chapter we saw that, by itself, a score is meaningless unless it is compared to a standard base or scale. Scores are often converted into units of the percentile rank scale in order to provide a readily understandable basis for their interpretation and comparison.

We saw that percentile ranks of scores and scores corresponding to a given percentile may be approximated from a cumulative percentage graph. Direct computational methods were demonstrated to permit a more precise location of the percentile rank of a score and the score corresponding to a given percentile. A percentile rank was shown to be meaningless in the abstract. It must always be expressed in relation to some reference group. A distribution can be divided into centiles, deciles, and quartiles, as discussed.

Terms to Remember

Centile   A percentage rank that divides a distribution in 100 equal parts. Same as a percentile.

Decile   A percentage rank that divides a distribution into 10 equal parts.

Percentile rank   Number that represents the percentage of cases in a distribution that had scores at or lower than the one cited.

Percentiles   Numbers that divide a distribution into 100 equal parts. Same as a centile.

Quartile   A percentage rank that divides a distribution into 4 equal parts.

Exercises

1. Estimate the percentile rank of the following ages, using Fig. 4.1.

   a) 1   **b)** 12   **c)** 34

2. Calculate the percentile rank of the ages in Exercise 1, using Table 4.1.

3. Estimate the ages corresponding to the following percentiles, using Fig. 4.1.

   a) 25   **b)** 50   **c)** 75

4. Calculate the ages corresponding to the percentiles in Exercise 3, using Table 4.1.

5. If your parents told you that they had just read in the newspaper that overall the public high school students in your county were at the 42nd percentile in a standardized reading test, how might you clearly explain the meaning of the test results to them?

6. Refer to Exercise 13 in Chapter 3.

   a) What is the percentile rank of an inmate who has been arrested six times?

   b) How many times would an inmate have to have been arrested to be at the 1st quartile? The 3rd quartile? The 2nd quartile?

7. A researcher administered a religious knowledge test to 36 persons, and their scores are presented here. Which specific scores would be above the 3rd quartile? What score is at the 50th percentile? The 50th centile? The 9th decile? Use $i = 3$ and assume that the lowest class interval is 0–2.

9	6	8	10
2	2	5	0
7	1	4	2
5	8	0	1
6	7	10	1
2	4	9	7
3	3	9	6
4	6	4	4
0	2	1	4

8. The following questions are based on Table 4.2.

   a) John H. proudly proclaims that he obtained a "higher" percentile ranking than his friend, Howard. Investigation of the fact reveals that his score on the test was actually lower. Must it be concluded that John H. was lying, or is some other explanation possible?

b) Jean obtained a percentile rank of 66 on the political science test. What was her score? What score would she have had to obtain to achieve the same percentile rank on the French test?

c) The Department of Sociology at Anomie University uses the Advanced Graduate Record Examination in Sociology as an element of the admissions procedure. No applicant obtaining a percentile rank below 74 is considered for admission, regardless of other qualifications. Thus the 74th percentile might be called a cutoff point. What score constitutes the cutoff point for this distribution?

9. Using the frequency distribution in Table 3.4, calculate the percentile ranks of the following ages.

  a) 30   **b)** 34   **c)** 51   **d)** 73

10. For Table 3.4, what age is at the

  a) 10th percentile?   **b)** 5th decile?

  c) 64th centile?   **d)** 99th percentile?

11. Given the frequency distribution you constructed for the social worker's case load in Exercise 5, Chapter 3, how many clients would a caseworker need to be placed in the

  a) 1st quartile?   **b)** 3rd quartile?

12. Given the accompanying table, (a) calculate the percentile rank for the following size families.

  i) 3   **ii)** 5   **iii)** 7

(b) In what way does this table violate the standard procedure outlined in Chapter 3 for constructing frequency distributions?

Size of U.S. families of Spanish origin in 1977

SIZE	NUMBER (IN THOUSANDS)
2	662
3	636
4	625
5	408
6	218
7 or more persons	216

Source: U.S. Bureau of the Census, Persons of Spanish origin in the United States: March 1978, *Current Population Reports,* Series P-20, No. 339, U.S. Government Printing Office, Washington, D.C., 1979.

13. Given Exercise 12, what size family lies at the 5th decile? The 2nd quartile? The 3rd quartile?

Measures of
Central
Tendency

5

5.1   Introduction

5.2   The Arithmetic Mean ($\overline{X}$)

5.3   The Median (Md_n)

5.4   The Mode (M_o)

5.5   Comparison of Mean, Median, and Mode

5.6   The Mean, Median, Mode, and Skewness

5.1 Introduction

The ambiguity in the use of the term *average* causes confusion among lay people and perhaps leads to their suspicion that statistics is more of an art than a science. Unions and management speak of average salaries and frequently cite numerical values that are in sharp disagreement with each other; television programs and commercials are said to be prepared with the "average viewer" in mind; politicians are deeply concerned about the views of the average American voter; the average family size is frequently given as a fractional value, a statistical abstraction that is ridiculous; the term *average* is commonly used as a synonym for the term *normal;* the TV weather reporter tells us we had an average day or that rainfall for the month is above or below average. Indeed, the term *average* has so many popular meanings that many statisticians prefer to drop it from the technical vocabulary and refer, instead, to **measures of central tendency.** We shall define a measure of central tendency as an *index of central location used in the description of frequency distributions.* Since the center of a distribution may be defined in different ways, there are a number of different measures of central tendency. In this chapter, we shall concern ourselves with three of the most frequently used measures of central tendency: the mean, the median, and the mode.

5.1.1 Describing Frequency Distributions

Through the first four chapters of the book, we dealt with organizing data into a meaningful and useful form. We want to go beyond that to describe our data so that meaningful statements can be made about them. A frequency distribution represents an organization of data, but it does not, in itself, permit us to make statements either describing the distribution or comparing two or more distributions.

Frequency distributions can be characterized by three different features: (1) Frequency data have a characteristic *form* or shape, as you learned in Section 3.8 where we discussed the forms of frequency data, including symmetrical and skewed distributions. (2) Frequency data cluster around a *central value* which lies between the two extreme values of the variable under study. This feature is the topic of this chapter. (3) Frequency data can be characterized by their degree of *dispersion,* spread, or variability about the central value. This last topic is discussed in Chapter 6.

Given these three features about a particular distribution, the social scientist can clearly convey to others the characteristics of a particular distribution. For example, the income distributions of the members of two church congregations might be summarized graphically as shown in Fig. 5.1. You see that in Fig. 5.1 the two distributions differ with respect to form, a central value, and variability. In this and the following chapter you will learn how to reduce considerable data to a few summary values. Being able to locate a single point of central tendency, particularly when combined with a summary description of the dispersion of scores about that point, can be very useful. While

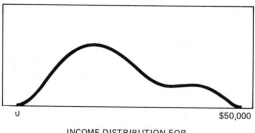

INCOME DISTRIBUTION FOR
MEMBERS OF CONGREGATION A

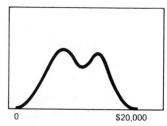

INCOME DISTRIBUTION FOR
MEMBERS OF CONGREGATION B

5.1 Hypothetical income distributions of two church congregations.

there is much to be gained by these summary values, the danger of course is that you lose considerable information that was available when the data were in their original form.

5.2 The Arithmetic Mean $(\overline{X})$

5.2.1 Methods of Calculation

You are already familiar with the arithmetic **mean,** for whenever you obtain an "average" of grades by summing the grades and dividing by the number of grades, you are calculating the arithmetic mean. In short, *the mean is the sum of the scores or values of a variable divided by the number of scores.* Stated in algebraic form:

$$\overline{X} = \frac{X_1 + X_2 + \cdots + X_N}{N} = \frac{\Sigma X}{N} = \frac{\text{sum of scores}}{\text{number of scores}} \quad (5.1)$$

where

$$X = \text{a raw score in a set of scores,}$$
$$\overline{X} = \text{the mean and is referred to as } X \text{ bar,*}$$
$$N = \text{the number of scores, and}$$
$$\Sigma = \text{sigma and directs us to sum all the scores.}$$

Note that we are using the shorthand version of X here. It is not necessary to include the subscript i because we are summing all of the X values.

During the period 1835 to 1968 a total of 81 political assassinations and assaults occurred in the following regions of the United States: Northeast—7, Southeast—28, North Central—11, South Central—25, and West—10.[†] Given Eq. (5.1) and the number of assassinations for each of the five regions of the United States, we can now compute the arithmetic mean number of assassinations per region.

---

* In Section 1.2 we indicated that italic letters would be used to represent sample statistics and Greek letters to represent population parameters. The Greek letter μ will be used to represent the population mean.

[†] *Source:* National Commission on Causes and Prevention of Violence, *Task Force Report on Assassinations,* 1969.

$$X_1 = 7, X_2 = 28, X_3 = 11, X_4 = 25, \text{ and } X_5 = 10$$

$$\overline{X} = \frac{7 + 28 + 11 + 25 + 10}{5} = 16.2.$$

The mean serves as a summary value for the distribution of the five scores. Note that interval level data are required before you can calculate the mean because both addition and division of the scores are required.

Obtaining the mean from an ungrouped frequency distribution   You will recall that we constructed a frequency distribution as a means of eliminating the constant repetition of scores that occur with varying frequency, in order to permit a single entry in the frequency column to represent the number of times a given score occurs. Thus in Table 5.1 we know from column f that the score of 8 occurred six times. In calculating the mean, then, it is not necessary to add 8 six times since we may multiply the score by its frequency and obtain the same value of 48. Since each score is multiplied by its corresponding frequency prior to summing, we may represent the mean for frequency distributions as follows:

$$\overline{X} = \frac{\Sigma fX}{N} . \tag{5.2}$$

Table 5.1   Computational procedures for calculating the mean with ungrouped frequency distributions

X	f	fX		
4	2	8		
5	2	10		
6	3	18	$\overline{X} = \dfrac{\Sigma fX}{N}$	
7	4	28		
8	6	48	$\overline{X} = \dfrac{232}{29}$	
9	4	36		
10	5	50	$\overline{X} = 8.00$	
11	2	22		
12	1	12		
	$N = 29$	$\Sigma fX = 232$		

Obtaining the mean from a grouped frequency distribution, raw score method
The calculation of the mean from a grouped frequency distribution involves essentially the same procedures that are used with ungrouped frequency distributions. To start, the midpoint of each interval is used to represent *all scores within that interval*. Thus the assumption is made that the scores in an interval are evenly distributed. The midpoint of each interval (X_m) is multiplied by its corresponding frequency, and the product is summed and divided by N. If you encountered a distribution such as in Exercise 12, Chapter 4 where an interval was not closed, it would be impossible to determine the midpoint and hence you could not calculate the mean. The procedures used to calculate the mean from a grouped frequency distribution are demonstrated in Table 5.2.

Table 5.2 Computational procedures for calculating the mean from a grouped frequency distribution

1 CLASS INTERVAL	2 FREQUENCY f	3 MIDPOINT X_m	4 FREQUENCY MULTIPLIED BY THE MIDPOINT fX_m	
75–79	3	77	231	
80–84	4	82	328	
85–89	8	87	696	
90–94	10	92	920	
95–99	15	97	1455	
100–104	20	102	2040	$\overline{X} = \dfrac{\Sigma fX_m}{N}$
105–109	15	107	1605	
110–114	10	112	1120	
115–119	8	117	936	$= \dfrac{10,195}{100}$
120–124	5	122	610	
125–129	2	127	254	$= 101.95$
	$N = 100$		$\Sigma fX = 10,195$	

5.2.2 Properties of the Arithmetic Mean

One of the most important properties of the mean requires that you understand the concept of deviation. A **deviation** is the distance and direction of a score from a reference point which is normally the mean. A deviation is positive when the score is larger than the mean and negative when the score is less than the mean.

The first important property of the mean is that it is the point in a distribution of measurements or scores about which the sum of the deviations are equal to zero. In other words, if we were to subtract the mean from each score and add the resulting deviations from the mean, this sum would equal zero. Symbolically,

$$\sum (X - \overline{X}) = 0. \qquad (5.3)$$

Therefore, the mean is the value that balances all the scores on either side of it. You may wish to think of it as a center of gravity. The arithmetic mean of the values $X_1 = 2, X_2 = 3, X_3 = 4, X_4 = 5, X_5 = 6$ is 4. Substituting these values into Eq. (5.3) we find that the sum of the deviations of these five values from their mean of 4 equals 0 as follows:

$$(2 - 4) + (3 - 4) + (4 - 4) + (5 - 4) + (6 - 4) =$$
$$(-2) + (-1) + (0) + (+1) + (+2) = 0.$$

A second important property of the mean is that *the mean is very sensitive to extreme values when these are not dispersed evenly on both sides of it.* Observe the two **arrays** of scores in Table 5.3. An *array is an arrangement of data according to their magnitude from the smallest to the largest value.* Note that all the scores in both distributions are the same except for the very large score of 33 in column X_2. This one extreme score is sufficient to double the size of the mean. The sensitivity of the mean to extreme scores is a characteristic that has important implications governing our use of it. These implications will be discussed in Section 5.5, in which we compare the three measures of central tendency.

Table 5.3   Comparison of the means of two arrays of scores, one of which contains an extreme value

GROUP 1 SCORE, X_1	GROUP 2 SCORE, X_2
2	2
3	3
5	5
7	7
8	33
$\Sigma X_1 = 25$	$\Sigma X_2 = 50$
$\overline{X}_1 = 5.00$	$\overline{X}_2 = 10.00$

A third important property of the mean is that the **sum of squares** *of deviations from the arithmetic mean is less than the sum of squares of deviations about any other score or potential score.* To illustrate this property of the mean, Table 5.4 shows the squares and the sum of squares when deviations are taken from the mean and various other scores in a distribution. It can be seen that the sum of squares is smallest in column 4, when deviations are taken from the mean. This property of the mean provides us with another definition: *the mean is that measure of central tendency that makes the sum of squared deviations around it minimal.*

Table 5.4   The squares and sum of squares of deviations taken from various scores in a distribution

1 X	2 $(X - 2)^2$	3 $(X - 3)^2$	4 $(X - \overline{X})^2$	5 $(X - 5)^2$	6 $(X - 6)^2$
2	0	1	4	9	16
3	1	0	1	4	9
4	4	1	0	1	4
5	9	4	1	0	1
6	16	9	4	1	0
Totals N = 5 X = 4	30	15	10	15	30

Each of these three properties of the arithmetic mean will be considered further in future chapters.

5.2.3 The Weighted Mean

Imagine that four classes in introductory sociology obtained the following mean scores on the final examination: 75, 78, 72, and 80. Could you sum these four means together and divide by four, to obtain an overall mean for all four classes? This could be done *only if* the N in each class is identical. What if, as a matter of fact, the mean of 75 is based on an N of 30, the second mean is based on 40 observations, the third on $N = 25$, and the fourth on $N = 50$?

The total sum of scores may be obtained by multiplying each mean by its respective N and summing.
Thus

$$\sum (N \cdot \overline{X}) = 30(75) + 40(78) + 25(72) + 50(80)$$
$$= 11,170.$$

The **weighted mean,** $\overline{X}_w$, can be expressed as the sum of the mean of each group multiplied by its respective weight (the N in each group) divided by the sum of the weights (that is, $\Sigma w = \Sigma N_i = N$).

$$\overline{X}_w = \frac{\Sigma(w \cdot \overline{X})}{\Sigma w} = \frac{\Sigma(N_i \cdot \overline{X})}{N}$$

$$\overline{X}_w = \frac{30(75) + 40(78) + 25(72) + 50(80)}{145}$$

$$= \frac{11,170}{145} = 77.03.$$

5.3 The Median (Md_n)

The **median** is defined as *that score or potential score in a distribution of scores, above and below which one-half of the frequencies fall.* If this definition sounds vaguely familiar to you, it is not by accident. The median is merely a special case of a percentile rank. Indeed, the median is the score at the 50th percentile.

Consider the following array of scores: 5, 19, 37, 39, 45. Note that the scores must be arranged in order of magnitude and that N is an odd number. A score of 37 is the median, since two scores fall above it and two scores fall below it.* If N is an *even* number, the median is the arithmetic mean of the two middle values. The two middle values in the array of scores 8, 26, 35, 43, 47, 73 are 35 and 43. The arithmetic mean

---

* When working with an array of numbers where N is odd, the definition of the median does not quite hold; that is, in the example above in which the median is 37, two scores lie below it and two above it, as opposed to one-half of N. Consider the score of 37 as falling one-half on either side of the median.

of these two values is (35 + 43)/2, or 39. Therefore the median is 39. Equation (5.4) provides a generalized rule for computing the median position. It should be clear at this point that the median is a positional measure that requires data measured on either an ordinal or interval scale.

$$\text{Median case number} = \frac{N + 1}{2}. \qquad (5.4)$$

In the first example with five values, the median case number would be 3 and the score of 37 is the median. When N is even, as in the case of the second example, the median case number is 3.5, hence we take the arithmetic mean of the 3rd and 4th values as we did in the example. Again we find the median to be 39.

5.3.1 The Median of a Grouped Frequency Distribution

The generalized procedures discussed in Chapter 4 for determining the score at various percentile ranks may be applied to the calculation of the median. The median of a grouped frequency distribution may be computed using a modification of Eq. (4.4) as shown in Eq. (5.5).

$$\text{Median} = X_{ll} + i\frac{(N/2) - \text{cum } f_{ll}}{f_i} \qquad (5.5)$$

where

$$X_{ll} = \text{score at lower real limit of the interval}$$
$$\text{containing the median,}$$
$$i = \text{width of the interval,}$$
$$\text{cum } f_{ll} = \text{cumulative frequency at the lower real}$$
$$\text{limit of the interval containing the}$$
$$\text{median,}$$
$$f_i = \text{number of cases within the interval}$$
$$\text{containing the median,}$$
$$N = \text{total number of cases in the frequency}$$
$$\text{distribution.}$$

Applied to the data appearing in Table 4.1 (Section 4.2.2) the median becomes:

$$\text{median} = 29.5 + 5\,\frac{(230.3/2) - 110.8}{19.3}$$

$$= 29.5 + 5\,\frac{(115.2 - 110.8)}{19.3}$$

$$= 29.5 + 1.14 = 30.64.$$

We are assuming that the scores within an interval are evenly distributed.

5.3.2 A Special Case

Occasionally the middle score in an array of scores is tied with other scores. How do we specify the median when we encounter tied scores?

Consider the following array of 20 scores: 2, 3, 3, 4, 5, 7, 7, 8, 8, 8, 8, 9, 10, 12, 14, 15, 17, 19, 19, 20. The easiest procedure is to convert the array to an ungrouped frequency distribution and apply Eq. (5.5). However, the i in the numerator may be eliminated since it is equal to 1.

$$\text{Median} = X_{\text{ll}} + \frac{(N/2 - \text{cum}\,f_{\text{ll}})}{f_i}$$

$$= 7.5 + \frac{(20/2 - 7)}{4}$$

$$= 7.5 + \frac{3}{4} = 8.25.$$

X	f	cum f	X	f	cum f
2	1	1	12	1	14
3	2	3	13	0	14
4	1	4	14	1	15
5	1	5	15	1	16
6	0	5	16	0	16
7	2	7	17	1	17
8	4	11	18	0	17
9	1	12	19	2	19
10	1	13	20	1	20
11	0	13			

[Handwritten margin notes:] Not correct (only for continuous interval data) grouped to whole numbers

5.3.3   Characteristics of the Median

An outstanding characteristic of the median is its *insensitivity* to extreme scores. Consider the following set of scores: 2, 5, 8, 11, 48. The median is 8. This is true in spite of the fact that the set contains one extreme score of 48. Had the 48 been a score of 97, the median would *remain the same.* This characteristic of the median makes it valuable for describing central tendency in certain types of distributions in which the *mean* is an unacceptable measure of central tendency due to its sensitivity to extreme scores. This point will be further elaborated in Section 5.5 when the uses of the three measures of central tendency are discussed.

5.4   The Mode (M_o)

Of all measures of central tendency, the **mode** is the most easily determined since it is obtained by inspection rather than by computation. *The mode is the score that occurs with greatest frequency.* For grouped data, the mode is designated as the midpoint of the interval containing the highest frequency count. In Table 5.2 the mode is a score of 102 since it is the midpoint of the interval (100–104) containing the greatest frequency.

If the data in Table 5.2 were represented as a frequency curve (polygon) or as a histogram, the mode would also be 102. The mode of a frequency curve is the highest value on the curve. In a histogram the mode is represented by the midpoint of the tallest column.

The mode can be determined for nominal-, ordinal-, or interval-level data and is therefore the most versatile of the three measures of central tendency discussed in this chapter. Consider a study in which the respondents were classified by their religious affiliation in the following manner: Protestant, 47%; Catholic, 42%; and Jewish, 11%. The modal category would be Protestant. Note that the mode is located in the category with the plurality of cases. A majority of cases, that is, over 50%, is not required.

The mode, from a statistical perspective, is also the most probable value. If the names of all the respondents classified by religious affiliation were placed in a large container, the name drawn first would most probably be that of a Protestant simply because Protestants are most numerous in this particular study.

In some distributions, such as Fig. 3.9(c), there are two high points which produce the appearance of two humps, as on a camel's back. Such distributions are referred to as being *bimodal*. A distribution with more than two humps is referred to as being multimodal. Figure 3.9(a) depicts a distribution with no mode.

5.5 Comparison of Mean, Median, and Mode

We have seen that the mean is a measure of central tendency in which the *sum* of the deviations on one side equals the sum of the deviations on the other side. The median divides a frequency distribution into two equal parts so that the *number* of scores below the median equals the *number* of scores above the median. The mode is the score that occurs with the greatest frequency.

In general, the arithmetic mean is the preferred statistic for representing central tendency because of several desirable properties. To begin with, the mean is a member of a mathematical system, which permits its use in more advanced statistical analyses because it is based on interval-level data. It is the preferred measure of central tendency if the distribution of scores is not skewed. We have used deviations from the mean to demonstrate two of its most important characteristics; that is, the sum of deviations is zero and the sum of squares is minimal. Deviations of scores from the mean provide valuable information about any distribution. We shall be making frequent use of deviation scores throughout the remainder of the text. In contrast, deviation scores from the median and the corresponding squared deviations have only limited applications to more advanced statistical considerations.

Another important feature of the mean is that it is the most stable or reliable measure of central tendency. Let's say that we draw 25 separate simple random samples of 500 persons who are registered voters in New York State (it is possible that a very few persons will be selected in more than one of the samples). If we computed the mean, the median, and the modal age for each of the 25 samples, we would find that in most instances the mean would show less fluctuation than either the median or mode. That is, the frequency distribution for the 25 mean ages would be more compact (show less variability) than would either the frequency distribution of the 25 median ages or the 25 modal ages. The reason that stability or reliability is particularly important stems from our effort to infer from a sample to the population. We learned in Chapter 1 that the population parameter is frequently unknown, but that we can calculate sample statistics. Thus, the sample is a convenient way to estimate population characteristics or parameters. Normally it is not possible to draw more than one sample so, because of its stability, we tend to use the sample mean as the best single estimate of its corresponding population parameter.

On the other hand, there are certain situations in which the median is preferred as the measure of central tendency. When the distribution is symmetrical, the mean, the median, and the mode are identical. Under these circumstances, the mean should be used. However, as we have seen, when the distribution is markedly skewed, the mean, which is analogous to a center of gravity, will provide a misleading estimate of central tendency because under these circumstances it is not a central score. The mean for group 2 in Table 5.3 is 10, even though four of the five scores are below this value. Annual family income is a commonly studied variable in which the median is preferred over the mean since the distribution of this variable is distinctly skewed in the direction of high salaries, with the result that the mean overestimates the income obtained by most families. You will find upon examination of most census information that the median is used more frequently than the mean or mode because the median is less sensitive to extreme values such as are encountered with age, educational attainment, and income distributions. Computation of the median requires that the data be measured at the ordinal level or higher.

The mode is used far less than the mean and median in the social sciences. It is particularly appropriate whenever a quick, rough estimate of central tendency is desired. The mode, however, is very versatile in that it can be used with all levels of measurement.

5.6 The Mean, Median, Mode, and Skewness

In Chapter 3, we demonstrated several forms of skewed distributions. We pointed out, however, that skew cannot always be determined by inspection. If you understand the differences between the mean and the median, you should be able to suggest a method for determining whether or not a distribution is skewed and, if so, the direction of the skew. The basic fact to keep in mind is that the mean is pulled in the direction of the skew, whereas the median and mode, unaffected by extreme scores, are not. When the mean is higher than the median, the distribution may be said to be positively skewed; when the mean is lower than the median, the distribution is negatively skewed. Figure 5.2 demonstrates the relation between the mean, the median, and the mode in positively skewed, negatively skewed, and symmetrical distributions.

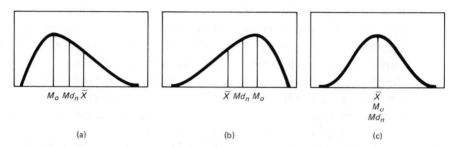

M_o Md_n $\bar{X}$ $\bar{X}$ Md_n M_o $\bar{X}$
M_o
Md_n

(a)   (b)   (c)

5.2 The relation between the mean, median and mode in (a) positively skewed (b) negatively skewed and (c) symmetrical distributions

Summary

In this chapter we discussed, demonstrated the calculation of, and compared three indices of central tendency that are frequently used to describe frequency distributions: the mean, the median, and the mode. We saw that the mean may be defined variously as the sum of scores divided by the number of scores, the point in a distribution that makes the summed deviations equal to zero, or the point in the distribution that makes the sum of the squared deviations minimal. The median divides the distribution in half, so that the number of scores below the median equals the number of scores above it. Finally, the mode is defined as the most frequently occurring score. We demonstrated the method for obtaining the weighted mean of a set of means when each of the individual means is based on a different N.

Because of special properties it possesses, the mean is the most frequently used measure of central tendency. However, because of the mean's sensitivity to extreme scores that are not balanced on both sides of the distribution, the median is usually the measure of choice when distributions are markedly skewed. The mode is rarely used in the social sciences.

Finally, we demonstrated the relationship between the mean, the median, and the mode in positively skewed, negatively skewed, and symmetrical distributions.

Terms to Remember

Array Arrangement of data according to their magnitude from the smallest to the largest value.

Deviation The distance and direction of a score from a reference point.

Mean Sum of the scores or values of a variable divided by their number.

Measure of central tendency Index of central location used in the description of frequency distributions.

Median Score in a distribution of scores, above and below which one-half of the frequencies fall.

Mode Score that occurs with the greatest frequency.

Sum of squares Deviations from the mean, squared and summed.

Weighted mean Sum of the mean of each group multiplied by its respective weight (the N in each group), divided by the sum of the weights (total N).

Exercises

1. Find the mean, the median, and the mode for each of the following sets of measurements. Show that $\Sigma(X - \overline{X}) = 0$.
   a) 10, 8, 6, 0, 8, 3, 2, 5, 8, 0
   b) 1, 3, 3, 5, 5, 5, 7, 7, 9
   c) 119, 5, 4, 4, 4, 3, 1, 0
2. In which of the sets of data in Exercise 1 is the mean a poor measure of central tendency? Why?

3. Which measure(s) of central tendency can be computed for Fig. 3.3 in Chapter 3? Why? Which measure(s) should not be computed? Why?

4. You have calculated measures of central tendency on family income and expressed your data in dollars. You decide to recompute after you have multiplied all the dollars by 100 to convert them to cents. How will this affect the measures of central tendency?

5. Calculate the measures of central tendency for the data in Table 3.3 in Chapter 3.

6. In Exercise 1(c), if the score of 119 were changed to a score of 19, how would the various measures of central tendency be affected?

7. On the basis of the following measures of central tendency, indicate whether or not there is evidence of skew and, if so, its direction.

    a) $\overline{X} = 56$,   median $= 62$,   mode $= 68$

    b) $\overline{X} = 68$,   median $= 62$,   mode $= 56$

    c) $\overline{X} = 62$,   median $= 62$,   mode $= 62$

    d) $\overline{X} = 62$,   median $= 62$,   mode $= 30$, mode $= 94$

8. What is the nature of the distributions in Exercises 7(c) and (d)?

9. Calculate the mean of the following array of scores: 3, 4, 5, 5, 6, 7.

    a) Add a constant, say 2, to each score. Recalculate the mean. Generalize: What is the effect on the mean of adding a constant to all scores?

    b) Subtract the same constant from each score. Recalculate the mean.
Generalize: What is the effect on the mean of subtracting a constant from all scores?

    c) Square all the scores. Recalculate the mean.
Generalize: What is the effect on the mean of squaring all the scores?

    d) Multiply each score by a constant, say 2. Recalculate the mean.
Generalize: What is the effect on the mean of multiplying each score by a constant?

    e) Divide each score by two. Recalculate the mean.
Generalize: What is the effect on the mean of dividing each score by a constant?

10. What is the mean, median, and mode for Fig. 3.4 in Chapter 3? How is each affected by the skew of the distribution?

Not correct!

11. If we know that the mean and median of a set of scores are equal, what can we say about the form of the distribution?

12. Given the information in the following table, calculate the appropriate measures of central tendency. (*Hint:* You may wish to exclude 1971–1977 for one or more of the measures of central tendency.)

Immigration to the United States, 1821 to 1977

PERIOD	NUMBER (IN THOUSANDS)
1821–1830	152
1831–1840	599
1841–1850	1713
1851–1860	2598
1861–1870	2315
1871–1880	2812
1881–1890	5247
1891–1900	3688
1901–1910	8795
1911–1920	5736
1921–1930	4107
1931–1940	528
1941–1950	1035
1951–1960	2515
1961–1970	3322
1971–1977	2797
Total	47,959

Source: U.S. Immigration and Naturalization Service, *Annual Report.*

13. In a departmental final exam, the following mean grades were obtained for classes of 25, 40, 30, 45, 50, and 20 students: 72.5, 68.4, 75.0, 71.3, 70.6, and 78.1.

    a) What is the total mean over all sections of the course?

    b) Draw a line graph showing the size of the class along the abscissa and the corresponding means along the ordinate. Does there appear to be a relationship between class size and mean grades on the final exam?

14. On the basis of examination performance, an instructor identifies the following groups of students:

    a) Those with a percentile rank of 90 or higher.

    b) Those with a percentile rank of 10 or less.

    c) Those with percentile ranks between 40 and 49.

d) Those with percentile ranks between 51 and 60.

Which group would the instructor work with if he or she wished to raise the *median* performance of the total group? Which group if he or she wished to raise the *mean* performance of the total group?

15. Which of the measures of central tendency is most affected by the degree of skew in the distribution? Explain.

16. Given the information in the following table, calculate the mean, median, and mode for marriages and divorces.

Marriages and Divorces in the United States

YEAR	MARRIAGES (1000s)	DIVORCES (1000s)
1962	1577	413
1963	1654	428
1964	1725	450
1965	1800	479
1966	1857	499
1967	1927	523
1968	2069	584
1969	2145	639
1970	2159	708
1971	2190	773
1972	2282	845
1973	2284	915
1974	2230	977
1975	2153	1036
1976	2155	1083
1977	2178	1091
1978	2243	1122

Source: U.S. Bureau of the Census, *Statistical Abstract of the United States, 1978* (99th edition), Washington, D.C., 1978, Table 78; and *Statistical Abstract of the United States, 1979* (100th edition),Washington, D.C., 1979, Table 117.

17. A student compiled the following academic record in college. What is her overall grade point average if $A = 4, B = 3, C = 2, D = 1, F = 0$?

CREDIT HOURS	GRADE
40	A
55	B
15	C
10	D
0	F

18. While backpacking on the Appalachian Trail you encounter 20 hikers and note the number of days they have been out. Given the following data, demonstrate that the sum of deviations about the mean equals zero.

5	1	15	9
4	42	19	9
39	6	22	8
14	18	14	1
12	14	8	19

19. A review of the files for 35 juvenile offenders presently held in a county detention home reveals that the majority have been arrested previously. Compute the mean, median, and modal number of arrests for the juvenile offenders.

NUMBER OF PREVIOUS ARRESTS	f
0	3
1	4
2	6
3	9
4	6
5	2
6	3
7	2
	$N = 35$

Measures of Dispersion

6

6.1 Introduction

6.2 The Range

6.3 The Interquartile Range

6.4 The Mean Deviation

6.5 The Variance (s^2) and Standard Deviation (s)*

6.6 Interpretation of the Standard Deviation

6.1 Introduction

In the introduction to Chapter 4, we saw that a score by itself is meaningless. A score takes on meaning only when it is compared with other scores or other statistics. If we know the mean of the distribution of a given variable, we can determine whether a particular score is higher or lower than the mean. But how much higher or lower? It is clear at this point that a measure of central tendency such as the mean provides only a limited amount of information. To more fully describe a distribution, or to more fully interpret a score, additional information is required concerning the **dispersion** of scores about our measure of central tendency.

Consider Fig. 6.1. In both examples of frequency curves, the mean of the distribution is exactly the same. However, note the difference in the interpretations of a score of 128. In Fig. 6.1(a), because the scores are widely dispersed about the mean, a score of 128 may be considered only moderately high. Quite a few individuals in the distribution scored above 128, as indicated by the area to the right of 128. In Fig. 6.1(b), on the other hand, the scores are compactly distributed about the same mean. This is a more *homogeneous* distribution. Consequently, the score of 128 is now virtually at the top of the distribution and it may therefore be considered a very high score.

In interpreting individual scores, we must find a companion to the mean or the median. This companion must in some way express the

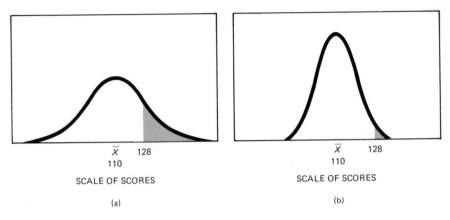

6.1 Two frequency curves with identical means but differing in dispersion or variability.

degree of dispersion of scores about the measure of central tendency. We will discuss five such measures of dispersion or variability: the **range,** the **interquartile range,** the **mean deviation,** the **variance,** and the **standard deviation.** Of the five, we will find the standard deviation to be our most useful measure of dispersion in both descriptive and inferential statistics. In advanced inferential statistics, as in analysis of variance (Chapter 16), the variance will become a most useful measure of variability.

6.2 The Range

When we calculated the various measures of central tendency, we located a *single point* along the scale of scores and identified it as the mean, the median, or the mode. When our interest shifts to measures of dispersion, however, we must look for an index of variability that indicates the *distance* along the scale of scores.

One of the first measures of distance that comes to mind is the **range.** The range is by far the simplest and the most straightforward measure of dispersion. It is the scale distance between the largest and smallest score. In a recent study conducted by one of the authors the youngest respondent was 28 years old and the oldest was 90. Thus the range was 90 − 28 or 62.

Although the range is meaningful, it is of little use because of its marked instability, particularly when it is based on a small sample. Note that if there is one extreme score in a distribution, the dispersion of scores will appear to be large when, in fact, the removal of that score may reveal an otherwise compact distribution. Several years ago, a resident of an institution for retarded persons was found to have an I.Q. score in the 140s. Imagine the erroneous impression that would result if the scores for the residents varied from 20 to 140 and the range was reported to be 120! Stated another way, the range reflects only the two most extreme scores in a distribution, and the remaining scores are ignored.

6.3 The Interquartile Range

In order to overcome the instability of the range as a measure of dispersion, the **interquartile range** is sometimes calculated. The interquartile range is calculated by subtracting the score at the 25th percentile (referred to as the first quartile or Q_1) from the score at the 75th percentile (the third quartile or Q_3).

$$\text{Interquartile range} = Q_3 - Q_1. \qquad (6.1)$$

This measure of score variation is more meaningful than the range because it is not based on two extreme scores. Rather, it reflects the middle 50% of the scores. It does, however, have two shortcomings: (1) Like the range, it does not by itself permit the precise interpretation of a score within a distribution, and (2) like the median, it does not enter into any of the higher mathematical relationships that are basic to inferential statistics. Figure 6.2 provides a comparison of the range and interquartile range.

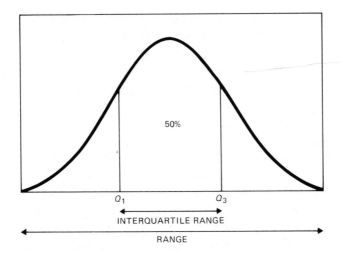

50%

Q_1 Q_3

INTERQUARTILE RANGE

RANGE

6.2 Comparison of range and interquartile range for a hypothetical distribution.

6.4 The Mean Deviation

In Chapter 5, we pointed out that, when we are dealing with data that are characterized by a bell-shaped distribution, the mean is our most useful measure of central tendency. We calculated the mean by adding together all the scores and dividing them by N. If we carried these procedures one step further, we could subtract the mean from each score, sum the differences (deviations from the mean), and thereby obtain an estimate of the amount of deviation from the mean. By dividing by N, we would have a measure that would be analogous to the arithmetic mean except that it would represent the dispersion of scores from the arithmetic mean.

If you think for a moment about the characteristics of the mean, which we discussed in the preceding chapter, you will encounter one serious difficulty. The sum of the deviations of all scores from the mean must add up to zero. If we defined the **mean deviation** (also referred to as the **average deviation**) as this sum divided by N, the mean deviation would always have to be zero. You will recall in Chapter 5 that we used the fact that $\Sigma(X - \overline{X}) = 0$ to arrive at one of several definitions of the mean.

Now if we were to add all the deviations *without regard to sign* and divide by N, we would still have a measure reflecting the mean deviation from the arithmetic mean. The resulting statistic would be based upon the **absolute value** of the deviations. The absolute value of a positive number or of zero is the number itself. The absolute value of a negative number can be found by changing the sign to a positive one. Thus the absolute value of $+3$ *or* -3 is 3. The symbol for an absolute value is $|\ |$. Thus $|-3| = 3$ and $|+3| = 3$.

The calculation of the mean deviation (MD) is shown in Table 6.1. The data in Tables 6.1 through 6.4 were compiled by the Joint Center for Political Studies in Washington, D.C. and include the number of black officials in 11 southern states holding an elected state position in July 1977.*

As a basis for comparison of the dispersion of several distributions, the mean deviation has some value. For example, the greater the mean deviation, the greater the dispersion of scores. However, for interpreting scores within a distribution, the mean deviation is less useful since there is no precise mathematical relationship between the mean deviation, as such, and the location of scores within a distribution.

Table 6.1   Computational Procedures for Calculating the Mean Deviation from an Array of Scores

STATE	NUMBER OF BLACK ELECTED OFFICIALS HOLDING STATE POSITIONS* X	$(|X - \overline{X}|)$	COMPUTATION
West Virginia	1	$|-5.55|$	
Virginia	2	$|-4.55|$	
Florida	3	$|-3.55|$	
Kentucky	3	$|-3.55|$	$MD = \dfrac{\Sigma(|X - \overline{X}|)^\dagger}{N}$ (6.2)
Mississippi	4	$|-2.55|$	
Arkansas	4	$|-2.55|$	
North Carolina	6	$|-0.55|$	
Louisiana	10	$|+3.45|$	$MD = \dfrac{45.65}{11} = 4.15.$
Tennessee	11	$|+4.45|$	
South Carolina	13	$|+6.45|$	
Alabama	15	$|+8.45|$	
	$\Sigma X = 72$	$(|X - \overline{X}|) = 45.65$	$N = 11$

$$\overline{X} = \frac{\Sigma X}{N} = \frac{72}{11} = 6.55$$

*Source: The Social and Economic Status of the Black Population in the United States: An Historical View, 1790–1978. Current Population Reports, Special Studies Series P-23, No. 80, Table 132. Reproduced by permission. Copyright by Joint Center for Political Studies, Washington, D.C. All rights reserved.

†For scores arranged in the form of a frequency distribution, the following equation for the mean deviation should be used:

$$MD = \Sigma f(|X - \overline{X}|)/N.$$

* These figures do not include black officials holding congressional, regional, city, county, or other elected positions.

You may wonder why we have bothered to demonstrate the mean deviation when it is of so little use in statistical analysis. As you shall see, the standard deviation and the variance, which have great value in statistical analysis, are very closely related to the mean deviation.

6.5 The Variance (s^2) and Standard Deviation (s)*

After looking at Table 6.1, you might pose this question, "We had to treat the values in column ($X - \overline{X}$) as absolute numbers because their sum was equal to zero. Why could we not square each ($X - \overline{X}$) and then add the squared deviations? In this way, we would legitimately rid ourselves of the minus signs, while still preserving the information that is inherent in these deviation scores."

The answer: We could if, by so doing, we arrived at a statistic of greater value in judging dispersion than those we have already discussed. It is fortunate that the standard deviation, based on the squaring of these deviation scores, is of considerable value in three different respects. (1) The standard deviation reflects dispersion of scores, so that the dispersion of different distributions may be compared using the standard deviation (s). (2) The standard deviation permits the precise interpretation of scores within a distribution. (3) The standard deviation, like the mean, is a member of a *mathematical system,* which permits its use in more advanced statistical considerations. We will use measures based upon s when we advance into inferential statistics, and will have more to say about the interpretive aspects of s after we have shown how it is calculated.

6.5.1 Calculation of Variance and Standard Deviation, Mean Deviation Method, with Ungrouped Scores

The **variance** is defined verbally as *the sum of the squared deviations from the mean divided by N.* Symbolically, it is represented as

---

* Italic letters will be used to represent sample statistics, and Greek letters to represent population parameters; for example, σ^2 represents the population variance and σ represents the population standard deviation. The problem of estimating population parameters from sample values will be discussed in Chapter 14.

$$s^2 = \frac{\Sigma(X - \overline{X})^2}{N} . \qquad (6.3)*$$

The **standard deviation** is the *square root* of the variance and is defined as

$$s = \sqrt{\frac{\Sigma(X - \overline{X})^2}{N}} . \qquad (6.4)$$

The computational procedures for calculating the standard deviation, utilizing the mean deviation method, are shown in Table 6.2.

You will recall from Section 5.2.2 that the sum of the $(X - \overline{X})^2$ column [that is, $\Sigma(X - \overline{X})^2$] is known as the **sum of squares** and that this sum is minimal when deviations are taken about the mean. From this point on in the text, we will encounter the sum of squares with regularity. It will take on a number of different forms, depending on the procedures that we elect for calculating it. However, it is important

Table 6.2 Computational procedure for calculating s, mean deviation method, from an array of scores

STATE	NUMBER OF BLACK ELECTED OFFICIALS HOLDING STATE POSITIONS* X	$(X - \overline{X})$	$(X - \overline{X})^2$	COMPUTATION
West Virginia	1	-5.55	30.80†	$s = \sqrt{\dfrac{\Sigma(X - \overline{X})^2}{N}}$
Virginia	2	-4.55	20.70	
Florida	3	-3.55	12.60	
Kentucky	3	-3.55	12.60	
Mississippi	4	-2.55	6.50	$= \sqrt{\dfrac{234.70}{11}}$
Arkansas	4	-2.55	6.50	
North Carolina	6	-0.55	0.30	
Louisiana	10	$+3.45$	11.90	$= \sqrt{21.34}$
Tennessee	11	$+4.45$	19.80	
South Carolina	13	$+6.45$	41.60	$= 4.62$
Alabama	15	$+8.45$	71.40	
	$\Sigma X = 72$	$\Sigma(X - \overline{X}) = -0.05$††	$\Sigma(X - \overline{X}) = 234.70$	

$$N = 11$$
$$\overline{X} = 6.55$$

†A comparison of the values in the $(X - \overline{X})$ column with those in the $(X - \overline{X})^2$ column demonstrates that the standard deviation weights extreme departures from the mean more heavily than does the mean deviation.

††Does not equal 0 due to rounding error.

* The important distinction between biased $\Sigma(X - \overline{X})^2/N$ and unbiased $\Sigma(X - \overline{X})^2/(N - 1)$ estimates of the population variance will be discussed in Chapter 14.

to remember that, whatever the form, the sum of squares represents the *sum of the squared deviations from the mean.*

The mean deviation method was shown to help you conceptually understand that the standard deviation is based on the deviation of scores from the mean. The mean deviation Eq. (6.4) is extremely unwieldy for use in calculation, particularly when the mean is a fractional value, which is usually the case.

6.5.2 Calculation of Standard Deviation, Raw Score Method with Ungrouped Scores

Equation (6.5) is the computational or raw score equation we will use for *s*. It is mathematically equivalent to the mean deviation Eq. (6.4)

Table 6.3 Computational procedures for calculating *s*, raw score method, from an array of scores

STATE	NUMBER OF BLACK ELECTED OFFICIALS HOLDING STATE POSITIONS* X	X^2	COMPUTATION
West Virginia	1	1	
Virginia	2	4	$s = \sqrt{\dfrac{\Sigma X^2}{N} - (\overline{X})^2}$
Florida	3	9	
Kentucky	3	9	
Mississippi	4	16	$= \sqrt{\dfrac{706}{11} - 6.55^2}$
Arkansas	4	16	
North Carolina	6	36	
Louisiana	10	100	$= \sqrt{64.18 - 42.90}$
Tennessee	11	121	
South Carolina	13	169	$= \sqrt{21.28} = 4.61$
Alabama	15	225	
	$\Sigma X = 72$	$\Sigma X^2 = 706$	$N = 11$
			$\overline{X} = 6.55$

*Source: The Social and Economic Status of the Black Population in the United States: An Historical View, 1790–1978. Current Population Reports, Special Studies Series P-23, No. 80, Table 132. Reproduced by permission. Copyright by Joint Center for Political Studies, Washington, D.C. All rights reserved.

used to compute s.

$$s = \sqrt{\frac{\Sigma X^2}{N} - (\overline{X})^2}.$$ *(6.5)*

You will note that the result using Eq. (6.5) agrees almost exactly with the answer we obtained by the mean deviation Eq. (6.4) in Table 6.2. Table 6.3 summarizes the computational procedure that is used to calculate s with Eq. (6.5).

6.5.3 Calculation of Standard Deviation, Raw Score Method, from an Ungrouped Frequency Distribution

If we take the data in Table 6.2 and arrange them into an ungrouped frequency distribution, we obtain the accompanying table.

X	f
1	1
2	1
3	2
4	2
6	1
10	1
11	1
13	1
15	1

To calculate s, square each score and multiply by its corresponding frequency. Sum these products to obtain $\Sigma f X^2$. Place this value in Eq. (6.6):

$$s = \sqrt{\frac{\Sigma f X^2}{N} - (\overline{X})^2}.$$ *(6.6)*

Table 6.4 summarizes the procedure for obtaining the standard deviation from an ungrouped frequency distribution.

Table 6.4   Procedures for calculating the standard deviation of scores from an ungrouped frequency distribution

X^*	f	fX	X^2	fX^2	COMPUTATION
1	1	1	1	1	$s = \sqrt{\dfrac{\Sigma fX^2}{N} - (\overline{X})^2}$
2	1	2	4	4	
3	2	6	9	18	
4	2	8	16	32	$= \sqrt{\dfrac{706}{11} - 6.55^2}$
6	1	6	36	36	
10	1	10	100	100	
11	1	11	121	121	$= \sqrt{64.18 - 42.90}$
13	1	13	169	169	
15	1	15	225	225	$= \sqrt{21.28} = 4.61$
	$N = 11$	$\Sigma fX = 72$		$\Sigma fX^2 = 706$	$\overline{X} = 6.55$

*Source: The Social and Economic Status of the Black Population in the United States: An Historical View, 1790–1978. Current Population Reports, Special Studies Series P-23, No. 80, Table 132. Reproduced by permission. Copyright by Joint Center for Political Studies, Washington, D.C. All rights reserved.

6.5.4 Errors to Watch For

In calculating the standard deviation using the raw score method, it is common for students to confuse the similar-appearing terms ΣX^2 (or ΣfX^2) and $(\Sigma X)^2$ (or $[\Sigma fX]^2$). It is important to remember the former represents the sum of the squares of each of the individual scores, whereas the latter represents the square of the sum of the scores. For example:

Given: $X_1 = 2,\ X_2 = 3,\ X_3 = 4,\ X_4 = 5,\ X_5 = 6$

$$\Sigma X^2 = 2^2 + 3^2 + 4^2 + 5^2 + 6^2 = 90$$

$$(\Sigma X)^2 = (2 + 3 + 4 + 5 + 6)^2 = 20^2 = 400$$

$$\Sigma X^2 \neq (\Sigma X)^{2*}$$

* Does not equal ($\neq$).

By definition, it is impossible to obtain a negative sum of squares or a negative standard deviation. In the event that you obtain a negative value under the square root sign, you have probably confused ΣX^2 and $(\Sigma X)^2$.

A rule of thumb for estimating the standard deviation is that the ratio of the range to the standard deviation is rarely smaller than 2 or greater than 6. Generally you will find that the more scores (larger the sample) you are dealing with, the larger the ratio. In our preceding example, the ratio was $14/4.62 = 3.03$. Had our sample been considerably larger, the ratio would have been closer to 6. This will become clearer after learning about the normal distribution in the next chapter. If you obtain a standard deviation that yields a ratio greater than 6 or smaller than 2 relative to the range, you have almost certainly made an error.

6.6   Interpretation of the Standard Deviation

You may have noticed that the size of the standard deviation is related to the variability in the scores. The more homogeneous the scores, the smaller the standard deviation and, conversely, the more heterogeneous the scores, the larger the standard deviation. The mean and sum of squares are equal in Arrays A and B in the accompanying table, for example, yet there is considerable difference in the variability of the scores and, therefore, the size of the standard deviations. You should also

	ARRAY A			ARRAY B	
X	$X - \overline{X}$	$(X - \overline{X})^2$	X	$X - \overline{X}$	$(X - \overline{X})^2$
4	0	0	2	−2	4
4	0	0	2	−2	4
4	0	0	3	−1	1
4	0	0	4	0	0
4	0	0	9	+5	25
$\Sigma X = 20$	$\Sigma(X - \overline{X}) = 0$	$\Sigma(X - \overline{X})^2 = 0$	$\Sigma X = 20$	$\Sigma(X - \overline{X}) = 0$	$\Sigma(X - \overline{X})^2 = 34$
$\overline{X} = 4$			$\overline{X} = 4$		
$N = 5$			$N = 5$		
	$s = \sqrt{0/5} = 0$			$s = \sqrt{34/5} = 2.6$	

note that extreme scores have a disproportionate effect on the standard deviation since we square the raw deviations from the mean. Examine the impact of the raw score 9 in Array B, for example, on the $(X - \overline{X})^2$ column. These points are best illustrated by the mean deviation method of computation.

A complete understanding of the meaning of the standard deviation requires a knowledge of the relationship between the standard deviation and the normal distribution. In order to be able to interpret the standard deviations that are calculated in this chapter, it will be necessary to explore the relationship between the raw scores, the standard deviation, and the normal distribution. This material is presented in the following chapter.

Summary

We have seen that to describe a distribution of scores fully, we require more than a measure of central tendency. We must be able to describe how these scores are dispersed about central tendency. In this connec-

Table 6.5   Summary procedures: Calculating the variance and the standard deviation from an array of scores

X	X^2	STEPS
0	0	1. Count the number of scores to obtain N. $N = 12$.
1	1	2. Sum the scores in the X column to obtain ΣX. $\Sigma X = 48$.
2	4	3. Find the mean. $\Sigma X/N = 48/12 = 4.00$.
3	9	4. Square each score and place it in the adjacent column.
4	16	5. Sum the X^2 column to obtain $\Sigma X^2 = 242$.
4	16	6. Substitute the values found in steps 3 and 5 in the
5	25	equation for s and solve:
5	25	
5	25	$$s = \sqrt{\frac{\Sigma X^2}{N} - (\overline{X})^2}$$
6	36	
6	36	
7	49	$$= \sqrt{\frac{242}{12} - 16}$$
$\Sigma X = 48$	$\Sigma X^2 = 242$	
		$$= \sqrt{20.17 - 16} = \sqrt{4.17} = 2.04.$$

Also,

$$s^2 = \frac{\Sigma X^2}{N} - (\overline{X})^2$$

$$= \frac{242}{12} - 16$$

$$= 20.17 - 16 = 4.17.$$

tion we discussed five measures of dispersion: the range, the inter-quartile range, the mean deviation, the standard deviation, and the variance. The summary procedure for calculating the variance and standard deviation from an array of scores is presented in Table 6.5 and from a grouped frequency distribution in Table 6.6.

Table 6.6   Summary procedures: Calculating the variance and standard deviation from a grouped frequency distribution

1	2	3	4	5	
		MIDPOINT OF			
CLASS INTERVAL	f	INTERVAL X_m	fX_m	fx_m^2	STEPS
0–2	1	1	1	1	1. Sum the f column to obtain $N = 63$.
3–5	4	4	16	64	
6–8	9	7	63	441	2. Prepare column 3, showing the midpoint of each interval.
9–11	10	10	100	1000	
12–14	16	13	208	2704	
15–17	11	16	176	2816	3. Multiply the f in each interval by the score at the midpoint of its corresponding interval. Values in column 2 are multiplied by corresponding values in column 3. Place in column 4.
18–20	8	19	152	2888	
21–23	3	22	66	1452	
24–26	1	25	25	625	
	$N = 63$		$\Sigma fX_m = 807$	$\Sigma fX_m^2 = 11{,}991$	

4. Sum column 4 to obtain ΣfX_m.

5. Divide ΣfX_m by N to obtain $\overline{X}$. Thus $\overline{X} = 807/63 = 12.81$.

6. Multiply values in column 4 by corresponding values in column 3 to obtain fX_m^2. Sum this column to obtain $\Sigma fX_m^2 = 11{,}991$.

7. Substitute the values in steps 4 and 5 in the equation for s^2 and s:

$$s^2 = \frac{\Sigma fX_m^2}{N} - (\overline{X})^2$$

$$= \frac{11{,}991}{63} - (12.81)^2$$

$$= 190.33 - 164.10$$

$$= 26.23.$$

Then

$$s = \sqrt{26.23}$$

$$= 5.12.$$

Terms to Remember

Absolute value of a number   The value of a number without regard to sign.

Dispersion   The spread or variability of scores about the measure of central tendency.

Interquartile range   A measure of variability obtained by subtracting the score at the 1st quartile from the score at the 3rd quartile.

Mean deviation (average deviation)   Sum of the deviation of each score from the mean, without regard to sign, divided by the number of scores.

Range   Measure of dispersion; the scale distance between the largest and the smallest score.

Standard deviation   Measure of dispersion defined as the square root of the sum of the squared deviations from the mean, divided by N. Also can be defined as the square root of the variance.

Sum of squares   Deviations from the mean, squared and then summed.

Variance   Sum of the squared deviations from the mean, divided by N.

Exercises

1. Calculate s^2 and s for the following array of scores: 3, 4, 5, 5, 6, 7.

   a) Add a constant, say, 2, to each score. Recalculate s^2 and s. Would the results be any different if you had added a larger constant, say, 200?
   Generalize: What is the effect on s and s^2 of adding a constant to an array of scores? Does the variability increase as we increase the magnitude of the scores?

   b) Multiply each score by a constant, say 2. Recalculate s and s^2. Generalize: What is the effect on s and s^2 of multiplying each score by a constant?

2. In the following table are listed the average sentence and number of months served by first-time federal prisoners for a variety of offenses. Calculate the standard deviation for both columns and discuss.

Average sentence and time served by prisoners released from Federal Institutions for the first time: 1965 to 1977

OFFENSE	AVERAGE SENTENCE (MONTHS)	AVERAGE TIME SERVED (MONTHS)
Counterfeiting	34.1	16.0
Drug Laws	39.5	19.5
Embezzlement	21.1	11.0
Forgery	29.5	16.2
Income Tax	13.7	8.3
Kidnapping	148.4	57.1
Robbery	123.1	45.1
White Slave Traffic	59.7	26.0

Source: U.S. Bureau of Prisons, Statistical Report, annual.

3. Why is it necessary to supplement measures of central tendency with measures of dispersion to describe a distribution completely?

4. A list follows of the greatest distances (in miles) 15 children have ever travelled from their homes. Calculate the mean, median, mode, standard deviation, and the variance. What are the advantages of each?

1000	100	400
600	3000	700
250	600	1100
400	1000	1300
500	50	200

5. What is the nature of a distribution if $s = 0$?

6. Calculate the standard deviations for the following sets of measurements:

a) 10, 8, 6, 0, 8, 3, 2, 2, 8, 0   b) 1, 3, 3, 5, 5, 5, 7, 7, 9
c) 20, 1, 2, 5, 4, 4, 4, 0   d) 5, 5, 5, 5, 5, 5, 5, 5, 5, 5

7. Why is the standard deviation in Exercise 6(c) so large? Describe the effect of extreme scores on s.

8. Determine the range for the sets of measurements in Exercise 6. For which of these is the range a misleading index of variability and why?

9. For the following table, find the standard deviation for females and for males. Which sex is more homogeneous in its number of physician visits per year?

Average number of physician visits by age and sex (hypothetical data).

AGE	MALES	FEMALES
Under 5	5.8	6.5
5–9	3.2	4.2
10–14	4.4	4.6
15–19	4.8	5.4
20–24	5.1	6.0
25–29	5.3	5.7
30–34	5.2	5.4

10. For the following grouped frequency distribution, find (a) the standard deviation and (b) the variance (hypothetical data).

Number of self-reported delinquent acts

DELINQUENT ACTS	f
0–2	13
3–5	6
6–8	3
9–11	2
12–14	4
15–17	2
	$N = 30$

11. Find (a) the standard deviation and (b) the variance for the following distribution. Which measure do you believe best reflects the distribution and why?

In how many political campaigns have you worked ten or more hours (hypothetical data)?

NUMBER OF CAMPAIGNS	f
1	22
2	15
3	11
4	9
5	4
6	2
7	1
8	1
	$N = 65$

12. Find the (a) mean, (b) standard deviation, and (c) range for each major in the following table. Which discipline has been most stable? Least stable? (*Hint:* Use the mean deviation method to ensure your calculator can accommodate the values.)

Earned Bachelor's Degrees in the United States

	POLITICAL SCIENCE	HISTORY	SOCIOLOGY
1969–1970	26,000	43,000	30,000
1970–1971	27,000	45,000	33,000
1971–1972	28,000	44,000	35,000
1972–1973	30,000	41,000	35,000
1973–1974	31,000	37,000	35,000
1974–1975	29,000	31,000	32,000
1975–1976	28,000	28,000	28,000

Source: U.S. Department of Health, Education, and Welfare, National Center for Education Statistics, reports on *Earned Degrees Conferred.* (Figures have been rounded to the nearest thousand.)

13. Compute the interquartile range and the mean deviation for the following scores:

    18, 2, 4, 6, 10, 7, 9, 11, 14, 8, 3, 16, 20, 22, 24, 25

The Standard Deviation and the Standard Normal Distribution

7

7.1   Introduction

7.2   The Concept of Standard Scores

7.3   The Standard Normal Distribution

7.4   Characteristics of the Standard Normal Distribution

7.5   Transforming Raw Scores to z Scores

7.6   Illustrative Problems

7.7   Interpreting the Standard Deviation

7.1 Introduction

We have emphasized previously that scores are generally meaningless by themselves. To take on meaning they must be compared to the distribution of scores from some reference group. Indeed, the scores derived from any index or scale become more meaningful when they are compared to some reference group of objects or persons. If we learned there were 491 violent crimes in San Antonio and 1475 violent crimes in Cleveland during a comparable time period, we would not be able to determine which city had the highest rate of violence until we knew the population of each city during the same time period. Simply noting that there is approximately a 1 to 3 ratio in actual numbers of violent acts between the two cities would prove misleading. We would need to calculate rates for a comparable number of persons, for example, a rate per 100,000 population for each city, before we could make a valid comparison.

7.2 The Concept of Standard Scores

In interpreting a single score, we want to place it in some position with respect to a collection of scores from a reference group. In Chapter 4, you learned to place a score by determining its percentile rank. Recall that the percentile rank of a score tells the percentage of scores that are of lower-scale value. Another approach for interpretation of a single score might be to view it with reference to some central point, such as the mean. Thus a score of 20 in a distribution with a mean of 23 might be reported as -3 because it is 3 units less than the mean. We can also express this deviation score as a **standard score** in terms of standard deviation units. If the standard deviation of a set of scores is 1.5 (that is, one standard deviation is 1.5), a score of 20 would be two standard deviations below the mean of 23 since the absolute difference between 20 and 23 is -3. Dividing -3 by our standard deviation of 1.5 yields -2 which is the number of standard deviations that the raw score of 20 is below the mean. This process of dividing the deviation of a score from the mean by the standard deviation is known as the transformation of a raw score to a z score. Symbolically, z is defined as

$$z = \frac{X - \overline{X}}{s} = \frac{X - \mu}{\sigma}$$

where

μ = the population mean, and
σ = the population standard deviation.

A z score represents the *deviation of a specific score from the mean expressed in standard deviation units.* For example, if the z score value of a raw score is $+2.1$, we know that the raw score lies 2.1 standard deviations to the right of the mean. A positive z score indicates the raw score is greater than the mean and a negative z score indicates the raw score is less than the mean.

In order to appreciate fully the value of transforming z scores, let us summarize three of their most important properties.*

1. The sum of the z scores is zero.
2. The mean of the z scores is zero.
3. The standard deviation and the variance of z scores is one.

Thus if we computed the z scores for all the raw scores in a bell-shaped frequency distribution using Eq. (7.1), we would find that the z scores would sum to zero, their mean would be zero, and their standard deviation and variance would be one.

What is the value of transforming a raw score to a z score? The conversion to z scores always yields a mean of 0 and a standard deviation of 1, but it does not normalize a nonnormal distribution. However, if the *population of scores* on a given variable is normal, we may express any score as a percentile rank by locating our z in the standard normal distribution. In addition, since z scores represent abstract numbers, as opposed to the concrete values of the original scores (inches, years of education, income, etc.), we may compare an individual's position on one variable with his or her position on a second. To understand these two important characteristics of z scores, we must make reference to the standard normal distribution.

---

* A complete mathematical explanation of several z score properties is provided here for those interested:

1. The sum of the z scores is zero. Symbolically stated,

$$\Sigma z = 0.$$

2. The mean of z scores is zero. Thus

$$z = \frac{\Sigma z}{N} = 0.$$

3. The sum of the squared z scores equals N. Thus

$$\Sigma z^2 = N.$$

This characteristic may be demonstrated mathematically.

$$\Sigma z^2 = \frac{\Sigma(X - \overline{X})^2}{s^2} = \frac{1}{s^2} \cdot \Sigma(X - \overline{X})^2$$

$$= \frac{N}{\Sigma(X - \overline{X})^2} \cdot \Sigma(X - \overline{X})^2$$

$$= N.$$

4. The standard deviation and the variance of z scores is one. Thus

$$S_z = S_z^2 = 1.$$

To demonstrate

$$S_z^2 = \frac{\Sigma(z - \overline{z})^2}{N}.$$

Since $\overline{z} = 0$, then

$$S_z^2 = \frac{\Sigma z^2}{N}.$$

Since $\Sigma z^2 = N$, then

$$S_z^2 = \frac{N}{N} = 1.$$

7.3 The Standard Normal Distribution

The **standard normal distribution** is a theoretical model that is based on the mathematical equation in the footnote.* There is a family of curves that may be called normal. A normal curve's exact shape is determined by its population mean (μ) and its population standard deviation (σ). Each unique combination of a normally distributed population's μ and σ yields a unique normally distributed curve. Students sometimes have the mistaken impression that the shape of all normal curves is as regular as the coat hangers one particular dry cleaner might use. This is not the case. While all coat hangers share some basic characteristics, we all know that there are several variations of their characteristic shape. Normal curves also vary; for example, some are flatter than others and some more peaked. All normal curves do, however, share the characteristics summarized in Section 7.4.

7.4 Characteristics of the Standard Normal Distribution

All standard normal distributions have a μ of 0, a σ of 1, and a total area equal to 1.00. They are also symmetrical. If you folded Fig. 7.1 along a line that would connect the point where the horizontal axis equals 0 and the highest point on the curve, two mirror images would result. A normal curve is sometimes referred to as a bell-shaped curve in

---

* It will be recalled that the Greek letters μ and σ represent the population mean and the standard deviation, respectively. The equation of the normal curve is:

$$Y = \frac{Ni}{\sigma\sqrt{2\pi}} \; e \; \frac{-(X - \mu)^2}{2\sigma^2}$$

in which

$Y =$ the frequency at a given value of X,
$\sigma =$ the standard deviation of the distribution,
$\pi =$ a constant equaling approximately 3.1416,
$e =$ approximately 2.7183,
$N =$ total frequency of the distribution,
$\mu =$ the mean of the distribution,
$i =$ the width of the interval,
$X =$ any score in the distribution.

It should be clear that there is a family of curves that may be called normal. By setting $Ni = 1$, a distribution is generated in which $\mu = 0$ and total area under the curve equals 1.

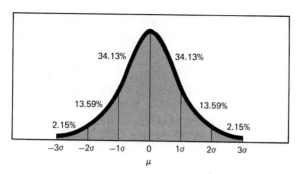

7.1 Areas between selected points under the normal curve.

that the mean, median, and modal values are equal and both sides of the curve slope gently away from its highest point. Note also that this theoretical curve never touches the X-axis (horizontal axis), hence the curve extends infinitely to the left and right without ever closing.*

There is also a fixed proportion of cases between a vertical line, or ordinate, erected at any one point on the horizontal axis and an ordinate erected at any other point. Taking a few reference points along the normal curve, we can make the following statements.

1. Between the mean and 1 standard deviation above the mean are found 34.13% of all cases. Similarly, 34.13% of all cases fall between the mean and 1 standard deviation below the mean. Stated in another way, 34.13% of the *area* under the curve is found between the mean and 1 standard deviation above the mean, and 34.13% of the *area* falls between the mean and 1 standard deviation below the mean.

2. Between the mean and 2 standard deviations above the mean are found 47.72% of all cases. Since the normal curve is symmetrical, 47.72% of the area also falls between the mean and 2 standard deviations below the mean.

3. Finally, between the mean and 3 standard deviations above the mean are found 49.87% of all the cases. Similarly, 49.87% of the cases fall between the mean and 3 standard deviations below the mean. Thus

---

* This characteristic of the standard normal distribution is called asymptosis.

99.74% (49.87% + 49.87%) of all cases fall between ±3 standard deviations. Here we see that a total of 6 standard deviation units (±3 standard deviations from the mean) account for nearly 100% of the cases and hence nearly 100% of the range of values. The remaining 0.26% (100% − 99.74%) of the cases lie beyond ±3 standard deviations from the mean. This leads us to another characteristic of the standard normal distribution: it is comprised of an infinite number of cases since the curve never closes.

7.5 Transforming Raw Scores to z Scores

Many variables in the real world such as the weight of a large number of persons, their attitudes toward the military, or their scores on a standardized examination approximate a normal distribution. As a result, we can use our theoretical distribution, the standard normal distribution, as a basis for comparison. To make such a comparison we must first transform the scores of the normally distributed variable to z scores. We are, in effect, creating a new variable with $\mu = 0$ and $\sigma = 1$, which is expressed in units of the standard normal curve. These units are referred to as standard deviation units or z scores. Figure 7.2 clarifies the relationships among raw scores on a standardized examination, z

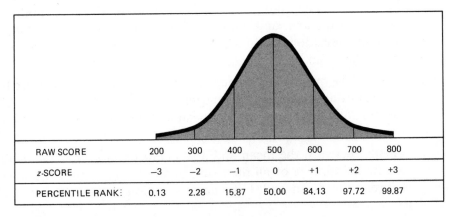

RAW SCORE	200	300	400	500	600	700	800
z-SCORE	−3	−2	−1	0	+1	+2	+3
PERCENTILE RANK:	0.13	2.28	15.87	50.00	84.13	97.72	99.87

7.2 Relationships among raw scores, z scores, and percentile ranks of a normally distributed variable in which $\mu = 500$ and $\sigma = 100$.

scores, and percentile ranks of a normally distributed variable. It assumes that $\mu = 500$ and $\sigma = 100$.

The mean of the raw scores in Fig. 7.2 is 500 and the corresponding mean of the standard normal distribution is 0. A raw score of 600 is one standard deviation to the right of the mean of 500 since one standard deviation equalled 100. The *z* score of $+1$, which is equivalent to 600 in raw units, is also located one standard deviation to the right of its *z* score mean of 0. The following calculations provide further clarification. Applying Eq. (7.1):

if $X = 500$, $\mu = 500$, and $\sigma = 100$

$$\text{then} \quad z = \frac{X - \mu}{\sigma} = \frac{500 - 500}{100} = 0$$

if $X = 600$, $\mu = 500$, and $\sigma = 100$

$$\text{then} \quad z = \frac{X - \mu}{\sigma} = \frac{600 - 500}{100} = +1.$$

You should carefully note that these relationships apply *only to scores from normally distributed populations*. When scores are from a population that is skewed (for example, the per capita income in the United States) rather than normally distributed, the standard normal distribution may not be used as a model for the data and *z* scores should not be computed. Transforming the raw scores to standard scores does not in any way alter the form of the original distribution nor does it, as we just saw, alter the relationship between the raw scores. The only change is to convert the mean to 0 and the standard deviation to 1. If the original distribution of scores is nonnormal, the *distribution* of *z* scores will be nonnormal. In other words, our transformation to *z*'s will *not* convert a nonnormal distribution to a normal distribution.

7.5.1 Finding the Area between Given Scores

Up to this point, we have confined our discussion of area under the standard normal curve to selected points. As a matter of fact, however, it is possible to determine the percent of areas between *any* two points by making use of the tabled values of the area under the normal curve (see Table A in Appendix C). You have learned to convert a raw distribution of scores into a standard normal distribution so that you can use

Table A, which was developed by a statistician who computed the area between the mean and every possible z score from 0 to 4.00. The left-hand column in Table A headed by z represents the deviation from the mean expressed in standard deviation units. *By referring to the body of the table, we can determine the proportion of total area between a given score and the mean (column B) and the area beyond a given score (column C).*

Thus if a woman had a score of 24.65 on a normally distributed variable with $\mu = 16$ and $\sigma = 5$, her z score would be $+1.73$ using Eq. (7.1).

$$z = \frac{X - \mu}{\sigma} = \frac{24.65 - 16}{5} = +1.73.$$

Referring to column B in Table A, we find that 0.4582 or 45.82%* of the area lies between her score and the mean. Since 50% of the area also falls below the mean in a symmetrical distribution, we may conclude that 50% + 45.82%, or 95.82%, of all the area falls below a score of 24.65. Note that we can now translate this score into a percentile rank of 95.82.

Suppose another individual obtained a score of 7.35 on the same normally distributed variable. His z score would be -1.73, as opposed to her z score of $+1.73$, because the raw scores (24.65 and 7.35) are both 8.65 units from the mean of 16. Using Eq. (7.1):

$$z = \frac{7.35 - 16}{5} = -1.73.$$

Since the normal curve is symmetrical, only the areas corresponding to the positive z values are given in Table A. The proportion of cases between a negative z value and the mean will equal the proportion of cases between a positive z value of equal value and the mean. Thus the area between the mean and a z of -1.73 is also equal to 0.4582 or 45.82% of the total area. The percentile rank of a score below the mean may be obtained either by subtracting 45.82% from 50%, or directly from column C. In either case, the percentile rank of a score of 7.35 is 4.18 because 4.18% of the scores were equal to or less than 7.35.

Before we consider some further uses of the normal curve, recall that one of its characteristics is that it contains an infinite number of

---

* The areas under the normal curve are expressed as proportions of area. To convert to percentage of area, multiply by 100 or move the decimal two places to the right.

cases. If you will look at the value in column C associated with z = 4.00, you will find that the proportion of cases beyond z = 4.00 is 0.00003. The area beyond z = 5.00 is 0.0000003 (not shown in Table A), and indeed there will always be cases regardless of how far we move to the left or right of the mean in this theoretical curve. Most social science texts do not show the area beyond 4 or 5 standard deviations; however, the precision of physics, for example, requires measurement up to ± 10 to 15 standard deviations from the mean!

7.6 Illustrative Problems

Let us examine several sample problems in which we assume that the mean of the general population, μ, is equal to 100 on a standardized test, and the standard deviation, σ, is 16. It is assumed that the variable is normally distributed.

Problem 1

Mary Jones obtains a score of 125 on a standardized test. What percent of cases fall between her score and the mean? What is her percentile rank in the general population?

At the outset, it is always wise to construct a crude diagram representing the relationships in question. Thus in the present example, the diagram would appear as shown in Fig. 7.3. To find the value of z

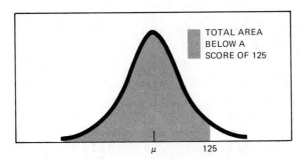

7.3 Proportion of area below a score of 125 in a normal distribution with μ = 100 and σ = 16.

corresponding to $X = 125$, we subtract the population mean from 125 and divide by 16. Thus using Eq. (7.1):

$$z = \frac{125 - 100}{16} = 1.56.$$

Looking up 1.56 in column B (Table A), we find that 44.06% of the area falls between the mean and 1.56 standard deviations above the mean. Mary Jones's percentile rank is, therefore, 50 + 44.06 or 94.06.

Problem 2

John Doe scores 93 on the same test. What is his percentile rank in the general population (Fig. 7.4)?

$$z = \frac{93 - 100}{16} = -0.44.$$

The minus sign indicates that the score is below the mean. Looking up 0.44 in column C, we find that 33.00% of the cases fall below his score. Thus his percentile rank is 33.00.

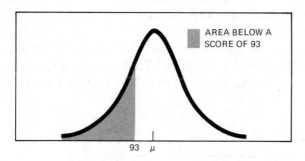

AREA BELOW A SCORE OF 93

93 μ

7.4 Proportion of area below a score of 93 in a normal distribution with $\mu = 100$ and $\sigma = 16$.

Problem 3

What proportion of the area falls between a score of 120 and a score of 88 (Fig. 7.5)?

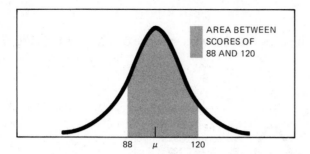

AREA BETWEEN
SCORES OF
88 AND 120

88 μ 120

7.5 Proportion of area between the scores 88
and 120 in a normal distribution with μ = 100
and σ = 16.

Note that to answer this question we do *not* subtract 88 from 120
and divide by σ. The areas in the normal probability curve are
designated in relation to the mean as a fixed point of reference. We
must, therefore, separately calculate the area between the mean and a
score of 120 and the area between the mean and a score of 88. We then
add the two areas to answer our problem.

Procedure

Find the z corresponding to X = 120:

$$z = \frac{120 - 100}{16} = 1.25.$$

Find the z corresponding to X = 88:

$$z = \frac{88 - 100}{16} = -0.75.$$

Find the required areas by referring to column B (Table A):

Area between the mean and z = 1.25 is 0.3944.

Area between the mean and z = -0.75 is 0.2734.

Add the two areas together.

Thus, the proportion of the area between 88 and 120 = 0.6678.

Problem 4

What percent of the area falls between a score of 123 and 135 (Fig. 7.6)?

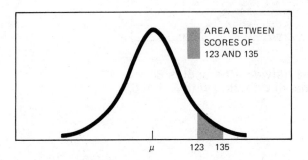

AREA BETWEEN
SCORES OF
123 AND 135

μ 123   135

7.6 Percent of area between the scores 123 and
135 in a normal distribution with $\mu = 100$ and
$\sigma = 16$.

Again, we cannot obtain the answer directly; we must find the
area between the mean and a score of 123 and subtract this from the
area between the mean and a score of 135.

Procedure

Find the z corresponding to $X = 135$.

$$z = \frac{135 - 100}{16} = 2.19.$$

Find the z corresponding to $X = 123$.

$$z = \frac{123 - 100}{16} = 1.44.$$

Find the required areas by referring to column B.

Area between the mean and $z = 2.19$ is 48.57%.

Area between the mean and $z = 1.44$ is 42.51%.

Subtract to obtain the area between 123 and 135. The result is

$$48.57 - 42.51 = 6.06\%.$$

Problem 5

We stated earlier that our transformation to z scores permits us to compare an individual's position on one variable with his or her position on another. Let us illustrate this important use of z scores.

On a standard aptitude test, John G. obtained a score of 245 on the verbal portion and 175 on the mathematics portion. The means and the standard deviations of each of these normally distributed tests are as follows: Verbal, $\mu = 220$, $\sigma = 50$; Math, $\mu = 150$, $\sigma = 25$. On which test did John score higher? We need to compare John's z score on each test. Thus

$$\text{verbal } z = \frac{245 - 220}{50} \qquad \text{math } z = \frac{175 - 150}{25}$$
$$= 0.50 \qquad\qquad\qquad = 1.00.$$

We may conclude, therefore, that John scored higher on the math portion because his math score was one full standard deviation ($z = 1.00$) above the mean of the math portion, whereas his verbal score was only one-half of a standard deviation ($z = 0.50$) above the mean of the verbal portion. We can also express these scores as percentile ranks by using Table A. John's percentile rank is 84.13 on the math test and only 69.15 on the verbal test.

Problem 6

Occasionally it is necessary to convert a z score to a raw score. The following equation is algebraically equivalent to Eq. (7.1) and can be used in this instance.

$$X = z(\sigma) + \mu.$$

For example, assume you drew a simple random sample of undergraduates at Ohio State and asked each of the students polled their current grade point average on a four-point scale ($A = 4$, $B = 3$, etc.). We will also assume that grade averages for the undergraduates are normally distributed with a mean of 2.1 and a standard deviation of 0.6. What is the grade average of a student who has a z score of 1.5?

$$X = z(\sigma) + \mu$$
$$X = 1.5(0.6) + 2.1$$
$$X = 0.90 + 2.1$$
$$X = 3.00$$

Thus we find that the student has a grade average of 3.00.

7.7 Interpreting the Standard Deviation

In Chapter 6 we noted that an interpretation of the standard deviation is dependent upon an understanding of the normal distribution. Imagine two random samples, *A* and *B,* from a normally distributed population with a mean age of 40. A total of 100 persons are included in sample *A* and 100 persons in sample *B*. We find that the ages of the persons in the two samples are distributed as shown in Fig. 7.7.

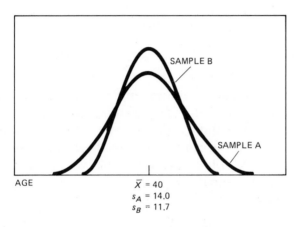

AGE $\bar{X} = 40$
           $s_A = 14.0$
           $s_B = 11.7$

7.7 Two frequency curves with identical means
but different standard deviations.

While both have a mean of 40, it is apparent that sample *A* exhibits more variability than sample *B,* as shown by their frequency curves and respective standard deviations of 14.0 and 11.7. But we can say much more about the two distributions in Fig. 7.7. For instance, we know from Table A that 68.26 (2 × 34.13) percent of the persons in sample *A* are between the ages of 26 and 54 since these values are − 1 and + 1 standard deviation units from the mean (40 + 14 = 54, 40 − 14 = 26 since $s_A = 14.0$). Recall that when we state that $s_A = 14.0$, we are saying that *one* standard deviation in sample *A* equals 14.0. By adding and subtracting one standard deviation (in this case it is 14.0 for sample *A*) to and from the mean of 40, we immediately know the ages equivalent to $z = + 1$ and $z = - 1$ since z scores are measured in standard deviation units. If you are interested in knowing the raw scores that are ± 2 standard deviations from the mean, move 28 (2×14.0) units to the right and left of the mean. Hence, the ages 12 and 68 are 2 standard deviations from the mean in sample A. Their z score equivalents are -2 and $+2$. Looking at Table A you will find that 95.44 (2×47.72) percent of the persons in sample A are between the ages of 12 and 68.

For comparative purposes, consider sample B with its standard deviation of 11.7. The ages equivalent to $z = \pm 1$ can be calculated by adding 11.7 to the mean of 40 and subtracting 11.7 from the mean.

$$40 + 11.7 = 51.7$$
$$40 - 11.7 = 28.3$$

Whereas in sample A we found that the ages corresponding to $z = \pm 1$ were 26 and 54, we find that for sample B the ages are 28.3 and 51.7. This finding supports our earlier contention that sample B is more homogeneous (shows less variability) than sample A.

Summary

In this chapter we demonstrated the value of the standard deviation for comparison of the dispersion of scores in different distributions of a variable, the interpretation of a score with respect to a single distribution, and the comparison of scores on two or more variables. We showed how to convert raw scores into units of the standard normal curve (transformation to z scores). Various characteristics of the standard normal curve were explained. A series of problems demonstrated the various applications of the conversion of normally distributed variables to z scores and z scores to raw scores.

Finally, we discussed the standard deviation as an interpretive tool.

Terms to Remember

Standard normal distribution   A frequency distribution that has a mean of 0, a standard deviation of 1, and a total area equal to 1.00.

Standard score (z) A score that represents the deviation of a specific score from the mean, expressed in standard deviation units.

Exercises

1. Given a normal distribution with a mean of 45.2 and a standard deviation of 10.4, find the standard score equivalents for the following scores.

   a) 55   **b)** 41   **c)** 45.2

   d) 31.5   **e)** 68.4   **f)** 18.9

2. Find the proportion of area under the normal curve between the mean and the following z scores.

   a) -2.05 **b)** -1.90 **c)** -0.25

   d) $+0.40$ **e)** $+1.65$ **f)** $+1.96$

   g) $+2.33$ **h)** $+2.58$ **i)** $+3.08$

3. Given a normal distribution based on 1000 cases with a mean of 50 and a standard deviation of 10, find:

   a) The proportion of area and the number of cases *between* the mean and the following scores:

   60   70   45   25

   b) The proportion of area and the number of cases *above* the following scores:

   60   70   45   25   50

   c) The proportion of area and the number of cases *between* the following scores:

   60–70   25–60   45–70   25–45

4. Below are student Spiegel's scores, the mean, and the standard deviation on each of three tests given to 3000 students.

TEST	μ	σ	SPIEGEL'S SCORE
Arithmetic	47.2	4.8	53
Verbal comprehension	64.6	8.3	71
Geography	75.4	11.7	72

a) Convert each of Spiegel's test scores to standard scores.

b) On which test did Spiegel stand highest? On which lowest?

c) Spiegel's score in arithmetic was surpassed by how many students? his score in verbal comprehension? in geography?

d) What assumption must be made in order to answer the preceding question?

e) Speigel's friend Sue had a z score of -1.1 on the geography test. What was her raw score?

5. On a normally distributed mathematics aptitude test, for females

$$\mu = 60, \qquad \sigma = 10,$$

and for males,

$$\mu = 64, \qquad \sigma = 8.$$

a) Arthur obtained a score of 62. What is his percentile rank relative to the male and female scores?

b) Helen's percentile rank is 73 relative to the female scores. What is her percentile rank relative to the male scores?

c) In comparison with other students of their own sex, who did better, Arthur or Helen?

6. If frequency curves were constructed for each of the following, which do you feel would approximate a normal curve?

a) Heights of a large representative sample of adult American males.

b) Means of a large number of samples with a fixed N (say, $N = 100$) drawn from a normally distributed population of scores.

c) Weights, in ounces, of ears of corn selected randomly from a cornfield.

d) Annual income, in dollars, of the breadwinner of a large number of American families selected at random.

e) Weight, in ounces, of all fish caught in a popular fishing resort in a season.

7. In a normal distribution with $\mu = 72$ and $\sigma = 12$:

a) What is the score at the 25th percentile?

b) What is the score at the 75th percentile?

c) What is the score at the 9th decile?

d) Find the percent of cases scoring above 80.

e) Find the percent of cases scoring below 66.

f) Between what scores do the middle 50% of the cases lie?

g) Beyond what scores do the most extreme 5% lie?

h) Beyond what scores do the most extreme 1% lie?

8. Answer Exercise 7(a) through (h), for

a) $\mu = 72$ and $\sigma = 8$.

b) $\mu = 72$ and $\sigma = 4$.

c) $\mu = 72$ and $\sigma = 2$.

9. Given the following information, determine whether Larry did better on Test I or Test II. On which test did Mindy do better?

	TEST I	TEST II
μ	500	24
σ	40	1.4
Larry's scores	550	26
Mindy's scores	600	25

10. Are all sets of z scores normally distributed? Why or why not?

11. You have just completed interviewing 75 ministers. Initial calculations indicate that the number of years of formal education they have completed is normally distributed with a mean of 13.4 and a standard deviation of 3.0.

a) How many years of education must a minister have completed to be in the top decile?

b) How many years of education must a minister have completed to be 1 standard deviation above the mean?

c) Approximately how many ministers will be within $\pm 2\ z$ scores of the mean?

d) Interpret the standard deviation for this problem. What does it tell you?

12. Your roommate has just received the results of her Graduate Record Examination in the mail. The mean of the verbal section was 510, the standard deviation 102, and the scores were normally distributed. Her reported score was 640, but the computer did not print her percentile rank. She has never had a statistics course and needs your assistance to calculate her percentile rank and help her understand the meaning of her score. State your explanation briefly. Feel free to use visual examples to supplement your answer.

An Introduction to Contingency Tables

8

8.1   Introduction

8.2   Dependent and Independent Variables

8.3   The Bivariate Contingency Table

8.4   Percentaging Contingency Tables

8.5   Existence, Direction, and Strength of a Relationship

8.6   Introduction to Measures of Association for Contingency Tables

8.7   Nominal Measures of Association: Lambda

8.8   Nominal Measures of Association: Goodman's and Kruskal's Tau

8.9   Ordinal Measures of Association

8.10 Goodman's and Kruskal's Gamma

8.11 Somer's *d*

8.1 Introduction

In Chapters 3–7 we have considered one variable at a time, focusing on the frequency distribution, percentiles, measures of central tendency, and measures of dispersion. We now turn to examining the statistical relationship between two variables.

Social scientists have a primary interest in determining the extent to which variables are related in order to understand and predict behavior. Consider the following examples: In what manner is the variable years of educational attainment related to annual income? Are additional years of education a wise investment in terms of a person's potential earning power? Is a person's choice of leisure activities determined by his or her social class? To what extent is group cohesiveness a function of work group size? Is interest in politics related to a woman's parents' political involvement? To what extent is the "welfare experience" shared by two or more generations within a family?

8.2 Dependent and Independent Variables

All of the questions can be answered by analyzing two variables simultaneously. A *bivariate* or two-variable analysis normally involves specifying a dependent and an independent variable. The **dependent variable** (symbolized by Y) is the variable that is being predicted or explained. It is dependent on the **independent variable** (symbolized by X) which is the predictor variable. In an experimental context, the dependent variable is referred to as the criterion variable and the independent variable as the experimental or treatment variable. In the videotape instructional example outlined in Chapter 1, the instruction technique (lecture versus videotape of the lecture) would be the independent or experimental variable (X) and the students' test scores would be the dependent or criterion variable (Y).

Consider the relationship between marital status and employment status of American women. Until relatively recently, it was fairly easy to predict the likelihood of a woman holding a job if we knew her marital status, as married women seldom worked outside the home. Such a relationship can be diagramed as shown in Fig. 8.1.

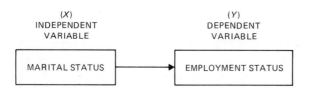

8.1 Relationship Between Marital and Marital Employment Status

Marital status is the independent or predictor variable and employment status is the dependent variable or the variable we wish to predict. The arrow indicates that marital status hypothetically exerts an influence on employment status. Whether a variable is designated the independent variable or the dependent variable is determined by its usage. Marital status could be a dependent variable if we wished to predict it from one's age. Or employment status could be used to predict an adult's self-esteem, in which case it would be the independent variable and self-esteem the dependent variable. The point is that a variable must be placed into a framework or research situation before it can be designated as either independent or dependent.

8.3   The Bivariate Contingency Table

In Chapter 3 you learned about frequency distributions. We will now return to that topic as we begin to consider the relationship between two variables. The U.S. Department of Labor conducted a national study in 1977 of employment patterns among American women.* Using data from that study, let us consider the relationship between employment status and marital status which was diagramed in Fig. 8.1. The frequency distributions for these variables are presented in Table 8.1.

Table 8.1   Tabulations and frequency distributions for employment status and marital status

RESPONDENT	EMPLOYMENT STATUS	MARITAL STATUS
1	Employed	Divorced
2	Not Employed	Never Married
3	Not Employed	Married
.	.	.
.	.	.
.	.	.
200	Employed	Married

EMPLOYMENT STATUS	FREQUENCY	%	MARITAL STATUS	FREQUENCY	%
Employed	(98)	49.0	Never married	(35)	17.5
Not Employed	(102)	51.0	Married	(125)	62.5
Total	(200)	100.0	Divorced	(15)	7.5
			Widowed	(25)	12.5
			Total	(200)	100.0

The percentages of cases within each category reflect the responses of the actual national sample; however, the distributions presented are based on only 200 hypothetical responses in order to simplify the presentation. The frequency distribution was constructed following a personal interview with each of our 200 hypothetical respondents. A portion of this information is presented with the fre-

---

*   U.S. Department of Labor, *U.S. Working Women: A Databook,* 1977.

quency distributions. At the time of the interview, 98 (49%) of the 200 women were employed, and 102 (51%) were not. Table 8.1 also shows that 35 respondents were never married, 125 are married, 15 are divorced, and 25 are widowed. Table 8.1 does not conveniently tell us the marital status of the 98 employed women nor of the 102 women not employed. We wish to know if a woman's marital status (the independent variable) might serve as a useful predictor of her employment status (the dependent variable).

Several other variables could have been considered as predictors of employment status, including the woman's financial needs, educational attainment, the number of children in her home, etc.; however, here we will seek only a partial understanding of why some women work and others do not by examining only the relationship between employment status and marital status.

Table 8.2 presents a bivariate distribution of the two variables. The 21 women in the upper left-hand cell of Table 8.2 have never been married *and* are presently employed. Had the tabulations of the individual respondents in Table 8.1 been shown in their entirety, we would be able to identify the 21 women whose pattern of responses was "employed–never married." We now have available the **conditional distribution** of the two variables and we can examine the distribution of the categories of one variable under the differing conditions or categories of another.

Table 8.2   Contingency table summarizing the relationship between employment status and marital status

(Y) EMPLOYMENT STATUS	(X) MARITAL STATUS				
	Never married	Married	Divorced	Widowed	Total
Employed	21	60	11	6	98
Not employed	14	65	4	19	102
Total	35	125	15	25	N = 200

Table 8.2 has been presented in the manner generally used by social scientists. Normally, the independent variable (X) is the column variable and is placed at the top of the table; the dependent variable (Y) is the row variable and is placed to the left of the table. However, this

convention is not followed by all social scientists, and it is important that you carefully examine tables to avoid misinterpretation. The boxes or cells contain the cell frequencies (for example, 21, 60, 11, 6, etc.) and/or the cell percentages. The column and row totals of the contingency table are referred to as *marginals*. **A contingency table** shows the joint distribution of two variables. There are as many column marginals as there are categories of the variable placed at the top of the table, and their values are the column totals (35, 125, 15, and 25). The row marginals are the row totals (98 and 102). Marginals provide a convenient summary of the distribution of cases for each of the variables. In fact, you have probably noticed that the marginals in Table 8.2 are equivalent to the frequency distributions in Table 8.1.

Table 8.2 is a 2 × 4 table (read "2 by 4 table") because there are two rows and four columns. Tables can have any number of rows and columns, depending on the number of categories of the variables.

8.4 Percentaging Contingency Tables

Contingency tables can be percentaged in three ways, depending on the base (denominator) that is used.* Table 8.3 illustrates the three ways of percentaging a table.

The most common way of percentaging a table (assuming that the independent variable is the column variable) is that of *percentaging down* as in Table 8.3(a) where you see that the percentages in each of the marital status columns total 100%. For example, 60% and 40% in the never married column total 100%. We typically percentage down because we are normally interested in assessing the effect of the independent variable on the dependent variable. When we percentage down, the column marginals (35, 125, 15, and 25) become the base (the denominator) with which the percentages are calculated. Focusing again on the upper left-hand cell, we can calculate that 60.0% of the never married respondents are employed by dividing 21 by 35 which is the column marginal for the never married respondents. *Percentaging down* is also referred to as *percentaging on the independent variable* when the independent variable is the column variable. Percentaging down allows

---

* In Chapter 2 you learned that a proportion is calculated by dividing a quantity in one category by the total of all the categories, and that a proportion can be converted to a percentage when multiplied by 100.

Table 8.3   Three ways of percentaging a contingency table.

Original table

(X)
MARITAL STATUS

(Y) EMPLOYMENT STATUS	Never married	Married	Divorced	Widowed	Total
Employed	21	60	11	6	98
Not employed	14	65	4	19	102
Total	35	125	15	25	N = 200

(a) Percentaging down using column marginals as the base

(X)
MARITAL STATUS

(Y) EMPLOYMENT STATUS	Never married	Married	Divorced	Widowed
Employed	60.0%	48.0%	73.3%	24.0%
Not employed	40.0%	52.0%	26.7%	76.0%
Total	100.0%	100.0%	100.0%	100.0%

(b) Percentaging across using the row marginals as the base

(X)
MARITAL STATUS

(Y) EMPLOYMENT STATUS	Never married	Married	Divorced	Widowed	Total
Employed	21.4%	61.2%	11.2%	6.1%	99.9%*
Not employed	13.7%	63.7%	3.9%	18.6%	99.9%*

*Does not add to 100.0% due to rounding error.

(c) Percentaging on the total using the total number of cases as the base

(X)
MARITAL STATUS

(Y) EMPLOYMENT STATUS	Never married	Married	Divorced	Widowed	Total
Employed	10.5%	30.0%	5.5%	3.0%	
Not employed	7.0%	32.5%	2.0%	9.5%	
Total					100.0%

us to determine the effect of the independent variable by comparing across the percentages within a row, that is, by comparing people in different categories of the independent variable. In Table 8.3(a) we see, by comparing across the employed row, that 60.0% of the never married respondents are employed, 48.0% of the married respondents are employed, as are 73.0% of the divorced respondents and 24.0% of the widowed respondents.

Percentaging across, as illustrated in Table 8.3(b), means that the row marginals are used as the base and that the percentages in both of the employment status rows total 100%. We see that 21.4% of the employed respondents have never married, whereas 13.7% of those respondents who are not employed have never married. Rather than percentaging down and comparing across as we did in Table 8.3(a), we now are percentaging across and comparing up and down. One value of percentaging across in Table 8.3(b) is that a profile of the employed versus the not employed respondents can be established in terms of their marital status. Such a usage does not provide us with any indication of the extent to which the independent variable affects the dependent variable.

A third method of percentaging uses the total number of cases as the base; therefore, the sum of all the cell percentages is 100%. Percentaging on the total (200) as in Table 8.3(c) offers yet another interpretation of the data. The 10.5% figure in the upper left-hand cell indicates that 10.5% or 21 of all the 200 respondents have never married *and* are employed. The joint percentage distribution that results from percentaging on the total, like the second method, also does *not* allow for determining the influence of the independent variable on the dependent variable and is rarely presented; however, it is an appropriate method of percentaging in certain instances.

8.5 Existence, Direction, and Strength of a Relationship

We are now ready to consider whether two variables are statistically related and, if so, in what manner. Table 8.4 illustrates three unique relationships between family income and marital satisfaction. Family income is the independent variable because it is used to predict marital satisfaction, the dependent variable.

Table 8.4   Illustration of three types of relationships in contingency tables

(a) No relationship		
	(X) FAMILY INCOME	
(Y) MARITAL SATISFACTION	Low	High
High	60%	60%
Low	40%	40%
Total	100%	100%

(b) Positive relationship		
	(X) FAMILY INCOME	
(Y) MARITAL SATISFACTION	Low	High
High	40%	60%
Low	60%	40%
Total	100%	100%

(c) Negative relationship		
	(X) FAMILY INCOME	
(Y) MARITAL SATISFACTION	Low	High
High	60%	40%
Low	40%	60%
Total	100%	100%

Labeling the categories of ordinal or interval scale variables normally follows a standard convention.* The lowest category of the independent variable (the column variable) is placed at the left, and the highest category of the independent variable is placed at the right. When labeling the dependent variable (the row variable), the highest category is placed at the top and the lowest category at the bottom.† You are cautioned that this convention is not followed by all social scientists; therefore, you should carefully examine all tables before interpreting the data.

Comparison of the conditional distribution of the two variables determines whether a relationship exists between the variables. In Table 8.4(a), for instance, all the conditional relationships are identical in that

---

* This placement of the categories of a nominal-level variable in a contingency table is determined by the user's personal preference because the categories are not ordered.

† This placement of the variables and variable categories corresponds to other statistical graphs including the scatter diagram presented in Chapters 9 and 10.

60% of the low-income marriages are characterized as highly satisfying, as are 60% of the high-income marriages. There is also no difference when we compare percentages across the low marital satisfaction row, and, therefore, knowing a couple's income does not help us predict their marital satisfaction. The difference in these percentages (0% in both instances), which was calculated by comparing the percentages within the categories of the dependent variable between the two extreme categories of the independent variable, is referred to as the **percentage difference** or **epsilon** (ϵ). A relationship is said to exist between two variables when none of the epsilons are equal to 0.

Table 8.4(b) has also been percentaged down due to the placement of the independent variable. Comparing the percentages across the high category of marital satisfaction we note a 20% percentage difference. Forty percent of the low-income marriages versus 60% of the high-income marriages reportedly are highly satisfying: Thus epsilon equals 20. The reverse is true upon comparing the percentages across the low category of marital satisfaction where we see that a larger percentage (60%) of the low-income marriages are characterized by low marital satisfaction than are high-income marriages (40%), and, therefore, epsilon equals 20. We can conclude that income is moderately associated with marital satisfaction and that it is a **positive relationship**. Relationships between two variables can either be positive or negative. A positive relationship is one in which high scores on one variable tend to be associated with high scores on the other variable, and conversely low scores on one variable tend to be associated with low scores on the other variable. Naturally, we cannot specify the direction of a relationship when one or both of the variables are nominal level, as in Table 8.2. If the categories of the variables are not rank ordered, we can only speak of the existence and strength of a relationship.

Table 8.4(c) provides an example of a **negative relationship.** Variables are said to be negatively related when high scores on one variable tend to be accompanied by low scores on the other. Conversely, low scores on one variable tend to be associated with high scores on the other. We see in Table 8.4(c) that high-income marriages tend to be less satisfying than low-income marriages. The variables have a moderate, negative relationship.

The strength of a relationship can range from nonexistent to perfect. A *nonexistent relationship* is one in which knowledge of the independent variable does not improve our prediction of the dependent variable or when the variables are unrelated, as in Table 8.4(a). A **perfect relationship** is one in which knowledge of the independent variable allows a perfect prediction of the dependent variable or vice versa.

8.6 Introduction to Measures of Association for Contingency Tables

Several measures of association appropriate for contingency tables are presented in the remainder of this chapter, none of which require that we make assumptions about the population distribution. Those measures that have been developed specifically for nominal-level data provide a measure of the strength of relationship. The measures for ordinal-level or interval level data include both a measure of the strength and the direction of the relationship.

All of the measures are either symmetric or asymmetric. **Symmetric measures** of association make no distinction between the independent and dependent variable. Symmetric measures provide a single summary value of mutual or two-way association. If we wished to establish the relationship between two items in a job satisfaction scale, a symmetric measure of association would be appropriate. We would be interested only in determining the relationship between the two items, neither of which would be classified as independent or dependent.

An **asymmetric measure** of association is a summary value of one-way association and requires that the independent and dependent variables be specified. Two summary values are available when an asymmetric statistic is calculated: one value results from designating the row variable as independent, and a second value results from designating the column variable as independent.* An ordinal-level asymmetric measure of association in which the column variable, family income, is designated as the independent variable would be an appropriate measure to use with a table like Table 8.4. At the same time, if you were interested in the effect of marital satisfaction on family income, the row variable, marital satisfaction, could serve as the independent variable, and a second summary value could be calculated.

---

* Unless noted otherwise, we will assume that X designates the independent variable and that it is the column variable.

8.7 Nominal Measures of Association: Lambda

Computing percentage differences, or epilson (ϵ), for data in contingency form is only one way of determining the existence of a relationship between two variables. In addition, social scientists are interested in the degree to which a variable is helpful in predicting an attitude or behavior as measured by a second variable. **Lambda (λ)** is a statistic used to evaluate the usefulness of one variable in predicting another. It helps answer the question: Do we merely need to know the marginal values of the dependent variable in order to predict, or does it help to partition the dependent variable into categories of the independent variable?

Lambda is a measure of association for nominal-level variables, based on the logic of **proportional reduction in error,** known as **PRE.***
It is useful for any size table. While computational procedures are available for asymmetrical and symmetrical lambda, only asymmetrical lambda will be presented.†

Let's consider the logic of PRE measures in general and lambda specifically using the data first presented in Table 8.2, repeated here as Table 8.5.

Table 8.5   Employment status by marital status

	(X) MARITAL STATUS				
(Y) EMPLOYMENT STATUS	Never married	Married	Divorced	Widowed	Total
Employed	21	60	11	6	98
Not employed	14	65	4	19	102
Total	35	125	15	25	$N = 200$

*   Lambda can be computed using ordinal- or interval-level data; however, lambda ignores ranks and intervals, and information would be lost. Lambda or another appropriate statistic such as Goodman's and Kruskal's tau (which is presented in the following section) must be computed whenever any of the variables are measured at the nominal level.

†   Symmetrical lambda measures the mutual influence of two variables on each other.

If you knew that 102 women in a sample of 200 women were not employed and you were asked to predict or guess the employment status of the 200 women in the sample without knowledge of their marital status, you would guess that none are employed, given the row marginals. Such a strategy would result in your correctly predicting 102 of the 200 women in the sample because 102 are not employed. You would have incorrectly predicted the employment status of 98 women. Therefore your guess of "not employed" would result in a slightly reduced number of errors.

If, on the other hand, you were asked to predict employment status while knowing marital status, you could make more refined guesses. If you knew a woman had been divorced *or* had never married, you would no longer want to guess "not employed" because the majority of women in both marital status categories are employed.

In order to analyze the usefulness of the partitioning of the data, social scientists calculate lambda, which expresses the relationship between the errors produced with the first type of guess and the errors produced with the second type of guess. To formalize this logic, statisticians refer to the predictions from tables as being based on one of two rules. Both rules govern procedures aimed at reducing the error in predictions. The first rule assumes no knowledge about the independent variable. *Rule I for lambda,* which is based on no knowledge of the independent variable, requires that we predict the modal category of the dependent variable.* In essence, we have only used the row marginals of the dependent variable to make our prediction.

Given additional information about our sample of women, we should be able to improve our prediction of their employment status. The data in Table 8.5 indicate, for example, that the women's employment status is related to their marital status. Now let us predict employment status given information about the respondent's marital status, which is the independent variable. We now use *Rule II for lambda,* which assumes knowledge of the independent variable. Rule II requires that we predict the within-category mode of the independent variable. Now we would predict that all of the never married respondents are employed, since the within-category mode of the never married group is the employed category of the dependent variable. Rule II also requires that we predict all of the married respondents are not employed, all of

---

* The mode of a category was defined in Chapter 5 as the score that occurs with the greatest frequency.

the divorced respondents are employed, and all of the widowed respondents are not employed.

Let's see how many errors in prediction we would make if we used the second rule. We would have correctly predicted the employment status of 21 out of the 35 never married respondents (therefore, 14 errors in prediction), 65 of the 125 married respondents (60 errors), 11 of the 15 divorced respondents (4 errors), and 19 of the 25 widowed respondents (6 errors). We have made a total of 84 prediction errors $(14 + 60 + 4 + 6)$ given the respondents' marital status (using Rule II), compared to 98 prediction errors *without* knowledge of the respondents' marital status (using Rule I). Thus we have reduced our prediction errors by 14 $(98 - 84 = 14)$ now that we have information on the independent variable. Lambda can be calculated using the following equation which is based on shifting from Rule I to Rule II once knowledge of the independent variable is available:

$$\lambda = \frac{(\text{errors using Rule I}) - (\text{errors using Rule II})}{\text{errors using Rule I}} . \quad \textbf{\textit{(8.1)}}^*$$

Substituting into Eq. (8.1) we find that lambda equals $+0.14$.

$$\lambda = \frac{98 - 84}{98} = \frac{14}{98} = +0.14.$$

While lambda is easily calculated using Eq. (8.1) in a table with few rows and cells, this is not the case with a larger table, and a computational equation is available for these purposes:

$$\lambda = \frac{\Sigma \text{ maximum frequency } (X) - \text{ maximum frequency } (Y)}{N - \text{ maximum frequency } (Y)} \quad \textbf{\textit{(8.2)}}$$

where

> maximum frequency (X) = the within-category mode for each category of the independent variable,
>
> maximum frequency (Y) = the modal frequency of the dependent variable, and
>
> N = sample size.

---

* Equation (8.1) assumes Y is the dependent variable. Under these circumstances, lambda is frequently written λ_{yx} with the first subscript designating the dependent variable. Subscript variables, for example, x and y, will appear in lowercase italics throughout the text.

In words, Eq. (8.2) can be read:

$$\text{lambda} = \frac{\frac{\begin{array}{c}\text{sum of the within-category}\\ \text{modes of the independent}\\ \text{variable}\end{array}}{\text{sample size}} - \begin{array}{c}\text{modal frequency of}\\ \text{the dependent variable}\end{array}}{- \begin{array}{c}\text{modal frequency of the}\\ \text{dependent variable}\end{array}} \cdot$$

Applying Eq. (8.2) to the data in Table 8.5:

$$\lambda = \frac{(21 + 65 + 11 + 19) - 102}{200 - 102}$$

$$= \frac{14}{98}$$

$$= +0.14.$$

8.7.1 Interpretation of Lambda

The value of lambda is determined by the proportional reduction in error when predicting the dependent variable upon shifting from prediction Rule I to prediction Rule II. Formally stated, the prediction rules for lambda are:

> Rule I: Guess the modal category of the dependent variable.

> Rule II: Guess the within-category mode of the independent variable.

Lambda measures the strength of the relationship and ranges from 0 when the variables are not related to 1 when they are perfectly related. If lambda equals 0, knowledge of the independent variable is of no value when predicting the dependent variable, and we cannot predict any better when shifting from Rule I to Rule II than we can by using only Rule I. When lambda equals 1, knowledge of the independent variable allows you to predict the dependent variable perfectly.

A lambda of 0.14 indicates a weak relationship between marital status and employment status and our reduction of the number of prediction errors by 14% when moving from Rule I to Rule II. Rule I resulted in an error rate of 98/200 or 49%, and Rule II resulted in an error rate of 84/200 or 42%. Shifting from Rule I to Rule II reduced the rate 7% (49% − 42% = 7%) which is the *absolute* reduction in error. To calculate *proportional* reduction in error we divide the 7% absolute reduction in error by the original error rate of 49%, yielding a lambda of +0.14.

8.7.2 Limitations of Lambda

In spite of lambda's strengths, which include use with nominal-level data, no assumptions concerning the distribution of the variables (a normal distribution is not necessary), and use with any size table, lambda does have two limitations. The first is that lambda is always positive in sign and hence gives us no indication of the direction of the relationship, a limitation that stems from its use with nominal-level data. The second limitation results from the prediction rules upon which lambda is based. Lambda equals 0 whenever *all* the within-category modes of the independent variable occur in the row containing the modal category of the dependent variable. In Table 8.6, lambda equals 0, yet it is clear that the variables are related as there is a 23.2% (85.7% − 62.5% = 23.2%) percentage difference between males and females who desire to have children, an 8.9% percentage difference be-

Table 8.6   Desire to have children by sex of respondent (hypothetical study of college students)

(Y) DESIRE TO HAVE CHILDREN?	(X) SEX OF RESPONDENT		
	Male	Female	Total
Yes	85.7 (60)	62.5 (50)*	110
No	8.6 (6)	17.5 (14)	20
Undecided	5.7 (4)	20.0 (16)	20
Total	100.0 (70)	100.0 (80)	N = 150

* While a variety of conventions are available, we prefer that when cell percentages *and* frequencies are presented in a contingency table, the frequencies be placed in parentheses. Contingency tables will be routinely percentaged on the independent variable unless stated otherwise.

tween males and females who have no desire for children and a 14.3% percentage difference between the undecided males and females. For males *and* females, the within-category modes of the independent variable (60 and 50) are both located in the row containing the modal category of the dependent variable; therefore, shifting from Rule I to Rule II will result in no reduction in prediction error; and lambda equals 0.

Using Eq. (8.2):

$$\lambda = \frac{\Sigma \text{ maximum frequency } (X) - \text{maximum frequency } (Y)}{N - \text{maximum frequency } (Y)}$$

$$= \frac{(60 + 50) - 110}{150 - 110}$$

$$= 0$$

While we have found that lambda equals 0, it is apparent that a relationship does exist between sex and desire to have children. As was noted in Section 8.5, we can only conclude that there is no relationship between two variables if the percentage differences between categories of the independent variable are zero. Goodman's and Kruskal's tau, which is based on a different set of prediction rules and does not share this limitation with lambda, should be calculated whenever the distribution of cases on the dependent variable is extremely skewed toward a single category, as was shown in Table 8.6.

8.8 Nominal Measures of Association: Goodman and Kruskal's Tau

Goodman's and Kruskal's tau (τ), which is based on a different set of prediction rules than is lambda, is an asymmetric measure of association for nominal-level variables that can be computed for any size table. Its value ranges from 0 (no reduction in error) when the variables are unrelated to 1 (total reduction in error) when there is a perfect relationship. We will again assume for instructional purposes that X is the independent variable.

Whereas the prediction rules for lambda were based on modal categories, the prediction rules for tau are based on the random assignment of cases. Lambda, you will recall from the previous section, equals 0 whenever all the within-category modes of the independent variable occur uniformly across a single row of the dependent variable. Because tau does not share this limitation with lambda, most social scientists prefer to use tau rather than lambda.

Tau is calculated in accordance with the following prediction rules:

Rule I: All cases are randomly assigned into the categories of the dependent variable while retaining the marginal distribution of the dependent variable.

Rule II: All cases are randomly assigned within the categories of the independent variable to a category of the dependent variable.

Rule I ignores information about the independent variable and is based only on the marginal distribution of the dependent variable. The data in Table 8.7 will serve as our example as we discuss tau. The respondents' attitude toward labor unions is the dependent variable and the respondents' region of orientation, that is, where they spent the first 15 years of their life, is the independent variable.

Table 8.7   Attitude toward labor unions by region of orientation

(Y) ATTITUDE TOWARD LABOR UNIONS	(X) REGION OF ORIENTATION Non-South	South	Total
Prounion	66.7 (40)	37.5 (15)	(55)
Antiunion	33.3 (20)	62.5 (25)	(45)
Total	100.0 (60)	100.0 (40)	N = 100

When operating under Rule I, we assume that we assign respondents to a category of the dependent variable based only on knowledge that 55 are prounion and 45 are antiunion and that we do not have information on individual cases. One by one we make the assignments, knowing we will make some errors. Due to the random assignment of cases, some of the 55 respondents who are placed in the prounion category could actually belong in the antiunion category. Hence, some of the cases will be assigned incorrectly with respect to their actual attitude toward unions, and some will be assigned correctly. For example, if a respondent were prounion, but the random assignment specified by Rule I resulted in his or her being placed in the antiunion category, an error in assignment would have been made. If the same respondent had been assigned to the prounion category, an error would not have been made. We can calculate the number of errors we would expect in each of the categories of the dependent variable in the following manner.

We can expect, in the long run, an error rate of 45% in classification using Rule I for the prounion category because 45 out of 100 of the respondents are antiunion and some of these respondents would be placed in the prounion category of the dependent variable. A 45% error rate would yield 24.75 errors for the prounion category because we are classifying 55 cases. Note that we must multiply the probability of error by the number of cases randomly assigned ($0.45 \times 55 = 24.75$). In essence, the probability of an antiunion respondent being randomly assigned to the prounion category is 45 out of 100 since the antiunion respondents constitute 45% of the sample.

An error rate of 55% would result from randomly assigning people to the antiunion category since 55 out of 100 of the respondents are prounion and some of them would be placed into the antiunion category. A total of 24.75 errors would also be made in the long run for the antiunion category because we are classifying 45 cases ($0.55 \times 45 = 24.75$). Overall, we would expect a total of 49.50 classification errors ($24.75 + 24.75$) using Rule I.

Rule II is based upon knowledge of the independent variable and requires that all cases be randomly assigned within categories of the independent variable to a category of the dependent variable. Antiunion cases constitute 33.3% of the non-South respondents and, therefore, we can expect 33.3% (or 20/60) of the non-Southern bred respondents to erroneously be assigned to the prounion category and 66.7% (or 40/60) of the non-Southern respondents who are prounion to erroneously be assigned to the antiunion category. Consequently, the non-South column would contribute 13.32 errors (0.333×40) plus 13.34 errors (0.667×20) or a total of 26.66 errors.*

---

*   The errors contributed by the two cells are not equal due to rounding.

Using the same procedure we can calculate the number of errors that will be made in the South column when we apply Rule II. We would have an error rate of 62.5% in the South prounion cell because 62.5% (or 25/40) of the Southern cases are antiunion and an error rate of 37.5% in the South/antiunion cell because 37.5% (or 15/40) of the Southern cases are prounion. The South column would contribute 9.375 errors (0.625 × 15) plus 9.375 errors (0.375 × 25) or a total of 18.75 errors. Employing Rule II, a total of 45.41 errors (26.66 + 18.75) can be expected, as opposed to 49.50 errors using Rule I. Tau can now be calculated using the following equation which involves shifting from Rule I to Rule II:

$$\text{tau} = \frac{(\text{errors using Rule I}) - (\text{errors using Rule II})}{\text{errors using Rule I}} = \frac{E_1 - E_2}{E_1}$$

$$= \frac{49.50 - 45.41}{49.50} \qquad (8.3)$$

$$= 0.08.$$

The procedure we have just discussed allows a conceptual understanding of tau; however, it is far more cumbersome than the computational approach using Eq. (8.6) below. The errors resulting from Rules I and II can be calculated using Eqs. (8.4) and (8.5), assuming X is the independent variable:

$$\text{errors using Rule I} = \sum \left(\frac{N - F_y}{N} (F_y) \right) \qquad (8.4)$$

$$\text{errors using Rule II} = \sum \frac{(F_x - f)f}{F_x} . \qquad (8.5)$$

The computational equation for tau assuming X is the independent variable is:

$$\tau = \frac{N \sum \frac{f^2}{F_x} - \sum F_y^2}{N^2 - \sum F_y^2} \qquad (8.6)$$

where

N = sample size,
F = marginal frequencies,
f = cell frequency,
Y = the dependent variable,
X = the independent variable.

Applying Eqs. (8.4) and (8.5) we can calculate the number of errors using Rules I and II for Table 8.7. Using Eq. (8.4):

$$\text{errors using Rule I} = \sum \left(\frac{N - F_y}{N} (F_y) \right)$$

$$= \frac{100 - 55}{100} (55) + \frac{100 - 45}{100} (45) = 49.50.$$

Note: There are as many terms to be summed when computing the number of errors using Rule I as there are categories of the dependent variable.

Using Eq. (8.5):

$$\text{errors using Rule II} = \sum \frac{(F_x - f)f}{F_x}$$

$$= \frac{(60 - 40)40}{60} + \frac{(60 - 20)20}{60} + \frac{(40 - 15)15}{40}$$

$$+ \frac{(40 - 25)25}{40}$$

$$= 45.41$$

Note: There are as many terms to be summed when computing the number of errors using Rule II as there are cells in the table.

Equation (8.3) was previously written symbolically in the following manner:

$$\tau = \frac{E_1 - E_2}{E_1},$$

where

$$E_1 = \text{errors using Rule I,}$$
$$E_2 = \text{errors using Rule II.}$$

Therefore, using the information from Eqs. (8.4) and (8.5) to calculate tau:

$$\tau = \frac{49.50 - 45.41}{49.50} = 0.08.$$

Or, using Eq. (8.6):

$$\tau = \frac{N\Sigma \frac{f^2}{F_x} - \Sigma f_y^2}{N^2 - \Sigma F_y^2}$$

$$= \frac{100\left(\frac{40^2}{60} + \frac{20^2}{60} + \frac{15^2}{40} + \frac{25^2}{40}\right) - (55^2 + 45^2)}{100^2 - (55^2 + 45^2)}$$

$$= 0.08.$$

8.8.1 Interpretation of Tau

Tau has a PRE interpretation that reflects the reduction in error when predicting the dependent variable upon shifting from prediction Rule I to Rule II. The tau of 0.08 we have calculated indicates an 8% reduction in error when predicting attitude toward labor unions, given knowledge about the respondents' region of orientation. We also can say that region of orientation is a weak predictor of attitude toward labor unions or that there is a weak relationship between the variables.

When tau equals zero we conclude that knowledge of the independent variable is of no value when predicting the dependent variable or that the variables are statistically unrelated. A tau of 1 indicates a perfect relationship between the variables or that shifting from Rule I to Rule II reduces all, that is, 100%, of the error associated with Rule I.

8.8.2 Limitations of Tau

Tau shares the limitation with all nominal-level statistics that its sign is always positive and its interpretation is limited to the strength and not the direction of the relationship. Tau can only equal 1 when the number of categories of the independent variable equals or exceeds the number of categories of the dependent variable because in a perfect relationship all cases associated with a particular column of an independent variable must occur in the same row of the dependent variable.

8.9 Ordinal Measures of Association*

Measures of association for contingency tables using ordinal-level (or interval-level) data are based on the logic of **pair-by-pair comparison.** We will present Goodman and Kruskal's gamma and Somer's *d*, two of the measures most frequently used by social scientists as ordinal measures of association.

8.9.1 The Logic of Pair-by-Pair Comparison

Each case in a contingency table can theoretically be paired with every other case in the table. In fact, we know the number of possible pairs in any table equals $N(N - 1)/2$. A table with 50 cases, for example, would contain 1225 possible pairs (50 × 49/2).

Table 8.8 contains four cases, *A, B, C,* and *D,* which can form six distinct pairs. Upon pairing cases *A* and *B* in Table 8.8 we see that *A* is lower on variable *X* than is case *B* (*A*'s *X* value is young and *B*'s is middle-aged), but *A* is higher on *Y* than is case *B*. Table 8.9 presents a summary of the relationships between all pairs of cases.

Table 8.8   Income by age

(Y) INCOME	Young	(X) AGE Middle-Aged	Old
High	A		D
Medium		B	
Low			C

A **discordant pair** is one in which the two cases are ranked in the opposite order on both variables. A **concordant pair** is one in which the two cases are ranked on the same order on both variables. A **tied pair** is one in which the two cases are ranked similarly on one or both of the variables. To the extent that a pattern emerges in which discordant pairs

---

*   Gamma and Somer's *d* can be used with interval-level data.

Table 8.9   Pairs of cases in Table 8.8

PAIR	SUMMARY	RELATIONSHIP
AB	A lower on X than is B A higher on Y than is B	Discordant
AC	A lower on X than is C A higher on Y than is C	Discordant
AD	A lower on X than is D A and D tied on Y	Tied
BC	B lower on X than is C B higher on Y than is C	Discordant
BD	B lower on X than is D B lower on Y than is D	Concordant
CD	C and D tied on X C lower on Y than is D	Tied

predominate, the relationship between the variables is negative as shown in Table 8.10(a). A situation in which concordant pairs predominate and which is characterized by a positive relationship is shown in Table 8.10(b).

Table 8.10   Concordant and discordant patterns in contingency tables

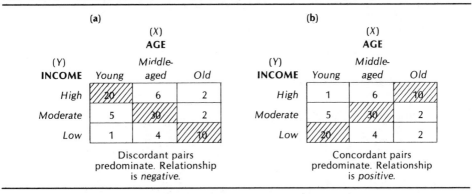

Discordant pairs predominate. Relationship is *negative.*

Concordant pairs predominate. Relationship is *positive.*

The diagonals have been shaded to highlight the predominant pattern in the tables. In Table 8.10(a) you will note that low scores on X *tend* to be paired with high scores on Y and vice versa. Gamma and Somer's d are sensitive to patterns such as these, and their values reflect the extent to which concordant and discordant pairs predominate. If gamma or Somer's d were calculated for Table 8.10(b), we would anticipate a positive relationship since the majority of cases lie on the positive diagonal; that is, low scores on X tend to be paired with low scores on Y, and high scores on X tend to be paired with high scores on Y.

8.10 Goodman and Kruskal's Gamma

Gamma (G)* is a symmetric measure of association for ordinal-level or interval-level data. No specification of which variable is independent and which is dependent is therefore necessary, and only one value of gamma is possible. Gamma ranges from -1 to $+1$, has a proportional reduction in error (PRE) interpretation, and is based on the logic of pair-by-pair comparison. Furthermore, gamma can be calculated for any size table. [†]

The equation for computing gamma compares the number of concordant pairs (C) to the number of discordant pairs (D). Gamma ignores ties, a weakness we will address later. The computation of gamma for Table 8.8 is shown in Eq. (8.7). Note that the sign of gamma in this instance is negative because discordant pairs outnumber concordant pairs 3 to 1.

$$G = \frac{C - D}{C + D}$$

(8.7)

where

C = number of concordant pairs, and
D = number of discordant pairs.

$$G = \frac{1 - 3}{1 + 3} = \frac{-2}{4} = \frac{-1}{2} = -0.5.$$

---

*   Most social scientists symbolize gamma by G, although some prefer the lowercase Greek letter (γ).

†   Yule's Q is a special case of gamma which was developed specifically for 2 × 2 tables. The values of G and Q are equivalent for 2 × 2 tables and are subject to the same interpretation and limitations. Social scientists generally prefer gamma because it can be used with any size contingency table containing ordinal-level or interval-level data.

8.10.1 Calculating Concordant and Discordant Pairs in Contingency Tables

Determining the number of concordant and discordant pairs for contingency tables with many cases is more complicated than the previous example which included only four cases. The data in Table 8.11 are hypothetical results of a study of the relationship between political liberalism and year in college for a sample of college students.

Table 8.11   Political liberalism by year in college

(Y) POLITICAL LIBERALISM	(X) YEAR IN COLLEGE Freshman	Sophomore	Junior	Senior	Total
Very liberal	a 2	b 3	c 7	d 10	22
Moderately liberal	e 8	f 12	g 9	h 7	36
Not liberal	i 10	j 5	k 4	l 3	22
Total	20	20	20	20	N = 80

Concordant pairs   The cells have been labeled with letters to facilitate the discussion. The lowest category of the independent variable, year in college, is freshman, and the highest category is senior. The not liberal category of the dependent variable is scored low, and the very liberal category scored high. The scores in Table 8.11, with obvious exceptions, tend to follow the *positive diagonal* connecting cells *i* and *d*.* We anticipate a positive relationship between *X* and *Y* due to the patterning of cases that are concentrated in cells *i, f, g,* and *d.* The positive diagonal in any contingency table connects the low-low cells of the dependent and independent variable (not liberal and freshman, therefore, cell *i* in Table 8.11) with the high-high cells of the dependent and independent variable (very liberal and senior, therefore, cell *d* in Table

---

*   Table 8.10(a) is characterized by a negative diagonal.

8.11). The negative diagonal always connects the low-high cells of the dependent and independent variable (not liberal and senior, therefore, cell *l* in Table 8.11) with the high-low cells of the dependent and independent variables (very liberal and freshman, therefore, cell *a* in Table 8.11).

The first step in identifying the concordant pairs in any contingency table is to locate the cell on the positive diagonal that represents a low score on *X* and *Y*. The procedure we will follow for computing gamma assumes that the table is laid out in the standard manner described in Section 8.5. In this example we must therefore begin at cell *i*. The cases that are concordant with the cases in cell *i* lie in those cells to the right and above cell *i*.* Cells concordant with cell *i* are cells *b, c, d, f, g,* and *h* because the cases in these cells are higher on both variables than the cases in cell *i*. The cells contain 3, 7, 10, 12, 9, and 7 cases, respectively, for a total of 48 cases. We can calculate the number of concordant *pairs* we have identified so far by multiplying the number of cases in cell *i* by the total number of cases in those cells above and to the right of cell *i;* therefore, we have identified 10 × 48 or 480 concordant pairs. We continue this procedure until we have counted all concordant pairs.

Our focal cell now is cell *j* because we must systematically move across the row, then up to the next row and across, etc., until we have identified all concordant pairs. We note that only cells *c, d, g,* and *h* are above and to the right of cell *j*. The cases in these four cells are concordant with the cases in cell *j* because all of the cases are higher on both variables than are the cases in cell *j*. A total of 33 cases (7 + 10 + 9 + 7) lie in these cells. Multiplying by the 5 cases in cell *j*, we now have identified 165 (5 × 33) additional concordant pairs. Continuing, we see that the 4 cases in cell *k* form concordant pairs with the cases in cells *d* and *h* for 68 (4 × 17) concordant pairs. The 8 cases in cell *e* are concordant with those in cells *b, c,* and *d,* and the 12 cases in cell *f* are concordant with the cases in cells *c* and *d*. Finally, the 9 cases in cell *g* are concordant with the cases in cell *d*. The calculations to determine the number of concordant pairs are summarized in Table 8.12.

---

* The initial focal cell for computing concordant pairs could have been cell *d*, which lies at the opposite end of the positive diagonal. If we began with cell *d*, we would locate cells to the left and below cell *d* because they are concordant with cell *d*. Hence, either cell *i* or *d* could have been chosen as the starting point.

Table 8.12   Calculation of concordant pairs

FOCAL CELL	PAIRED CELLS	CALCULATION OF CONCORDANT PAIRS	
i	b, c, d, f, g, h	10(3 + 7 + 10 + 12 + 9 + 7) =	480
j	c, d, g, h	5(7 + 10 + 9 + 7) =	165
k	d, h	4(10 + 7) =	68
e	b, c, d	8(3 + 7 + 10) =	160
f	c, d	12(7 + 10) =	204
g	d	9(10) =	90
		Total concordant pairs =	1167

Discordant pairs   Identifying the discordant pairs of cases involves a procedure similar to identifying the concordant pairs. Our focal cell lies on the *negative diagonal* that runs from cell *a* to cell *l* in Table 8.11. We begin with cell *a* and work down the diagonal locating those cells that contain cases whose value is less than the cases in cell *a* on both the independent and dependent variable. Therefore, cells *f, g, h, j, k,* and *l,* all of which lie below and to the right of cell *a,* contain cases discordant with those in cell *a.** We would then move systematically across and down to identify the remaining focal cells which include cells *b, c, e, f,* and *g.* The calculations necessary to determine the discordant pairs are presented in Table 8.13.

Table 8.13   Calculation of discordant cells

FOCAL CELL	PAIRED CELLS	CALCULATION OF CONCORDANT PAIRS	
a	f, g, h, j, k, l	2(12 + 9 + 7 + 5 + 4 + 3) =	80
b	g, h, k, l	3(9 + 7 + 4 + 3) =	69
c	h, l	7(7 + 3) =	70
e	j, k, l	8(5 + 4 + 3) =	96
f	k, l	12(4 + 3) =	84
g	l	9(3) =	27
		Total discordant pairs =	426

---

*   The discordant pairs could also be calculated by beginning with cell *l,* which is located at the opposite end of the negative diagonal, and locating cells above and to the left. The choice of either starting point is a matter of personal preference.

We have found that concordant pairs predominate and can now calculate gamma, making use of Eq. (8.7).

$$G = \frac{C - D}{C + D} = \frac{1167 - 426}{1167 + 426} = +0.465.$$

8.10.2 Interpreting Gamma

Gamma ranges from -1.00 when there is a perfect negative relationship between the variables (all untied pairs are discordant) to $+1.00$ where there is a perfect positive relationship (all untied pairs are concordant).* When the variables are unrelated, gamma equals 0 because there are an equal number of concordant and discordant pairs.

The gamma of $+0.465$ indicates a moderate, positive relationship between year in school and political liberalism, or in other words, the seniors at the college we surveyed are more politically liberal than are the freshmen. Our data of course do not allow us to know if the observed relationship is due to the liberalizing effect of a college education or if natural aging by four years would have the same liberalizing effect. Because gamma has a PRE interpretation, we can also say that using knowledge of the independent variable (year in college) to predict the order of the cases on the dependent variable (political liberalism), we have reduced our errors by 46.5%. However, because gamma is a symmetric measure of association, we can also say that, given knowledge of the dependent variable, we can reduce our errors by 46.5% when predicting the independent variable.

8.10.3 Limitations of Gamma

Gamma ignores tied pairs, and hence a gamma value may reflect far less than all the possible pairs. There are $N(N - 1)/2$ or 3160 possible pairs of cases in Table 8.11. Of these pairs, 1593 (426 + 1167) are discordant *or* concordant. Tied pairs total 1567 (3160 − 1593), and therefore, nearly 50% of all the possible pairs have been ignored in the calculation

---

*   Recall that a tied pair is one in which two cases are ranked similarly on one or both of the variables.

of gamma. As a result, gamma tends to overstate the actual relationship. Indeed, it is possible for gamma to equal -1 or $+1$ when all but one pair of cases is tied! Somer's *d,* which was developed specifically to overcome the problem, is discussed in the following section.

Gamma, along with all other ordinal-level statistics, can only detect a straight-line relationship between the variables. If a horseshoe-shaped curve best described the relationship between the variables, for example, the value of gamma would be very low. Naturally you must learn to identify situations such as this in which a nonlinear* relationship exists, and not compute gamma.

In spite of these drawbacks, gamma is the most frequently used measure of association for ordinal-level and interval-level data in contingency form.

8.11 Somer's *d*

Somer's *d* is an asymmetrical measure of association for ordinal-level data that incorporates tied ranks. It is similar to gamma in that it is based on pair-by-pair comparison. Two values of *d* (d_{yx} where *Y* is the dependent variable and d_{xy} in which *X* is treated as the dependent variable) are possible; however, here we will assume *Y* is dependent. The computational equation requires that we determine the number of concordant pairs (*C*), the number of discordant pairs (*D*) and T_y which is the number of pairs with different *X* values, but with the same *Y* value. Thus the cases in cell *a* are tied with those in cells *b, c,* and *d* because all are ranked very liberal on *Y* yet are in differing years in college. The computational procedure for Somer's *d* (with *Y* dependent) is presented in Eq. (8.8):

$$d_{yx} = \frac{C - D}{C + D + T_y} \qquad (8.8)$$

where

T_y = pairs with different *X* values but with the same *Y* values,
C = number of concordant pairs, and
D = number of discordant pairs.

---

* *Nonlinear* means not characterized by a straight line.

The logic of including ties on the dependent variable when calculating a measure of association for contingency data is that a tie on the dependent variable given differing X values suggests that X exerts no influence on Y.* Rather than ignoring ties as does gamma, Somer's d, which is a more conservative measure, includes ties on the dependent variable to insure a more accurate reflection of the true strength of association. Somer's d must always be less than or equal to gamma when calculated for the same contingency table due to the inclusion of the T_y term in the denominator. While Somer's d does not have a direct PRE interpretation, it has become very popular with social scientists in the past decade. Somer's d_{yx} has been calculated for the data in Table 8.11 (repeated here as Table 8.14).

Table 8.14 Computation of d_{yx} for the data in Table 8.11

		(X) YEAR IN COLLEGE				
(Y) POLITICAL LIBERALISM		Freshman	Sophomore	Junior	Senior	Total
Very liberal		a 2	b 3	c 7	d 10	22
Moderately liberal		e 8	f 12	g 9	h 7	36
Not liberal		i 10	j 5	k 4	l 3	22
Total		20	20	20	20	N = 80

$$d_{yx} = \frac{C - D}{C + D + T_y}$$

$$C = 1167$$
$$D = 426 \left.\right\} \text{(From Section 8.10.1)}$$

$$T_y = 2(3 + 7 + 10) + 3(7 + 10) + 7(10) +$$
$$8(12 + 9 + 7) + 12(9 + 7) + 9(7) +$$
$$10(5 + 4 + 3) + 5(4 + 3) + 4(3) = 807$$

---

* Somer's d_{yx} ignores ties on the independent variable. If we wished to compute d_{xy} where X is dependent, the T_y term in Eq. (8.8) would be replaced by T_x which includes pairs with different Y values but which share the same X value. For example, cell a is tied with cells e and i because all are freshmen, yet they are ranked differently on political liberalism.

Using Eq. (8.8)

$$d_{yx} = \frac{1167 - 426}{1167 + 426 + 807} = +0.309$$

You will note that this value is considerably less than the gamma of $+0.465$ we computed for the same set of data. Somer's d ranges from -1 (perfect negative relationship) to $+1$ (perfect positive relationship) and assumes a linear or straight-line relationship between the variables.

Summary

In this chapter we have learned about the presentation and interpretation of contingency tables. We have also discussed four distribution-free measures of association appropriate for nominal- or ordinal-level data.

Lambda and tau are measures of association for nominal data based on the logic of proportional reduction in error. Lambda can serve as either a symmetric or asymmetric measure, whereas tau is an asymmetric statistic. The values of both range from 0 to 1, and they can be used with tables of any dimension. Lambda equals 0 when the distribution of cases on the dependent variable is extremely skewed, and under this circumstance tau should be calculated.

Gamma and Somer's d are measures of association for ordinal data based on the logic of pair-by-pair comparison. Gamma is a symmetric measure of association with a PRE interpretation, while Somer's d is an asymmetric measure and does not have a PRE interpretation. While gamma ignores ties, Somer's d is a more conservative measure that incorporates tied ranks on the dependent variable. Gamma and Somer's d range from -1 to $+1$, can be used with tables of any dimension, and assume the variables are linearly related.

Terms to Remember

Asymmetric measure of association   A measure of the one-way effect of one variable upon another.

Concordant pair   Two cases are ranked similarly on two variables.

Conditional distribution   The distribution of the categories of one variable under the differing conditions or categories of another.

Contingency table   A table showing the joint distribution of two variables.

Dependent variable   The variable that is being predicted. Also referred to as the criterion variable.

Discordant pair   Two cases are ranked on the opposite order on two variables.

Epsilon (ϵ)   The percentage difference within a category of the dependent variable between the two extreme categories of the independent variable.

Gamma (*G*)   A symmetric measure of association for ordinal-level data based upon pair-by-pair comparison with a PRE interpretation.

Independent variable   The predictor variable. Referred to as the experimental variable in an experiment.

Lambda (λ)   An asymmetric or symmetric measure of association for nominal-level data.

Marginals   The column and row totals of a contingency table.

Negative relationship   High scores on one variable tend to be associated with low scores on the other, and conversely low scores on one variable tend to be associated with high scores on the other variable.

Pair-by-pair comparison   An approach to prediction for ordinal-level data in contingency form.

Percentage difference   Differences in percentages normally measured between the two extreme categories of the independent variable within a category of the dependent variable. Also referred to as epsilon (ϵ).

Perfect relationship   A relationship in which knowledge of the independent variable allows a perfect prediction of the dependent variable or vice versa.

Positive relationship   High scores on one variable tend to be associated with high scores on the other variable, and conversely low scores on one variable tend to be associated with low scores on the other variable.

Proportional reduction in error (PRE)   A ratio of the prediction errors without information about the independent variable to the prediction errors having information about the independent variable.

Somer's _d_   An asymmetrical measure of association for ordinal-level data that incorporates tied ranks.

Symmetric measure of association   A measure of the mutual association between two variables.

Goodman's and Kruskal's tau (τ)   An asymmetric measure of association for nominal-level data with a PRE interpretation.

Tied pair   Two cases are ranked similarly on one or both of two variables.

Exercises

1. Percentage the following table down, across, and on the total. Interpret the data for each of the ways you have percentaged.

(Y) DRUG USE	(X) HIGH SCHOOL GRADE POINT AVERAGE		
	Low	_Moderate_	_High_
High	0	5	15
Moderate	15	10	10
Low	10	5	5

$N = 75$

2. Identify and defend your choice of independent and dependent variables for the table in Exercise 1. Which direction would you percentage, given your selection?

3. Discuss the existence, direction, and strength of a relationship for the data in Exercise 1 by examining the percentages and distribution of cases. Calculate an appropriate measure of association, defend your choice, and interpret the results.

4. Compute lambda and tau. Interpret your answers.

	(X) RELIGIOUS AFFILIATION		
(Y) EDUCATIONAL ATTAINMENT	Protestant	Catholic	Jewish
College	10	15	15
High School	10	20	10

N = 80

5. Compute gamma and Somer's *d*. Compare and interpret your answers.

	(X) SOCIAL ISOLATION		
(X) PERCEIVED HEALTH	Low	Moderate	High
Excellent	10	4	3
Average	5	6	7
Poor	2	4	9

N = 50

6. In 1978, over 5 million adults in the United States reported that they wanted a job, yet were not seeking employment. Assuming that sex is the independent variable and that reason for not seeking work is the dependent variable, percentage the following table and compute an appropriate measure of association. Interpret the results.

(Y) REASON FOR NOT SEEKING WORK	(X) SEX	
	Male	Female
In school	693	681
Ill or disabled	326	394
Keeping house	32	1226
Think cannot get a job	305	540
Total	1356	2841

N = 4197

(In thousands)

Source: U.S. Bureau of Labor Statistics, *Employment and Earnings,* monthly.

7. Using a contingency table of your choice for illustrative purposes, discuss the logic underlying proportional reduction in error. Discuss the logic of pair-by-pair comparison using a different example.

8. For the following research situations construct a hypothetical table and select an appropriate measure of association:

   a) You have conducted a health survey and wish to examine the relationship between the number of close friends a person has and whether or not the respondent has seriously considered suicide.

   b) You wish to explore the relationship between the length of time U.S. senators have served in the Senate and the region of the country in which their home state is located.

   c) You wish to determine if lower-income persons are more likely to be the victim of a street assault than are upper-income persons.

9. A weakness of gamma is that it overstates the actual relationship between variables when there are ties in the data. Why is this the case, and how does Somer's *d* avoid this weakness?

10. Interracially married couples in the United States totaled approximately 319,000 in 1970 and 421,000 in 1977. The following table indicates the racial composition of interracial marriages by year. Percentage the table, and discuss the relationship between year and racial composition of the marital pair.

RACIAL COMPOSITION OF MARRIAGE PARTNERS	YEAR		
	1970	1977	
Husband black, wife white	41	95	
Wife black, husband white	24	30	
Husband black, wife other*	8	20	
Wife black, husband other*	4	2	
Husband white, wife other*	139	177	
Wife white, husband other*	94	97	
Total	310	421	N = 731
	(In thousands)		

Source: U.S. Bureau of the Census, Current Population Reports, p. 23, No. 77.
*"Other" indicates spouse is neither black nor white but is Oriental or another unspecified racial category.

11. Discuss the strengths and weaknesses of lambda and tau, gamma and Somer's *d*.

12. A study was conducted to examine the possible precipitating influences or causes of a person having a religious experience. Calculate lambda and tau for the following table, and interpret your results.

(Y) RELIGIOUS EXPERIENCE?	(X) POSSIBLE PRECIPITATOR			
	Death of a loved one	Romantic breakup of a relationship	Loss of a job	Severe physical pain
Yes	10	10	5	15
No	10	40	10	10

N = 110

13. Construct a table with ordinal-level variables in which there is a strong relationship between two variables, but because it is not linear, that is, it does not follow either the positive or negative diagonal, neither gamma nor Somer's *d* will accurately reflect the strong pattern that is evident. Calculate gamma or *d* and discuss. (*Hint:* Choose two variables that have several categories to allow for the demonstration of a nonlinear relationship.)

14. Construct tables in which:

    a) Lambda = 1, lambda = 0.

    b) Tau = 1, tau = 0.

    c) Gamma = 1, gamma = 0.

    d) Somer's d = 1, Somer's d = 0.

Correlation

9

9.1   The Concept of Correlation

9.2   Calculation of Pearson's r

9.3   Pearson's r and z Scores

9.4   The Correlation Matrix

9.5   Interpreting Correlation Coefficients

9.6   Spearman's Rho (r_s)

9.1 The Concept of Correlation

We are frequently interested in determining the relationship between two interval- or ratio-scaled variables. For example, college admissions personnel are very concerned with the relationship between SAT scores and performance at college. Do students who score high on the SAT also perform well in college? Conversely, do high school students who score low on the SAT perform poorly in college? Is there a relationship between socioeconomic status and recidivism in crime? Is a spouse's power in a marriage related to his or her earnings? What is the relationship between the size of a law firm and beginning salaries for new members? In order to express the extent to which two variables are related, it is necessary to compute a measure of association called a **correlation coefficient.** There are many types of measures of association, several of which were presented in Chapter 8.

The decision as to which measure of association to use with a specific set of data depends on such factors as (1) the type of scale or measurement in which each variable is expressed (nominal, ordinal, interval, or ratio), (2) the nature of the underlying distribution (continuous or discrete), and (3) the characteristics of the distribution of scores (linear or nonlinear). Table 9.1 summarizes the measures of association that are included in this text.

Table 9.1   A summary of measures of association presented in this text

SCALE	SYMBOL	USED WITH	CHAPTER
Nominal	λ (lambda)	Two nominal-level variables.	8
	τ (Goodman's and Kruskal's tau)	Two nominal-level variables.	8
	φ (phi coefficient)	Two dichotomous variables.	18
	C (contingency coefficient)	Two nominal-level variables with equal rows and columns.	18
	V (Cramer's V)	Two nominal-level variables.	18
Ordinal	G (Goodman's and Kruskal's gamma)	Two ordinal-level variables.	8
	d (Somer's d)	Two ordinal-level variables.	8
	r_s (Spearman's rho)	Two ordinal-level variables. If one variable is ordinal level and one is interval level, both must be expressed as ranks prior to calculating r_s.	9
Interval/Ratio	r (Pearson's correlation coefficient)	Two interval- and/or ratio-level variables.	9
	R (multiple correlation coefficient)	Three or more interval- and/or ratio-level variables.	11

The Pearson correlation coefficient (*r*) (also referred to as **Pearson's *r*,** the *Pearson product-moment correlation coefficient,* and *r*) will be discussed in this chapter. Pearson's *r* is a measure of the linear or straight-line relationship between two interval-level variables.

The form of the relationship between two interval- or ratio-level variables can be presented visually in a *scatter diagram* (also referred to as a *scattergram* or *scatterplot*). A **scatter diagram** is a graphic device used to visually summarize the relationship between two variables. The *X*-axis is traditionally the horizontal axis and represents the independent variable. The vertical axis normally represents *Y*, the dependent variable.* The axes are drawn perpendicular to each other, approximately equal in length and marked to accommodate the full range of scores. Each coordinate represents two values: a score on the *X* variable and a score on the *Y* variable. Figure 9.1 displays the relationship between the desired number of children and the actual number of children reported by six hypothetical couples. Couple *A* preferred two children and have two children; therefore, their coordinate has been located at

---

* As with all statistical conventions, they are not always followed; therefore, always determine the independent and dependent variable before interpreting the data.

the point at which X and Y both equal 2. The scatter diagram allows for a visual inspection of the relationship between the two variables, and it is evident that three of the couples (B, D, and E) do not have the number of children that they desired. The Pearson correlation coefficient (r) for the data in Fig. 9.1 equals $+0.21$. Let us consider the meaning of this value along with the correlation coefficients below the scatter diagrams in Fig. 9.2.

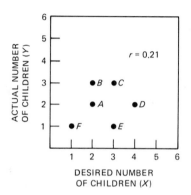

COUPLE	DESIRED NUMBER OF CHILDREN (X)	ACTUAL NUMBER OF CHILDREN (Y)
A	2	2
B	2	3
C	3	3
D	4	2
E	3	1
F	1	1

9.1 Scatter diagram showing the relationship between the desired and actual number of children for 6 couples (hypothetical data)

The values of the Pearson correlation coefficient vary between $+1.00$ [see Fig. 9.2(a)] and -1.00 [see Fig. 9.2(f)]. Both extremes represent perfect relationships between variables, and 0.00 represents the absence of a relationship [see Fig. 9.2(d)]. The *size* of the correlation coefficient indicates the strength of the relationship. The closer the points in the scatter diagram approach the form of a straight line, the stronger the relationship between X and Y. We ignore the sign of the coefficient when interpreting the strength of a relationship; hence, the relationships in Fig. 9.2(a) and (f) are equally strong in spite of their opposite signs.

The direction of a relationship is indicated by the *sign* of the correlation coefficient. A **positive relationship** (or direct relationship) indicates that high scores on one variable tend to be associated with high scores on a second variable and, conversely, low scores on one variable

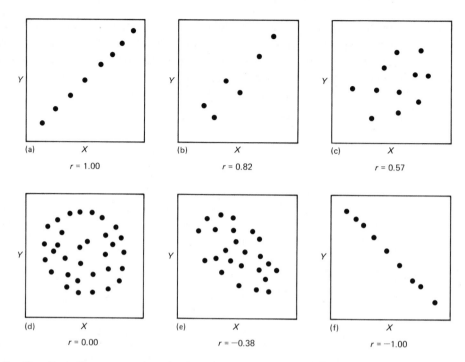

9.2 Scatter diagrams showing various degrees of relationship between two variables.

tend to be associated with low scores on the second variable [see Fig. 9.2(a), (b), and (c)]. A **negative relationship** (also referred to as an *inverse or indirect relationship*) indicates that low scores on one variable tend to be associated with high scores on a second variable. Conversely, high scores on one variable tend to be associated with low scores on the second variable [see Fig. 9.2(e) and 9.2(f)].

9.2 Calculation of Pearson's r

9.2.1 Mean Deviation Method

The mean deviation method for calculating Pearson's r is not often used by social scientists because it involves more time and effort than other computational techniques. It is being presented here primarily because it sheds further light on the characteristics of Pearson's *r*. However, with small samples, it is as convenient a computational equation as any, unless a calculator is available. The mean deviation computational equation for *r* is

$$r = \frac{\Sigma(X - \overline{X})(Y - \overline{Y})}{\sqrt{[\Sigma(X - \overline{X})^2][\Sigma(Y - \overline{Y})^2]}}. \qquad (9.1)$$

Equation (9.1) provides a conceptual understanding of the Pearson correlation coefficient. The numerator of the equation is the **covariation** of *X* and *Y* or the extent to which *X* and *Y* vary together. It is the numerator that determines the *sign* of the correlation coefficient. A positive *r* results from high values of *X* tending to be paired with high values on *Y* and, conversely, low values of *X* tending to be paired with low values of *Y*. Substitute the values from Fig. 9.1 into the numerator of Eq. (9.1) to ensure you understand covariation and how it determines the sign of *r*. Also note that the covariation has an unrestricted upper range and that, by itself, it is not a good measure of association.

The denominator is always positive because the values are squared. It consists of the square root of the product of the standard deviation of *X* and *Y* since, in the more complex version of the equation upon which Eq. (9.1) is based, both the numerator and denominator are divided by *N*. Thus it is the denominator that restricts the range of *r* to ±1 and results in *r* being independent of the measurement units of *X* and *Y*. The data in Table 9.2, which summarize educational attainment scores for seven female respondents and for their mothers, will be used to illustrate the mean deviation method for computing *r*.

Table 9.2   Computational procedures for *r* using the mean deviation method (hypothetical data)

RESPON-DENT	MOTHER'S EDUCATION X	$(X - \overline{X})$	$(X - \overline{X})^2$	DAUGHTER'S EDUCATION Y	$(Y - \overline{Y})$	$(Y - \overline{Y})^2$	$(X - \overline{X})(Y - \overline{Y})$
A	1	-6	36	7	-6	36	36
B	3	-4	16	4	-9	81	36
C	5	-2	4	13	0	0	0
D	7	0	0	16	3	9	0
E	9	2	4	10	-3	9	-6
F	11	4	16	22	9	81	36
G	13	6	36	19	6	36	36
	$\overline{X} = 7$		$\Sigma(X - \overline{X})^2 = 112$	$\overline{Y} = 13$		$\Sigma(Y - \overline{Y})^2 = 252$	$\Sigma(X - \overline{X})(Y - \overline{Y}) = 138$

$$r = \frac{\Sigma(X - \overline{X})(Y - \overline{Y})}{\sqrt{[\Sigma(X - \overline{X})^2][\Sigma(Y - \overline{Y})^2]}} = \frac{138}{\sqrt{(112)(252)}} = \frac{138}{168.00} = 0.82$$

The computational procedures for the mean deviation method should be familiar to you. The symbols $\Sigma(X - \overline{X})^2$ and $\Sigma(Y - \overline{Y})^2$ were used to compute standard deviation in Chapter 6. In fact, in calculating r only one step has been added, that of obtaining the sum of the cross-products $\Sigma(X - \overline{X})(Y - \overline{Y})$, which was identified earlier as the covariation of X and Y. This is obtained by multiplying the deviation of each individual's score from the mean of the X variable by its corresponding deviation on the Y variable and then summing all of the cross-products. Notice that, if maximum deviations in X had lined up with maximum deviations in Y and so on down through the array, $\Sigma(X - \overline{X})(Y - \overline{Y})$ would have been equal to 168.00, which is the same as the value of the denominator, and would have produced a correlation of 1.00. In this instance, we found the correlation between the mother's and daughter's education to be 0.82.

9.2.2 Raw Score Method

In calculating Pearson's r by the raw score method, we will use Eq. (9.2):

$$r = \frac{N\Sigma XY - (\Sigma X)(\Sigma Y)}{\sqrt{[N\Sigma X^2 - (\Sigma X)^2][N\Sigma Y^2 - (\Sigma Y)^2]}} \, . \qquad (9.2)$$

The procedure for calculating r by the raw score method is summarized in Table 9.3. Note that we calculate all of the quantities separately and substitute them into the equation. Here you find exactly the same coefficient ($r = 0.82$) as you did using the mean deviation method. As with the mean deviation method, all the procedures are familiar to you. The quantity ΣXY is obtained by multiplying each X value by its corresponding Y and then summing these products. Also, be sure to make the distinction between ΣX^2 and $(\Sigma X)^2$ as well as between ΣY^2 and $(\Sigma Y)^2$.

9.3 Pearson's r and z scores

The Pearson correlation coefficient can be also computed using z scores and, furthermore, z scores provide an intuitive interpretation that the raw score method is unable to offer. A high positive Pearson's r in-

Table 9.3 Computational procedures for *r* using the raw score method (hypothetical data)

RESPON- DENT	MOTHER'S EDUCATION X	X^2	DAUGHTER'S EDUCATION Y	Y^2	XY
A	1	1	7	49	7
B	3	9	4	16	12
C	5	25	13	169	65
D	7	49	16	256	112
E	9	81	10	100	90
F	11	121	22	484	242
G	13	169	19	361	247
$N = 7$	$\Sigma X = 49$ $\overline{X} = 7$	$\Sigma X^2 = 455$	$\Sigma Y = 91$ $\overline{Y} = 13$	$\Sigma Y^2 = 1435$	$\Sigma XY = 775$

$$r = \frac{N\Sigma XY - (\Sigma X)(\Sigma Y)}{\sqrt{[N\Sigma X^2 - (\Sigma X)^2][N\Sigma Y^2 - (\Sigma Y)^2]}}$$

$$= \frac{7(775) - (49)(91)}{\sqrt{[7(455) - (49)^2][7(1435) - (91)^2]}}$$

$$= \frac{5425 - 4459}{\sqrt{[3185 - 2401][10,045 - 8281]}}$$

$$= \frac{966}{\sqrt{[784][1764]}}$$

$$= \frac{966}{1176}$$

$$= 0.82$$

dicates that each individual obtains approximately the same *z* score on both variables. In a *perfect* positive correlation ($r = +1.00$), each individual obtains exactly the same *z* score on both variables.

With a high negative *r*, each individual obtains approximately the same *z* score on both variables, but the *z* scores are opposite in sign.

Remembering that the *z* score represents a measure of relative position in standard deviation units on a given variable (that is, a high positive *z* represents a high score relative to the remainder of the distribution, and a high negative *z* represents a low score relative to the remainder of the distribution), we may now generalize the meaning of Pearson's *r*.

Pearson's r represents the extent to which the same individuals, events, etc., occupy the same relative position on two variables.

In order to explore the fundamental characteristics of the Pearson correlation coefficient, let us slightly modify the mother's and daughter's educational attainment scores example to demonstrate a perfect positive correlation. In Table 9.4, we find the new scores for the seven pairs of mothers and daughters on the two variables, X and Y, where X represents the educational attainment of the mother and Y represents the educational attainment of the daughter.

Table 9.4   Raw scores and corresponding z scores for seven pairs of mothers and daughters on two variables (hypothetical data)

RESPON-DENT	MOTHER'S EDUCATION X	$X - \bar{X}$	$(X - \bar{X})^2$	z_x	DAUGHTER'S EDUCATION Y	$Y - \bar{Y}$	$(Y - \bar{Y})^2$	z_y	$z_x z_y$
A	1	-6	36	-1.5	4	-9	81	-1.5	2.25
B	3	-4	16	-1.0	7	-6	36	-1.0	1.00
C	5	-2	4	-0.5	10	-3	9	-0.5	0.25
D	7	0	0	0.0	13	0	0	0.0	0.00
E	9	2	4	0.5	16	3	9	0.5	0.25
F	11	4	16	1.0	19	6	36	1.0	1.00
G	13	6	36	1.5	22	9	81	1.5	2.25
	$\Sigma X = 49$		$\Sigma(X - \bar{X})^2 = 112$		$\Sigma Y = 91$		$\Sigma(Y - \bar{Y})^2 = 252$		$\Sigma z_x z_y = 7.00$
	$\bar{X} = 7$		$s_x = \sqrt{\frac{112}{7}} = 4.00$		$\bar{Y} = 13$		$s_y = \sqrt{\frac{252}{7}} = 6.00$		

The scale values of X and Y do not need to be the same for the calculation of r. In this example, we see that X ranges from 1 through 13, whereas Y ranged from 4 through 22. This independence of r from specific scale values, which was initially noted in Section 9.2.1, permits us to investigate the relationships among an unlimited variety of variables. If we measure, for example, a person's yearly income in dollars *or* hundreds of dollars and separately correlate both measures of income with a second variable such as occupational status, the resulting correlation between income and status will be the same regardless of which income measure we use.

The z scores of each respondent on each variable are identical in the event of a perfect positive correlation.* Had we reversed the order of either variable, that is, paired 1 with 22, paired 3 with 19, etc., the z

scores would still be identical, but would be opposite in sign. In this latter case, the correlation would be a maximum *negative* ($r = -1.00$).

If we multiply the paired z scores and then sum the results, we will obtain maximum values only when the correlation is 1.00. Indeed, as the correlation approaches zero, the sum of the products of the paired z scores also approaches zero. Note that when the correlation is perfect, the sum of the products of the paired z scores is equal to N, where N equals the number of pairs. These facts lead to an equation for r using z scores that is algebraically equivalent to Eqs. (9.1) and (9.2).

$$r = \frac{\Sigma(z_x z_y)}{N}. \qquad (9.3)$$

We suggest that you take the data in Table 9.1 and calculate r, using Eq. (9.3). You will arrive at a far more thorough understanding of r in this way.

Equation (9.3) is unwieldy because it requires calculation of separate z's for each score of each individual. Imagine the task of calculating r when N exceeds 50 cases, as it often does. For this reason, computational Eq. (9.2) is normally used.

9.4 The Correlation Matrix

Researchers frequently present a **correlation matrix** which is a summary of the statistical relationships between all possible pairs of variables. Table 9.5 presents the Pearson correlation coefficients for all the possible relationships between five variables for 420 hypothetical respondents. Computation of the coefficients required that five measures, that is, one for each of the five variables in Table 9.5, be available for all of the respondents. The diagonal of a correlation matrix represents the correlation of a variable with itself; therefore, all the values on the diagonal are 1.00. Each row in the matrix represents the correlation of a specific variable with other variables, as does each column. The variable age, for example, is presented in row 1 and in column 1. If you wish to

---

* Recall that z scores were discussed in Chapter 7.

know the relationship between age and education, read down the age column until you come to the education row where you find $r = -0.23$. Notice that the scoring procedure for the attitude toward the federal funding for abortion variable has been stated explicitly to avoid misunderstanding. In this case we see that a person who is opposed to federal funding for abortion would receive a low score and a pro-federal funding person would receive a high score. Since it would be possible to score this item in the opposite manner (that is, negative attitude scored high), it is always important to specify the scoring procedure in a potentially ambiguous situation such as this. If the scoring of the abortion item were reversed, all of the signs of the correlation coefficients associated with the abortion variable would also be reversed.

Table 9.5   Pearson correlation coefficients between five variables ($N = 420$) (hypothetical data)

	1	2	3	4	5
1. Age	1.00				
2. Occupational status	0.02	1.00			
3. Income	−0.19	0.33	1.00		
4. Education	−0.23	0.50	0.40	1.00	
5. Attitude toward federal funding of abortion (negative scored low)	−0.47	0.01	−0.19	0.14	1.00

Some social scientists prefer to display the correlation coefficients below the diagonal as shown in Table 9.5 and others above the diagonal. Either approach is equally appropriate because the coefficients above and below the diagonal are a mirror image of each other. Hence, you will seldom see a matrix with coefficients above and below the diagonal (referred to as a *square matrix*).

9.5 Interpreting Correlation Coefficients

The correlation between age and attitude toward the federal funding of abortion in Table 9.5 is -0.47, which may also be written $r_{15} = -0.47$ where the subscripts 1 and 5 refer to the variables as numbered in Table 9.5. The negative coefficient indicates that there is a negative, inverse, or indirect relationship between the variables. Younger respondents tend to favor federal funding of abortion and, conversely, older respondents tend to be opposed to federal funding of abortion. While there is no established rule that specifies what constitutes a weak, moderate, or strong relationship, these very general guidelines may be useful: a weak relationship, $r = \pm 0.01$ to ± 0.30; a moderate relationship, $r = \pm 0.31$ to ± 0.70; and a strong relationship, $r = \pm 0.71$ to ± 0.99. A perfect relationship is ± 1.00, and no relationship is indicated when $r = 0$.

Students frequently ask how large a coefficient must be before it is respectable. The magnitude and sign of a coefficient should be interpreted relative to previous inquiries whenever possible. The correlation between occupational status and educational attainment for a national sample of adult males normally falls between 0.50 and 0.60.* If an investigator found a marked departure from these values when studying similar variables, he or she would closely examine the computational procedures and data for errors.

Those social scientists who investigate the relationship between attitudes and behaviors, on the other hand, routinely expect the relationship to be weaker, generally between 0.20 and 0.40. Thus previous studies can serve as guidelines in interpreting the relationship you choose to investigate.

When low correlations are found, one is strongly tempted to conclude that there is little or no relationship between the two variables under study. However, it must be remembered that r reflects only the linear relationship between two variables. The failure to find evidence of a relationship may be due to one of three possibilities: (1) the variables are, in fact, unrelated, (2) the range of values on one or both of the variables is restricted, or (3) the variables are not related linearly.† In the latter instance, Pearson's r would not be an appropriate measure of the degree of relationship between the variables because it assumes a linear relationship.

To illustrate, if we were plotting the relationship between age and church attendance, we might obtain a scatter diagram like that shown in Fig. 9.3. It is usually possible to determine whether there is a substantial departure from a straight line relationship by examining the scatter diagram. If the distribution of points in the scatter diagram is egg

---

*   The sign of a coefficient is frequently omitted when it is positive.

†   A fourth possibility, that of a third variable suppressing the relationship, will not be discussed.

shaped or elliptical, without the decided bending that occurs in Fig. 9.3, it may safely be assumed that the relationship is linear. Any small departures from linearity will not greatly influence the size of the correlation coefficient. On the other hand, where there is marked curvilinearity, as in Fig. 9.3, a curvilinear coefficient of correlation would better reflect the relationship between the two variables under investigation. Although it is beyond the scope of this text to investigate nonlinear coefficients of correlation, you should be aware of this possibility and, as a matter of course, should construct a scatter diagram prior to calculating Pearson's *r*.

9.3   Scatter diagram of two variables that are related in a nonlinear fashion (hypothetical data)

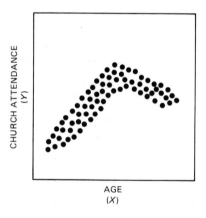

The assumption of linearity of relationship is the most important requirement to justify the use of Pearson's *r* as a measure of relationship between two variables. Pearson's *r* need not be calculated only with normally distributed variables, particularly when the sample size exceeds 50. So long as the distributions of the two variables are unimodal and relatively symmetrical, Pearson's *r* may legitimately be computed.

Another situation giving rise to artificially low correlation coefficients results from restricting the range of values of one or both of the variables. For example, if we were interested in the relationship between age and height for children from 3 to 16 years of age, we would undoubtedly obtain a rather high correlation coefficient between these two variables. However, suppose that we were to restrict the range of one of our variables? What effect would this have on the size of the coefficient? Consider the same relationship between age and height, but only for those children between the ages of 9 and 10. We would probably end up with a low coefficient.

You will note that the overall relationship illustrated in Fig. 9.4 indicates a moderately strong negative relationship between age and life satisfaction ($r = -0.67$), but when we focus only on those respondents 40 years of age and less ($r = -0.51$) or those over 40 ($r = -0.38$), we see that the relationship is attenuated or weakened. **Restriction of range** (sometimes referred to as *truncated range*) can normally be attributed to one of two causes: (1) a group of respondents might be inadvertently excluded from the study, and (2) the respondents might be naturally homogeneous with respect to a particular variable. The exclusion of potential respondents can occur, for example, when home interviews are conducted only during daylight hours. Such a procedure results in interviewing very few respondents in the work force and hence introduces the possibility of a restricted range in the respondents' ages.

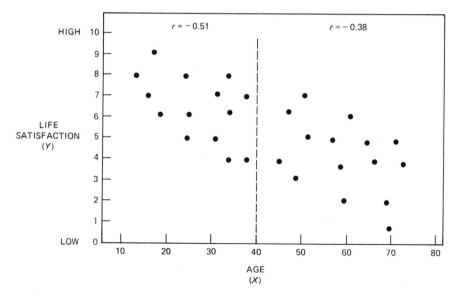

9.4 Scatter diagram illustrating high correlation over entire range of X and Y values ($r = -.67$) but lower correlation when the range is restricted above or below 40 years of age. (Hypothetical Data).

The second contributing factor, that of a naturally homogeneous group of respondents, is not uncommon in social science research, for much of this research is conducted in colleges and universities where respondents have been preselected for attitude and related variables. Thus they represent a fairly homogeneous group with respect to these variables. Consequently, when an attempt is made to demonstrate the relationship between variables like SAT scores and college grades, the resulting coefficient may be low because of the restricted range. Furthermore, the correlations would be expected to be lower for colleges that select their students from within a narrow range of SAT scores.

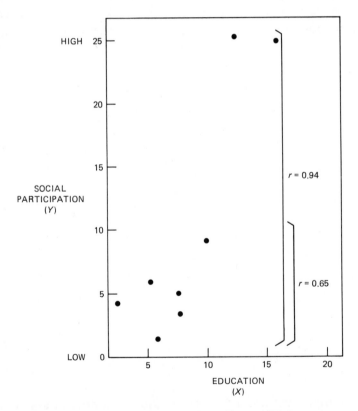

9.5 Scatter diagram illustrating the effect of extreme scores on the correlation coefficient. (Hypothetical data).

Correlations can be artificially high when we are dealing with extremely small samples or when the data set contains extreme scores.* In the former instance, the correlation coefficient calculated using the sample data may not accurately reflect the actual correlation in the population from which the sample was drawn. This possibility will be considered further in Chapter 12.

The impact of extreme scores on a correlation coefficient is shown in Fig. 9.5 where the inclusion of the two extreme scores inflates the initial correlation of education and social participation from $r = 0.65$ to $r = 0.94$.

9.6 Spearman's rho (r_s)

The second measure of correlation we shall discuss, r_s, is computed for ranked data on two variables. Imagine you are a sociologist, and after long years of observation you have developed a strong suspicion that the social class of husbands and wives prior to marriage is very similar. In an effort to test your idea, you obtain a measure of the social class background for a sample of 15 couples. You are now able to use the social class measure to rank the wives from highest to lowest relative to the other wives, and you can rank the husbands in the same manner as shown in Table 9.6.

Spearman's rho requires that you obtain the differences in the ranks, square each difference, sum the squared differences, and substitute the resulting values into Eq. (9.3),

$$r_s = 1 - \frac{6\Sigma D^2}{N(N^2 - 1)}, \qquad (9.3)$$

where

$$D = \text{rank } X - \text{rank } Y, \text{ and}$$
$$N = \text{the number of ranked pairs.}$$

As a matter of course, ΣD should be obtained even though it is not used in any of the calculations. It constitutes a useful check on the accuracy of your calculations up to this point since ΣD must equal zero.

---

* Also referred to as *deviant cases* or *outliers*.

Table 9.6   Computational procedures for calculating r_s from ranked variables (hypothetical data)

WIFE'S RANK (X)	HUSBAND'S RANK (Y)	D	D²	
1	4	-3	9	$r_s = 1 - \dfrac{6\Sigma D^2}{N(N^2 - 1)}$
2	2	0	0	
3	9	-6	36	
4	1	3	9	$= 1 - \dfrac{6(204)}{15(224)}$
5	7	-2	4	
6	10	-4	16	
7	8	-1	1	$= 1 - \dfrac{1224}{3360}$
8	13	-5	25	
9	5	4	16	$= 1 - 0.36$
10	3	7	49	
11	11	0	0	$= 0.64$
12	6	6	36	
13	12	1	1	
14	15	-1	1	
15	14	1	1	
		$\Sigma D = 0$	$\Sigma D^2 = 204$	

If you obtain any value other than zero, you should recheck your original ranks and the subsequent subtractions.

The value of r_s will equal $+1.00$ when the two sets of ranks are in perfect agreement, -1.00 when the two sets of ranks are in perfect disagreement, and 0 when there is no agreement; hence, r_s is a measure of the extent to which the two sets of ranks are in agreement or disagreement. The computations in Table 9.6 in which r_s was found to equal 0.64 indicate that there is considerable difference in the social class rankings prior to marriage of our sample of 15 wives and their husbands.

Occasionally, upon converting scores to ranks, you will find two or more tied scores. In this event, assign the mean of the tied ranks to each of the tied scores. The next score in the array receives the rank normally assigned to it. Thus the ranks of the scores 128, 122, 115, 115, 115, 107, 103 would be 1, 2, 4, 4, 4, 6, 7, and the ranks of the scores 128, 122, 115, 115, 107, 103 would be 1, 2, 3.5, 3.5, 5, 6.

When there are tied ranks on either or both the X and the Y variables, the Spearman equation yields an inflated value, particularly when the number of tied ranks is high. The Pearson's r equation should be applied to the ranked data when there are numerous ties.

Summary

In this chapter we discussed the concept of correlation and demonstrated the calculation of the Pearson correlation coefficient (r) using the mean deviation, raw score, and z score computational procedures. The scatter diagram was presented as an interpretive and summary device.

Correlation is concerned with determining the extent to which two variables are related or tend to vary together. The quantitative expression of the extent of the relationship is given in terms of the magnitude of the correlation coefficient. Correlation coefficients vary between values of -1.00 and $+1.00$; both extremes represent perfect relationships. A coefficient of zero indicates the absence of a relationship between two variables. The correlation matrix is used to summarize the correlations between all possible variable pairs.

Pearson's r is appropriate only for variables that are linearly related. Several situations that can contribute to artificially low or high correlations were discussed.

Spearman's rho (r_s) is a measure of association for ordinal data that provides a measure of the extent to which two sets of ranks are in agreement or disagreement. The value of r_s will equal $+1.00$ when the two sets of rankings are in perfect agreement and -1.00 when the two sets of rankings are in perfect disagreement.

Terms to Remember

Correlation coefficient   A measure that expresses the extent to which two variables are related.

Correlation matrix   Summary of the statistical relationships between all possible pairs of variables.

Covariation   Extent to which two variables vary together.

Negative relationship   Variables are said to be negatively related when a high score on one variable tends to be associated with a low score on the other variable. Conversely, low scores on one variable tend to be associated with high scores on the other.

Pearson's *r* (product-moment correlation coefficient)   Correlation coefficient used with interval- or ratio-scaled variables.

Positive relationship   Variables are said to be positively related when a high score on one variable tends to be associated with a high score on the other. Conversely, low scores on one variable tend to be associated with low scores on the other.

Scatter diagram   Graphic device used to summarize visually the relationship between two variables.

Spearman's rho (r_s)   A measure of association for ordinal-level data that provides a measure of the extent to which two sets of ranks are in agreement or disagreement.

Restricted range   Truncated range on one or both variables, resulting in a deceptively low correlation between these variables.

Exercises

1. You have hypothesized that membership growth in large city churches is positively associated with distance from the central business district (CBD). Compute *r* for the following data set of 20 churches and determine if the data support your hypothesis.

CHURCH	% GROWTH IN PAST YEAR	DISTANCE FROM CBD (IN MILES)	CHURCH	% GROWTH IN PAST YEAR	DISTANCE FROM CBD (IN MILES)
1	14	6	11	3	0
2	8	3	12	1	6
3	9	21	13	14	12
4	2	0	14	6	3
5	4	3	15	2	1
6	0	4	16	4	7
7	15	9	17	7	4
8	22	9	18	8	2
9	7	4	19	11	9
10	9	3	20	6	7

2. Using distance from the CBD as the independent variable and percentage of growth in the past year as the dependent variable, construct a scattergram for the data in Exercise 1. Does the relationship appear to be reasonably linear?

3. Twenty-five high school students were randomly selected from public high schools in the Boston area and asked to report anonymously the number of rock concerts they attended during the past year and the average number of days per week that they smoke marijuana. Given the accompanying data, compute r and interpret your answer. What does it tell you?

NUMBER OF ROCK CONCERTS ATTENDED	DAYS PER WEEK MARIJUANA SMOKED	NUMBER OF ROCK CONCERTS ATTENDED	DAYS PER WEEK MARIJUANA SMOKED
4	2	4	2
7	4	4	2
0	0	3	0
1	2	11	7
6	3	14	6
2	1	6	3
2	0	3	0
3	1	1	2
2	4	4	5
3	3	0	0
2	1	1	1
0	0	3	1
		0	1

4. Explain in *your own words* the meaning of correlation.

5. In each of the following examples, identify a potential problem in the interpretation of the results of a correlational analysis.

   a) The relationship between age and frequency of drinking for an adolescent sample between the ages of 13 and 18.

   b) The correlation between aptitude and grades for honor students at a university.

   c) The relationship between vocabulary and reading speed among children in a "culturally deprived" community.

6. For a group of 50 individuals $\Sigma z_x z_y$ is 41.3. What is the correlation between the two variables?

7. The following scores were made by five students on two tests. Calculate the Pearson r (using $r = \Sigma z_x z_y / N$). Convert to ranks and calculate r_s.

STUDENT	TEST X	TEST Y
A	5	1
B	5	3
C	5	5
D	5	7
E	5	9

Generalize: What is the effect of tied ranks on r_s?

8. What effect does departure from linearity have on Pearson r?

9. How does the range of scores sampled affect the size of the correlation coefficient?

10. What is the effect of reversing the scoring of a variable included in a correlation matrix?

11. A recent study focusing on political activism among black ministers reported that the Pearson r between age and an "attitude toward the police" scale (negative scored low) was -0.14. Interpret this finding. Church size and the minister's education were correlated 0.41. What does this finding tell you and how might it be interpreted? (*Source:* William M. Berenson, Kirk W. Elifson and Tandy Tollerson III, Preachers in politics: political activism among the black ministry, *Journal of Black Studies,* Volume 6, June 1976, pp. 373–392).

12. The following data show scores on advanced sections of the Graduate Record Examination and college grade point averages. What is the relationship between these two variables?

GRE SCORES	GRADE POINT AVERAGES	GRE SCORES	GRADE POINT AVERAGES
440	1.57	528	2.08
448	1.83	550	2.15
455	2.05	582	3.44
460	1.14	569	3.05
473	2.73	585	3.19
485	1.65	593	3.42
489	2.02	620	3.87
500	2.98	650	3.00
512	1.79	690	3.12
518	2.63		

13. Explain the difference between $r = -0.76$ and $r = +0.76$.

14. Construct a scatter diagram for each of the following four sets of data.

a)		b)		c)		d)	
X	Y	X	Y	X	Y	X	Y
1.5	0.5	0.5	5.0	0.5	0.5	0.5	1.0
1.0	0.5	0.5	4.5	1.0	1.0	0.5	2.5
1.0	2.0	1.0	3.5	1.0	1.5	0.5	4.5
1.5	1.5	1.5	4.0	1.5	2.5	1.0	3.5
1.5	2.0	1.5	2.5	1.5	3.5	1.5	1.0
2.0	2.0	2.0	3.0	2.0	2.5	1.5	2.5
2.5	2.5	2.5	2.0	2.0	3.5	1.5	4.0
2.5	3.2	2.5	3.5	2.5	4.5	2.0	1.0
3.0	2.5	3.0	2.5	3.0	3.5	3.0	2.0
3.0	3.5	3.0	2.0	3.5	3.0	3.0	3.5
3.5	3.5	3.5	2.0	3.5	2.5	3.0	4.5
3.5	4.5	3.5	2.5	3.5	2.0	3.5	1.0
4.0	3.5	4.0	1.5	4.0	2.5	3.5	1.0
4.0	4.5	4.0	0.7	4.0	2.0	3.5	3.5
4.5	4.5	5.0	0.5	4.5	1.0	4.0	3.5
5.0	5.0			5.0	1.0	4.0	4.5
				5.0	0.5	4.5	2.5
						4.5	1.0

15. By inspecting the scatter diagrams for the data in Exercise 14, determine which one represents:

   a) A curvilinear relation between X and Y.
   b) A positive correlation between X and Y.
   c) Little or no relation between X and Y.
   d) A negative correlation between X and Y.

16. Discuss the following correlation matrix, which is based on interviews with 275 lawyers (hypothetical data).

	1	2	3	4
1. Size of law firm	1.00			
2. Salary	0.40	1.00		
3. Age	-0.11	0.67	1.00	
4. Education	0.03	0.00	0.07	1.00

17. The following data show the scores obtained by a group of 20 students on a college entrance examination and a verbal comprehension test. Prepare a scatter diagram and calculate a Pearson r for these data.

STUDENT	COLLEGE ENTRANCE EXAM (X)	VERBAL COMPRE-HENSION TEST (Y)	STUDENT	COLLEGE ENTRANCE EXAM (X)	VERBAL COMPRE-HENSION TEST (Y)
A	52	49	K	64	53
B	49	49	L	28	17
C	26	17	M	49	40
D	28	34	N	43	41
E	63	52	O	30	15
F	44	41	P	65	50
G	70	45	Q	35	28
H	32	32	R	60	55
I	49	29	S	49	37
J	51	49	T	66	50

18. Two consulting agencies have independently ranked ten counties in terms of the quality of their public schools. Compute r_s to determine the extent to which the two sets of rankings correspond.

COUNTY	RANK OF SCHOOLS BY AGENCY A	RANK OF SCHOOLS BY AGENCY B
A	1	1
B	4	2
C	3	3
D	2	4
E	7	6
F	5	7
G	6	5
H	8	9
I	9	10
J	10	8

19. The data in the table below represent scores obtained by 10 students on a statistics examination, and their final grade point average. Prepare a scatter diagram and calculate Pearson's r for these data.

STUDENT	SCORE ON STATISTICS EXAM, X	GRADE POINT AVERAGE, Y	STUDENT	SCORE ON STATISTICS EXAM, X	GRADE POINT AVERAGE, Y
A	90	2.50	F	70	1.00
B	85	2.00	G	70	1.00
C	80	2.50	H	60	0.50
D	75	2.00	I	60	0.50
E	70	1.50	J	50	0.50

20. A sociological study involved the rating of teachers. In order to determine the reliability of the ratings, the ranks given by two different observers were tabulated. Are the ratings reliable? Explain your answer.

ANIMAL	RANK BY OBSERVER A	RANK BY OBSERVER B	ANIMAL	RANK BY OBSERVER A	RANK BY OBSERVER B
A	12	15	I	6	5
B	2	1	J	9	9
C	3	7	K	7	6
D	1	4	L	10	12
E	4	2	M	15	13
F	5	3	N	8	8
G	14	11	O	13	14
H	11	10	P	16	16

Regression and Prediction

10

10.1 Introduction to Prediction

10.2 Linear Regression

10.3 Residual Variance and Standard Error of Estimate

10.4 Explained and Unexplained Variation

10.5 Correlation and Causation

10.1 Introduction to Prediction

Is length of courtship a good predictor of marital success? What role does sexual compatibility play in a successful marriage? Knowing a person's employment record, what can we say about the likelihood of her receiving a promotion? Knowing a student's mathematics aptitude score, can we estimate how well he will do in a statistics course?

Let us look at an example. Suppose we are trying to predict student Jones's score on the final exam. If the only information available was that the class mean on the final was 75 ($\overline{Y} = 75$), the best guess we could make is that he would obtain a score of 75 on the final.* However, far more information is usually available; for example, Mr. Jones obtained a score of 62 on the midterm examination. How can we use this information to make a better prediction about his performance on the final exam? If we know that the class mean on the midterm examination was 70 ($\overline{X} = 70$), we could reason that since he scored below the mean on the midterm, he would probably score below the mean on the final. At this point, we appear to be closing in on a more accurate prediction of his performance. How might we further improve the accuracy of our prediction? Simply knowing that he scored below the

---

* See Section 5.2.2, in which we demonstrated that the sum of the deviations from the mean is zero and that the sum of squares of deviations from the arithmetic mean is less than the sum of squares of deviations about any other score or potential score.

mean on the midterm does not give us a clear picture of his relative standing on this exam. If, however, we know the standard deviation on the midterm, we could express his score in terms of his relative position, that is, his z score. Let us imagine that the standard deviation on the midterm was 4 ($s_x = 4$). Since he scored 2 standard deviations below the mean ($z_x = -2$), would we be justified in guessing that he would score 2 standard deviations below the mean on the final ($z_y = -2$)? That is, if $s_y = 8$, would you predict a score of 59 on the final? No! You will note that an important piece of information is missing: the correlation between the midterm and the final. You may recall from our discussion of correlation[*] that Pearson's r represents the extent to which the same individuals or events occupy the same relative position on two variables. Thus we are justified in predicting a score of exactly 59 on the final only when the correlation between the midterm and the final exam is perfect (that is, when $r = +1.00$). Suppose that the correlation is equal to zero. Then it should certainly be obvious that we are not justified in predicting a score of 59; rather, we are once again back to our original prediction of 75 (that is, $\overline{Y}$).

In summary, when $r = 0$, our best prediction is 75 ($\overline{Y}$); when $r = +1.00$, our best prediction is 59 ($z_y = z_x$). It should be clear that predictions from intermediate values of r will fall somewhere between 59 and 75.[†]

An important advantage of a correlational analysis, then, stems from its application to problems involving predictions from one variable to another. Sociologists, educators, political scientists, and economists are constantly being called upon to make predictions. To provide an adequate explanation of r and to illustrate its specific applications, it is necessary to proceed with an analysis of linear regression.

10.2 Linear Regression

To simplify our discussion, let us start with an example of two variables that are usually perfectly or almost perfectly related: monthly salary and yearly income. In Table 10.1 we have listed the monthly income of

---

[*]   See Section 9.2.

[†]   We are assuming that the correlation is positive. If the correlation were -1.00, our best prediction would be a score of 91, that is, $z_y = -z_x$.

eight employees of a small welfare office. These data are shown graphically in Fig. 10.1. It is customary to refer to the horizontal axis as the X-axis, and to the vertical axis as the Y-axis. If one variable occurs earlier in time (or before) the other, the earlier one (or the independent variable) is represented on the X-axis. It will be noted that all salaries are represented on a straight line extending diagonally from the lower left-hand corner to the upper right-hand corner.

Table 10.1 Monthly salaries and annual income of eight employees of a welfare office (hypothetical data)

EMPLOYEE	MONTHLY SALARY	ANNUAL INCOME
A	600	7,200
B	700	8,400
C	725	8,700
D	750	9,000
E	800	9,600
F	850	10,200
G	900	10,800
H	975	11,700

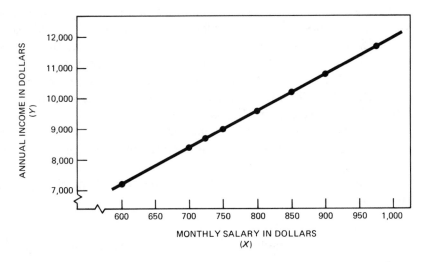

10.1 Monthly salaries and annual income of eight employees of a welfare office.

10.2.1 Equation for Linear Relationships

The equation relating monthly salary to annual salary may be represented as

$$Y' = 12X$$

where the prime mark on Y indicates Y is being predicted.

You may substitute any value of X (the independent variable) into the equation and obtain the value of Y (the dependent variable) directly. For example, if another employee's monthly salary were $1000, her annual income would be

$$Y' = 12(1000) = 12,000.$$

Let us add one more factor to this linear relationship. Suppose all of the employees received a $500 raise. The equation would now read

$$Y' = 500 + 12X.$$

Thinking back to your high school days of algebra, you will probably recognize this equation as a special case of the general equation for a straight line, that is,

$$Y' = a + b_y X \qquad (10.1)$$

in which Y and X represent variables that change from individual to individual, and a and b_y represent constants for a particular set of data. More specifically, b_y represents the *slope of a line* relating values of Y to values of X. Thus Y is the dependent variable, and X is the independent variable. This is referred to as the *regression of Y on X*. In the present example, the slope of the line is 12, which means that Y changes by a factor of 12 for each unit change in X. The letter a represents the value of Y when $X = 0$ and is referred to as the *Y-intercept* or the point at which the line crosses the Y-axis.

Equation (10.1) may be regarded as a method for predicting Y from known values of X. When the correlation is 1.00 (as in the present case), the predictions are perfect.

The slope is frequently defined as the ratio of the change in Y to the change in X.

$$b_y = \frac{\text{change in } Y}{\text{change in } X} = \frac{\Delta Y}{\Delta X} = \frac{Y_2 - Y_1}{X_2 - X_1}. \qquad (10.2)$$

Using any two coordinates on the line we can quickly calculate the slope. Looking at Table 10.1 and Fig. 10.1, we can arbitrarily use employee A's monthly salary and annual income as our X_1 and Y_1 values and employee C's monthly salary and annual income as our X_2 and Y_2 values because the coordinates for employee's A and C lie on the line. Substituting into Eq. (10.2) we confirm that the slope is 12.

$$b_y = \frac{Y_2 - Y_1}{X_2 - X_1} = \frac{8700 - 7200}{725 - 600} = \frac{1500}{125} = 12.$$

The reason we can arbitrarily choose any two coordinates on the line is because the slope is constant and unchanging for any given linear regression example.

The slope can also be negative. Figure 9.1(e) and (f) would have negative slopes, for example, because as X increases, Y decreases or as X decreases, Y increases. When $b = 0$, the line parallels the X-axis and a equals the mean of Y because the line will cross the Y-axis at $\overline{Y}$.

10.2.2 Predicting Y from Knowledge of X

In social science research the correlations we obtain are almost never perfect. Therefore we must find a straight line that *best fits* our data and make predictions from that line. But what do we mean by "best fit"? When discussing the mean and the standard deviation, we defined the mean as that point in a distribution that makes the sum of squares of deviations from it minimal, that is, less than the sum of squares of deviations about any other score or potential score. When applying this principle to correlation and regression, the *best-fitting straight line* is defined as that line that makes the squared deviations around it minimal. This straight line is referred to as a **regression line.**

The term *prediction,* as used in statistics, does not necessarily carry with it any implication about the future. The term *predict* simply refers to the fact that we are using information about one variable to obtain information about another. Thus if we know a woman's occupational status, we may use this information to predict how much education she has completed (which we would assume preceded her having been hired for the job).

At this point we will highlight a new symbol you recently encountered: Y'.* This may be read as "Y prime," "the predicted Y," or

---

* The symbol $\hat{Y}$ (read "Y cap"), an alternative symbol that also designates the predicted Y, is preferred by some social scientists.

"estimated Y." We use this symbol with the regression line or the regression equation to estimate or predict a score on one variable from a known score on another variable.

Returning to the equation for a straight line, we are faced with the problem of determining b and a for a particular set of data so that Y' may be obtained. One equation for obtaining the slope of the line relating Y to X, which is known as the line of regression of Y on X, is the mean deviation equation.

$$b_y = \frac{\Sigma(X - \overline{X})(Y - \overline{Y})}{\Sigma(X - \overline{X})^2}. \qquad (10.3)$$

You may have noticed the numerator for this equation and for the mean deviation computational Eq. (9.1) for r are identical. The numerator in Eq. (9.1) determined the sign of r, and the numerator of Eq. (10.3) also determines the sign of b. It follows that the sign of r will always be the same as the sign of b for a given data set. Equation (10.3) is mathematically equivalent to Eq. (10.4) which clearly shows the relationship between r and b where b is equal to r times the ratio of the standard deviation of Y and the standard deviation of X. Hence, r equals b only when the standard deviations of X and Y are equal.

$$b_y = r\frac{s_y}{s_x}. \qquad (10.4)$$

Equations (10.3) and (10.4) are mathematically equivalent to Eq. (10.5) which is the raw score computational equation preferred by many social scientists. You will note that it is very similar to the raw score Eq. (9.2) used to compute r.

$$b_y = \frac{N(\Sigma XY) - (\Sigma X)(\Sigma Y)}{N(\Sigma X^2) - (\Sigma X)^2}. \qquad (10.5)$$

The constant a is given by Eq. (10.6).

$$a = \overline{Y} - b_y\overline{X}. \qquad (10.6)$$

10.2.3 An Illustrative Regression Problem

We now can use Eq. (10.1) to predict Y. Let us solve a sample problem using the data introduced in Section 10.1.

Mr. Jones, you will recall, scored 62 on the midterm examination. What is our prediction concerning his score on the final examination? The following table lists the relevant statistics.

MIDTERM (X VARIABLE)	FINAL (Y VARIABLE)
$\overline{X} = 70$	$\overline{Y} = 75$
$s_x = 4$	$s_y = 8$
$X = 62$	
$r = 0.60$	

We can use Eq. (10.1) to predict Mr. Jones's score on the final examination in the following manner. Using Eq. (10.4) we can calculate b_y.

$$b_y = r\frac{s_y}{s_x} = 0.60\frac{8}{4} = 1.20.$$

We see that $b = 1.20$ or that, as X increases one unit on the X-axis, Y increases 1.20 units on the Y-axis. Now we can calculate a, or our Y-intercept, using Eq. (10.6).

$$a = \overline{Y} - b_y\overline{X} = 75 - (1.20)70 = -9.$$

We now know that $a = -9$ or that the regression line crosses the Y-axis at the value $Y = -9$. Using the values we have just calculated for b and a and his score of 62 on the midterm, we can now use Eq. (10.1) to predict Mr. Jones's score on the final exam.

$$Y' = a + b_yX = -9 + 1.20(62) = 65.40.$$

Our prediction is that Mr. Jones will receive a score of 65.40 on the final examination. This example should begin to convey to you the interrelatedness of the various approaches to calculating a and b and predicting Y.

A reasonable question at this point is, "Since we know $\overline{Y}$ and s_y in the preceding problem, we presumably have all the observed data at hand. Therefore, why do we wish to predict Y from X?" It should be pointed out that the purpose of this example was to acquaint you with a prediction equation. In actual practice, however, correlational techniques are most commonly used to make predictions about future samples where Y is unknown.

For example, let us suppose that you are a demographer and over a period of years have accumulated much information concerning the relationship between the state of the economy and average family size. You find that it is now possible to use economic information (the independent variable, X) to predict average family size (the dependent variable, Y), and then use this information to establish population projections.

Since we discussed the relationship between Pearson's r and z scores in Chapter 9, it should be apparent that the prediction equation may be expressed in terms of z scores. Mathematically, it can be shown that

$$z_{y'} = rz_x \qquad (10.7)$$

where $z_{y'} = Y'$ (the predicted final exam score) is expressed in terms of a z score.

Mr. Jones's score of 62 on the midterm can be expressed as a $z = -2.00$, since it is two standard deviations to the left of the mean, and we know the correlation between the midterm and final exam scores is 0.60. Thus $z_{y'} = 0.60(-2.00) = -1.20$.

10.2.4 The Mean Deviation and Raw Score Computational Procedures for Linear Regression

Table 10.2 illustrates the mean deviation method of computation to provide you with a conceptual understanding of regression analysis. The computational procedures and the figures will be familiar to you as they were used to compute r in Table 9.2. We find b to equal 1.232 which indicates that, as X increases one unit, Y increases 1.232 units. The regression line intersects the Y-axis at 4.38 since $a = 4.38$. These two values, a and b, determine the regression line and also provide us with a general prediction equation for these data. For example, using Eq. (10.1) we can predict respondent C's educational attainment knowing only her mother's educational attainment (5 years of formal education) in the following manner. Because

$$Y' = 4.38 + 1.232X$$
$$Y'_c = 4.38 + 1.232(5)$$
$$Y'_c = 10.54.$$

Table 10.2   Computational procedures for linear regression using the mean deviation method (hypothetical data)

RESPON-DENT	MOTHER'S EDUCATION X	$(X - \overline{X})$	$(X - \overline{X})^2$	DAUGHTER'S EDUCATION Y	$(Y - \overline{Y})$	$(X - \overline{X})(Y - \overline{Y})$
A	1	-6	36	7	-6	36
B	3	-4	16	4	-9	36
C	5	-2	4	13	0	0
D	7	0	0	16	3	0
E	9	2	4	10	-3	-6
F	11	4	16	22	9	36
G	13	6	36	19	6	36

$N = 7$ $\quad$ $\Sigma X = 49$ $\quad\quad\quad$ $\Sigma(X - \overline{X})^2 = 112$ $\quad$ $\Sigma Y = 91$ $\quad\quad$ $\Sigma(X - \overline{X})(Y - \overline{Y}) = 138$

$\overline{X} = 7$ $\quad\quad\quad\quad\quad\quad\quad\quad\quad\quad$ $\overline{Y} = 13$

Using Eq. (10.3)

$$b_y = \frac{\Sigma(X - \overline{X})(Y - \overline{Y})}{\Sigma(X - \overline{X})^2} = \frac{138}{112} = 1.232.$$

Using Eq. (10.6)

$$a = \overline{Y} - b_y(\overline{X}) = 13 - 1.232(7) = 4.38.$$

Since

$$Y' = a + b_y X$$
$$Y' = 4.38 + 1.232X.$$

Also

$$r = 0.82 \text{ (see Table 9.2).}$$

Our prediction is that respondent C has completed 10.54 years of education; however, we know she has actually completed 13 years, or 2.46 more years than we predicted (see Table 10.2).

Table 10.3 illustrates raw score computational Eq. (10.5). The raw score procedure provides us with the same results as did the mean deviation method presented in Table 10.2. Note that the correlation coefficient is positive ($r = +0.82$), as is the sign of the slope ($b = +1.232$).

Table 10.3 Raw score procedure for linear regression (hypothetical data)

RESPON-DENT	MOTHER'S EDUCATION X	X^2	DAUGHTER'S EDUCATION Y	Y^2	XY
A	1	1	7	49	7
B	3	9	4	16	12
C	5	25	13	169	65
D	7	49	16	256	112
E	9	81	10	100	90
F	11	121	22	484	242
G	13	169	19	361	247
$N = 7$	$\Sigma X = 49$ $\overline{X} = 7$	$\Sigma X^2 = 455$	$\Sigma Y = 91$ $\overline{Y} = 13$	$\Sigma Y^2 = 1435$	$\Sigma XY = 775$

Using Eq. (10.5)

$$b_y = \frac{N(\Sigma XY) - (\Sigma X)(\Sigma Y)}{N(\Sigma X^2) - (\Sigma X)^2}$$

$$= \frac{7(775) - (49)(91)}{7(455) - (49)^2}$$

$$= \frac{5425 - 4459}{3185 - 2401}$$

$$= \frac{966}{784}$$

$$b_y = 1.232.$$

Also, using Eq. (10.6)

$$a = \overline{Y} - b_y\overline{X}$$
$$a = 13 - 1.232(7)$$
$$a = 4.38.$$

Using Eq. (10.1)

$$Y' = a + b_y X$$
$$Y' = 4.38 + 1.232X.$$

Also

$$r = 0.82 \text{ (see Table 9.2).}$$

10.2.5 Constructing a Regression Line

Let us return to the problem of constructing a regression line for predicting scores on the Y variables. As we have already pointed out, the regression line seldom passes through all the paired scores. It will, in fact, pass among the paired scores in such a way as to minimize the squared deviations between the regression line (the predicted scores) and the obtained scores. Earlier we pointed out that the mean is the point in a distribution that makes the squared deviations around it minimal. The regression line is analogous to the mean since, as we shall demonstrate, the sum of deviations of scores around the regression line is zero and the sum of squares of these deviations is minimal. Some students find it helpful to think of the regression line as a "generalization" of the data.

The regression line always crosses the coordinate $(\overline{X}, \overline{Y})$; therefore, a convenient method of constructing our regression line involves locating the Y-intercept (a) on the Y-axis (in this instance, $a = 4.38$) and the coordinate determined by the means of X and Y. Connecting this coordinate with the Y-intercept determines the regression line that must pass through the coordinate determined by the means of X and Y.

It is helpful to remember that the closer the data points (coordinates) are to the line, the larger the magnitude of r. When $r = 1.00$, all of the data points will lie on the line, and the predicted Y values will equal the actual Y values.

The coordinate for respondent C has been highlighted in Fig. 10.2 to help you further understand prediction. Recall that in the previous section we predicted that respondent C would complete 10.54 years of education, given that her mother had completed the fifth grade. By examining Fig. 10.2 you should see that the predicted Y value (Y') lies on the line predicting Y from known values of X. Indeed, the regression line is also referred to as the *prediction line*. Projecting a perpendicular line from any point on the X-axis (assuming we are predicting Y from known values of X) to the prediction line allows us to determine the approximate predicted Y value for the corresponding X value from which the perpendicular line originated. As is shown in Fig. 10.2, the predicted Y value of 10.54 years (see Section 10.2.4), given that X equals 5 years, is 2.46 years lower than the actual Y value (13) for respondent C. *That is, respondent C's Y score is 2.46 years higher than* we expected, given her mother's education.

The small triangle in Fig. 10.2 confirms the slope that we calculated earlier for the data in Tables 10.2 and 10.3. You should see that, as Y increases 2.46 units, X increases 2 units. Substituting into Eq. (10.2)

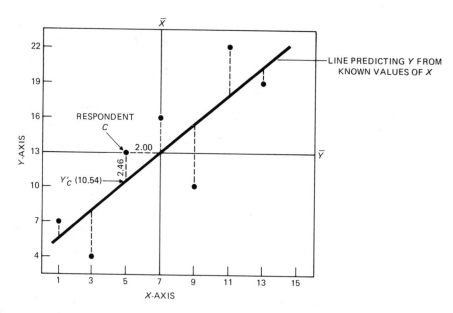

10.2 Scatter diagram representing paired scores on two variables and regression line for predicting *Y* from *X*. Note that the regression line minimizes the vertical or *Y*-axis deviations of the data points.

$$b_y = \frac{\Delta Y}{\Delta X} = \frac{2.46}{2.00} = 1.23$$

or

$$b_y = \frac{Y_2 - Y_1}{X_2 - X_1} = \frac{13 - 10.54}{7 - 5} = \frac{2.46}{2} = 1.23.$$

10.3 Residual Variance and Standard Error of Estimate

Figure 10.3 shows a series of scatter diagrams (reproduced from Fig. 10.2) and the regression line for predicting *Y* from known values of *X*. The regression line represents our best basis for predicting *Y* scores from known values of *X*. As we can see, not all the *Y* values fall on the regression line. However, if the correlation had been 1.00, all the *Y*

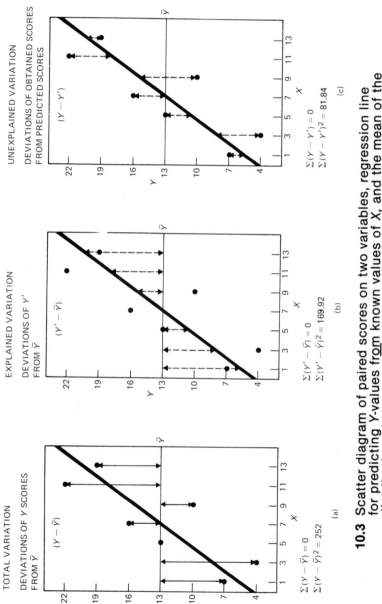

10.3 Scatter diagram of paired scores on two variables, regression line for predicting Y-values from known values of X, and the mean of the distribution of Y-scores ($\bar{Y}$): $r = 0.82$ (from data in Table 10.2). (a) Deviations of scores $(Y - \bar{Y})$ from the mean of Y, total variation. (b) Deviations of predicted scores $(Y' - \bar{Y})$ from the mean of Y, explained variation. (c) Deviations of scores $(Y - \bar{Y})$ from the regression line, unexplained variation.

scores (the years of education completed by the seven respondents) would have fallen right on the regression line. The deviations $(Y - Y')$ in Fig. 10.3(c) and Table 10.4 represent our errors in prediction. The $(Y - Y')$ column in Table 10.4 shows the errors in prediction for each of the seven respondents. Earlier we calculated respondent C's predicted Y score (Y') to be 10.54. Substituting this value and the actual Y score for her, we can calculate the size *and* direction of the error in prediction.

$$Y_{\text{respondent } C\text{'s mother}} - Y'_{\text{respondent } C} = 13 - 10.54 = +2.46.$$

The value $+2.46$ tells us that the coordinate for respondent C lies 2.46 years above the regression line. The size and direction of the error in prediction (referred to as a **residual**) for the remaining six respondents is shown in the $(Y - Y')$ column of Table 10.4. Note that there will be as many residuals as cases (respondents, etc.), although a residual will equal 0 if an obtained Y score lies on the prediction line.

Table 10.4   Computational procedures for total, explained, and unexplained variation (data from Table 10.2)

RESPON- DENT	X	Y	Y'	$(Y - \overline{Y})$	$(Y - \overline{Y})^2$	$(Y' - \overline{Y})$	$(Y' - \overline{Y})^2$	$(Y - Y')$	$(Y - Y')^2$
A	1	7	5.61	-6	36	-7.39	54.61	1.39	1.93
B	3	4	8.07	-9	81	-4.93	24.30	-4.07	16.56
C	5	13	10.54	0	0	-2.46	6.05	2.46	6.05
D	7	16	13.00	3	9	0	0	3.00	9.00
E	9	10	15.46	-3	9	2.46	6.05	-5.46	29.81
F	11	22	17.93	9	81	4.93	24.30	4.07	16.56
G	13	19	20.39	6	36	7.39	54.61	-1.39	1.93
	$\overline{X} = 7$	$\overline{Y} = 13$			$\Sigma(Y - \overline{Y})^2 = 252$		$\Sigma(Y' - \overline{Y})^2 = 169.92*$		$\Sigma(Y - Y')^2 = 81.84*$

*Do not add to 252 due to rounding $(169.92 + 81.84 \neq 252)$.

Note the similarity of $Y - Y'$ (the deviation of obtained scores from the predicted scores, which always lie on the regression line) to $Y - \overline{Y}$ (the deviation of Y scores from $\overline{Y}$); the algebraic sum of the deviations around the regression line is equal to zero. Earlier, we saw that the algebraic sum of the deviations around the mean is also equal to zero.

You will recall that in calculating the variance (s^2) we squared the deviations from the mean, summed, and divided by N. The square root of the variance was the standard deviation. Now, if we were to square and sum the deviations of the scores from the regression line,

$\Sigma(Y - Y')^2$, we would have a basis for calculating another variance and standard deviation. The variance around the regression line is known as the **residual variance** and is defined as:

$$s^2_{\text{est } y} = \frac{\Sigma(Y - Y')^2}{N}.$$ *(10.8)*

The standard deviation around the regression line (referred to as the **standard error of estimate**) is the square root of the residual variance. The standard error of estimate provides a measure of how well the regression equation predicts Y, given information about X. Thus

$$s_{\text{est } y} = \sqrt{\frac{\Sigma(Y - Y')^2}{N}}.$$ *(10.9)*

Fortunately there is a simplified method for calculating $s_{\text{est } y}$.

$$s_{\text{est } y} = s_y\sqrt{1 - r^2}.*$$ *(10.10)*

You will note that when $r = \pm 1.00$, $s_{\text{est } y} = 0$, which means that there are no deviations from the regression line, and therefore no errors in prediction. On the other hand, when $r = 0$, the errors of prediction are maximal for that given distribution, that is, $s_{\text{est } y} = s_y$.

With the data in Table 10.3, the following statistics were calculated:

$$\overline{X} = 7 \qquad\qquad \overline{Y} = 13$$
$$s_x = 4.32 \qquad\qquad s_y = 6.48$$
$$r = 0.82$$

Using Eq. (10.10),

$$s_{\text{est } y} = 6.48\sqrt{1 - 0.82^2}$$
$$= 6.48(0.5724)$$
$$= 3.71$$

As we have already indicated, the standard error of estimate has properties that are similar to those of the standard deviation. The smaller the standard error of estimate, the lesser the dispersion of scores around the regression line.

---

* The values of $\sqrt{1 - r^2}$ may be obtained directly from Table G. Thus for $r = 0.82$, we have $\sqrt{1 - r^2} = 0.5724$.

For each value of X, there is a distribution of scores on the Y variable. The mean of each of these distributions is Y', and the standard deviation of each distribution is $s_{est\ y}$. When the distribution of Y scores for each value of X has the same variability, we refer to this condition as **homoscedasticity.** In addition, if the distribution of Y scores for each value of X is normally distributed, we may state relationships between the standard error of estimate and the normal curve (see Section 7.3). We can, for example, predict a value of Y from any given X, and then describe an *interval* within which it is likely that the true value of Y will be found. For normally distributed variables, approximately 68% of the time the value of Y (that is, Y_T) will be within the following interval:

$$\text{Interval including } Y_T = Y' \pm s_{est\ y} \sqrt{1 + \frac{1}{N} + \frac{(X - \overline{X})^2}{\Sigma(X - \overline{X})^2}} \quad (10.11)*$$

where Y_T equals the true value of Y.

Thus if we predicted a Y value of 10.54 from $X = 5$ (see respondent C), given $s_{est\ y} = 3.71$, $Y' = 10.54$, $\overline{X} = 7$, $\Sigma(X - \overline{X})^2 = 112$ (see Table 9.2), and $N = 7$.

$$\text{Interval including } Y_T = 10.54 \pm 3.71 \sqrt{1 + \frac{1}{7} + \frac{(5 - 7)^2}{112}}$$

$$= 10.54 \pm 3.71(1.09)$$

$$= 10.54 \pm 4.04.$$

In other words, when $X = 5$, about 68% of the time the true Y (Y_T) will be found between the scores of 6.50 and 14.58.[†]

Using these data, we draw two lines, one above and one below the regression line for predicting Y from X (Fig. 10.4). For normally distributed variables, approximately 68% of the time the true values will be found between these lines when we predict Y from various values of X. Note that the lines are more spread out when predicting Y from extreme values of X. The result is two slightly bowed lines, with minimum distance between them at the mean of X. (It should be noted that these relationships refer to the distributions of *both* X and Y.)

---

[*]   Note that, if we wanted either the 95% or 99% confidence interval, we would multiply $s_{est\ y}$ in equation (10.11) by 1.96 or 2.58, respectively.

[†]   A technical interpretation of confidence intervals is presented in Section 14.7.

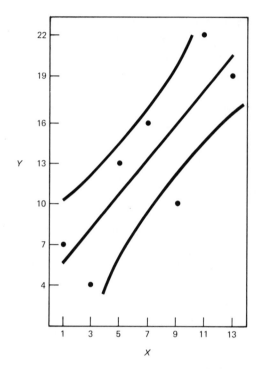

10.4 Line of regression for predicting Y from X with curved lines one $s_{est\,y}$ above and below the regression line (from data in Table 10.3).

10.4 Explained and Unexplained Variation*

If we look again at Fig. 10.3, we can see that there are three separate sums of squares that can be calculated from the data. These are:

1. Variation of scores around the sample mean [Fig. 10.3(a)]. This variation is given by $\Sigma(Y - \overline{Y})^2$ and is referred to as the **total variation** or the **total sum of squares.** It is basic to the determination of the variance and the standard deviation of the sample.

2. Variation of scores around the regression line (or predicted scores) [Fig. 10.3(c)]. This variation is given by $\Sigma(Y - Y')^2$ and is referred to as

---

* Although analysis of variance is not covered until Chapter 16, much of the material in this section will serve as an introduction to some of the basic concepts of analysis of variance.

unexplained variation or **residual sum of squares.** The reason for this choice of terminology should be clear. If the correlation between two variables is ± 1.00, all the scores fall on the regression line. Consequently, we have in effect explained *all* the variation in Y in terms of the variation in X and, conversely, all the variation of X in terms of the variation in Y. In other words, in the event of a perfect relationship, there is no unexplained variation. However, normally the correlation is less than perfect, and many of the scores will not fall right on the regression line, as we have seen. The deviations of these scores from the regression line represent variation that is not accounted for in the correlation between two variables. Hence, the term *unexplained variation* is used.

3. Variation of predicted scores about the mean of the distribution [Fig. 10.3(b)]. This variation is given by $\Sigma(Y' - \overline{Y})^2$ and is referred to as **explained variation** or **regression sum of squares.** The reason for this terminology should be clear from our discussion in the preceding paragraph and our prior reference to predicted deviation (Section 10.3). You will recall our previous observation that the greater the correlation, the greater the predicted deviation from the sample mean. It further follows that the greater the predicted deviation, the greater the explained variation. When the predicted deviation is maximum, the correlation is perfect, and 100% of the variation of the scores around the sample mean is explained.

It can be shown mathematically that the total variation consists of two components which may be added together. These two components represent explained variation and unexplained variation, respectively. Thus

$$\Sigma(Y - \overline{Y})^2 = \Sigma(Y - Y')^2 + \Sigma(Y' - \overline{Y})^2. \quad \textbf{\textit{(10.12)}}$$

Total   Unexplained   Explained
variation   variation   variation

These calculations are shown in Fig. 10.3. You will note that the sum of the explained variation (169.92) and the unexplained variation (81.84) is equal to the total variation. The slight discrepancy found in this example is due to rounding r to 0.82 prior to calculating the predicted scores.

Now, when $r = 0.00$, then $\Sigma(Y' - \overline{Y})^2 = 0.00$. (Why? See Section 10.2.) Consequently, the total variation is equal to the unexplained variation. Stated another way, when $r = 0$, all the variation is unexplained. On the other hand, when $r = 1.00$, then $\Sigma(Y - Y')^2 = 0.00$, since all the scores are on the regression line. Under these circumstances, total variation is the same as explained variation. In other words, all the variation is explained when $r = 1.00$.

The ratio of the explained variation to the total variation is referred to as the **coefficient of determination** and is symbolized by r^2. The equation for the coefficient of determination is:

$$r^2 = \frac{\text{explained variation}}{\text{total variation}} = \frac{\Sigma(Y' - \overline{Y})^2}{\Sigma(Y - \overline{Y})^2}. \qquad (10.13)$$

Referring to Fig. 10.3, we see that the proportion of explained variation to total variation is:

$$r^2 = \frac{169.92}{252} = 0.67.$$

It can be seen that the coefficient of determination indicates the proportion of total variation that is explained in terms of the magnitude of the correlation coefficient. When $r = 0$, the coefficient of determination, r^2, equals 0. When $r = 0.5$, the coefficient of determination is 0.25. In other words, 25% of the total variation is accounted for. Finally, when $r = 1.00$, then $r^2 = 1.00$ and all variation is accounted for or explained.

Figure 10.5 depicts graphically the proportion of variation in one variable that is accounted for by the variation in another variable when r takes on different values.

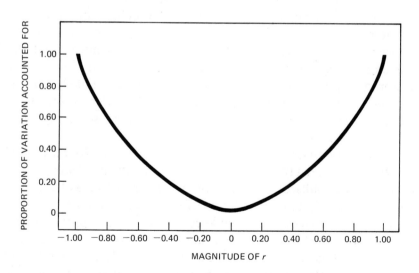

10.5 The proportion of the variation on one variable accounted for in terms of variations of a correlated variable at varying values of r^2.

We have previously shown that the coefficient of determination, r^2, is equal to 0.67 for the data summarized in Fig. 10.3. Note that the square root of this number (that is, $\sqrt{0.67}$) equals 0.82—which is r for these data.

Since r^2 represents the proportion of variation accounted for, $(1 - r^2)$ represents the proportion of variation that is *not* explained in terms of the correlation between X and Y. In this instance, $1 - r^2 = 1 - 0.67 = 0.33$.

10.5 Correlation and Causation

You have seen that when two variables are related it is possible to predict one from your knowledge of the other. This relationship between correlation and prediction often leads to a serious error in reasoning; that is, the relationship between two variables frequently carries with it the implication that one has caused the other. This is especially true when there is a temporal relationship between the variables in question, that is, when one precedes the other in time. What is often overlooked is the fact that the variables may not be causally connected in any way, but that they may vary together by virtue of a common link with a third variable. Thus if you are a bird watcher, you may note that, as the number of birds increases in the spring, the grass becomes progressively greener (see Fig. 10.6). However, recognizing that the extended number of hours and the increasing warmth of the sun is a third factor influencing both of these variables, you are not likely to conclude that the birds cause the grass to turn green, or vice versa. However, there are many occasions, particularly in the social sciences, when it is not so easy to identify the third factor.

Suppose that you have demonstrated that there is a high positive correlation between the number of hours students spend studying for an exam and their subsequent grades on that exam. You may be tempted to conclude that the number of hours of study causes grades to vary. This seems to be a perfectly reasonable conclusion, and is probably in close agreement with what your parents and instructors have been telling you for years. Let us look closer at the implications of a causal relationship. On the assumption that a greater number of hours of study causes grades to increase, we would be led to expect that *any* student who

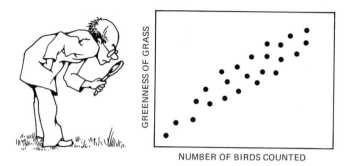

CORRELATION IS NOT CAUSATION

10.6 Note that when the census count of birds is low, the grass is not very green. When there are many birds, the grass is very green. Therefore, the number of birds determines how green the grass will become. What is wrong with this conclusion?

devotes more time to study is guaranteed a high grade and that one who spends less time on studying is going to receive a low grade. This is not necessarily the case. We have overlooked the fact that it might be that the better student (by virtue of higher intelligence, stronger motivation, better study habits, etc.) who devotes more time to study, performs better simply because he or she has a greater capacity to do so.

What we are saying is that correlational studies simply do not permit inferences of causation. Correlation is a *necessary* but not a *sufficient* condition to establish a causal relationship between two variables.

Huff's book* includes an excellent chapter devoted to the confusion of correlation with causation. He refers to faulty causal inferences from correlational data as the *post hoc* **fallacy.** The following excerpt illustrates a common example of the *post hoc* fallacy.

> Reams of pages of figures have been collected to show the value in dollars of a college education, and stacks of pamphlets have been published to bring these figures—and conclusions more or less based on them—to the attention

---

* Reprinted from D. Huff and I. Geis, *How to Lie with Statistics,* New York: W. W. Norton & Co., Inc., 1954, with permission.

of potential students. I am not quarreling with the intention. I am in favor of education myself, particularly if it includes a course in elementary statistics. Now these figures have pretty conclusively demonstrated that people who have gone to college make more money than people who have not. The exceptions are numerous, of course, but the tendency is strong and clear.

The only thing wrong is that along with the figures and facts goes a totally unwarranted conclusion. This is the *post hoc* fallacy at its best. It says that these figures show that if *you* (your son, your daughter) attend college you will probably earn more money than if you decide to spend the next four years in some other manner. This unwarranted conclusion has for its basis the equally unwarranted assumption that since college-trained folks make more money, they make it because they went to college. Actually we don't know but that these are the people who would have made more money even if they had *not* gone to college. There are a couple of things that indicate rather strongly that this is so. Colleges get a disproportionate number of two groups of kids—the bright and the rich. The bright might show good earning power without college knowledge. And as for the rich ones—well money breeds money in several obvious ways. Few sons of rich men are found in low-income brackets whether they go to college or not.

Summary

Let us briefly review what we have learned in this chapter. We have seen that it is possible to "fit" a straight line to a bivariate distribution of scores for predicting Y scores from known X values.

We saw that, when the correlation is perfect, all the scores fall upon the regression line. There is, therefore, no error in prediction. The lower the degree of relationship, the greater the dispersion of scores around the regression line, and the greater the errors of prediction. Finally, when $r = 0$, the mean of the sample provides our best predictor for a given variable.

The regression line was shown to be analogous to the mean: the summed deviations around it are zero and the sum squares are minimal. The standard error of estimate was shown to be analogous to the standard deviation.

We saw that three separate sums of squares, reflecting variability, may be calculated from correlational data. Variation about the mean of the distribution for each variable: This variation is referred to as the *total variation* or *total sum of squares*. Variation of each score about the regression line: This variation is known as *unexplained variation* or *residual sum of squares*. Variation of each predicted score about the mean of the distribution for each variable: This variation is known as *explained variation* or *regression sum of squares*. We saw that the sum of the explained variation and the unexplained variation is equal to the total variation.

Finally, we saw that the ratio of the explained variation to the total variation provides us with the proportion of the total variation that is explained. The term applied to this proportion is *coefficient of determination*.

Terms to Remember

Coefficient of determination (r^2)   The ratio of the explained variation to the total variation.

Explained variation   Variation of predicted scores about the mean of the distribution. Same as the regression sum of squares.

Homoscedasticity   Condition that exists when the distribution of Y scores for each value of X has the same variability.

Post hoc **fallacy**   Faulty causal inferences from correlational data.

Regression line (line of "best fit")   Straight line that makes the squared deviations around it minimal.

Regression sum of squares   Variation of predicted scores about the mean of the distribution. Same as the explained variation.

Residual   Size and direction of the error in prediction or the vertical distance from the observed data coordinate to the regression line.

Residual sum of squares   Variation of scores around the regression line. Same as the unexplained variation.

Residual variance   Variance around the regression line.

Standard error of estimate   Standard deviation of scores around the regression line.

Total sum of squares   Variation of scores around the sample mean. Same as the total variation.

Total variation   Variation of scores around the sample mean. Same as the total sum of squares.

Unexplained variation   Variation of scores around the regression line. Same as the residual sum of squares.

Exercises

1. Find the equation of the regression line for the following data

X	1	2	3	4	5
Y	5	3	4	2	1

2. In a study concerned with the relationship between two variables, X and Y, the following was obtained.

$$\overline{X} = 119 \qquad \overline{Y} = 1.30$$
$$s_x = 10 \qquad s_y = 0.55$$
$$r = 0.70$$
$$N = 100$$

a) Sally B. obtained a score of 130 on the X variable. Predict her score on the Y variable.

b) A score of 1.28 on the Y variable was predicted for Bill B. What was his score on the X variable?

c) Determine the standard error of estimate of Y. Interpret your answer.

3. A study was undertaken to find the relationship between family stability and performance in college. The following results were obtained.

FAMILY STABILITY	COLLEGE AVERAGE
$\overline{X} = 49$	$\overline{Y} = 1.35$
$s_x = 12$	$s_y = 0.50$
$r = 0.36$	
$N = 60$	

a) Norma obtained a score of 65 on the X variable. What is your prediction of her score on the Y variable?

b) Determine the standard error of estimate of Y. Interpret your answer.

c) What proportion of the total variation in Y is accounted for by the X variable?

4. Assume that $\overline{X} = 30$, $s_x = 5$; $\overline{Y} = 45$, $s_y = 8$. Draw a separate graph showing the regression line for the following values of r.

    a) 0.00   **b)** 0.20   **c)** 0.40

    d) 0.60   **e)** 0.80   **f)** 1.00

5. For the following data (assume enrollment is the independent variable):

a) Calculate a and b_y. Interpret your answer; that is, what does it tell you?

b) Plot the data and draw the regression line.

c) Calculate r and r^2. Interpret your answer.

Public elementary and secondary schools enrollment and expenditures data for selected U.S. cities, 1978*

CITY	ENROLLMENT (1,000) 1978	TOTAL EXPENDITURES (MILLION $'S) 1978
Baltimore, Md.	152	310
Boston, Mass.	77	261
Chicago, Ill.	511	1,285
Cleveland, Ohio	115	292
Dallas, Tex.	135	259
Detroit, Mich.	238	459
Houston, Tex.	207	378
Indianapolis, Ind.	78	117
Los Angeles, Calif.	587	1,335
Memphis, Tenn.	116	150
Milwaukee, Wis.	101	227
New Orleans, La.	91	139
New York, N.Y.	1,046	2,522
Philadelphia, Pa.	254	575
Phoenix, Ariz.	184	281
St. Louis, Mo.	78	144
San Antonio, Tex.	64	98
San Diego, Calif.	119	270
San Francisco, Calif.	65	176
Washington, D.C.	120	293

* *Source:* U.S. National Center for Education Statistics, *Statistics of Public Elementary and Secondary Day Schools,* annual.

6. For the following data (assume percentage registering is the independent variable):

    a) Calculate a, b_y, r, and the standard estimate of Y.

    b) Interpret your answers.

Percent reported registered to vote and percent voting, 1968–1978*

YEAR	PERCENTAGE OF U.S. ADULTS REPORTING THEY REGISTERED TO VOTE	PERCENTAGE OF U.S. ADULTS REPORTING THEY VOTED
1968	74.3	67.8
1972	72.3	63.0
1974	62.2	44.7
1976	66.7	59.2
1978	62.6	45.9

Source: U.S. Bureau of the Census, *Current Population Reports,* series P-20, No. 344, and earlier reports.

7. Assume that students take two tests for entrance into the college of their choice. Both are normally distributed tests. The college-entrance examination has a mean of 47.63 and a standard deviation of 13.82. The verbal comprehension test has a mean equal to 39.15 and a standard deviation equal to 12.35. The correlation between the two tests equals 0.85.

a) Estelle obtained a score of 40 on her college entrance examination. Predict her score on the verbal comprehension test.

b) How likely is it that she will score at least 40 on the verbal comprehension test?

c) Howard obtained a score of 40 on the verbal comprehension test. Predict his score on the college entrance examination.

d) How likely is it that he will score at least 40 on the college entrance examination?

e) REC University finds that students who score at least 45 on the verbal comprehension test are most successful. What score on the college entrance examination should be used as the cutoff point for selection?

f) Harris obtained a score of 55 on the college entrance examination. Would he be selected by REC University? What are his chances of achieving an acceptable score on the verbal comprehension test?

g) Rona obtained a score of 60 on the college entrance examination. Would she be accepted by REC University? How likely is it that she will *not* achieve an acceptable score on the verbal comprehension test?

8. For the following data (assume size of work group is the independent variable):

a) What is your prediction of the measure of group cohesiveness for a work group numbering 15?

b) What percent of the total variation in Y is explained by X?

GROUP	SIZE OF WORK GROUP	MEASURE OF GROUP COHESIVENESS (LOW VALUE INDICATES A COHESIVE WORK GROUP)
A	10	1
B	18	4
C	4	2
D	20	5
E	12	3
F	13	3
G	11	4

9. The per capita gross national product (GNP) is widely recognized as an estimate of the living standard of a nation. It has been claimed that per capita energy consumption is, in turn, a good predictor of per capita GNP. The accompanying table shows the GNP (expressed in dollars per capita) and the per capita energy consumption (expressed in millions of BTU's per capita) of various nations.

    a) Construct a scatter diagram from the accompanying data.

    b) Determine the correlation between GNP and energy expenditures.

    c) Construct the regression line for predicting per capita GNP from per capita energy consumption.

    d) The following nations were not represented in the original sample. Their per capita energy expenditures were: Chile, 21; Ireland, 49; Belgium, 88. Calculate the predicted per capita GNP for each country. Compare the predicted values with the actual values which are, respectively, 400; 630; 1400.

COUNTRY	ENERGY CONSUMPTION (IN MILLIONS OF BTU'S)	GNP (DOLLARS)
India	3.4	55
Ghana	3.0	270
Portugal	7.7	240
Colombia	12.0	290
Greece	12.0	390
Mexico	23.0	310
Japan	30.3	550
USSR	69.0	800
Netherlands	75.0	1100
France	58.0	1390
Norway	67.0	1330
West Germany	90.0	1410
Australia	88.0	1525
United Kingdom	113.0	1400
Canada	131.0	1900
United States	180.0	2900

10.

X	3	4	5	6	7	8	9	10	11
Y	4	3	5	6	8	7	9	9	11

    a) Determine the correlation for the accompanying scores.

    b) Given this correlation, and the computed mean for Y, predict the value of Y' for each X.

    c) Calculate $\Sigma(Y - \overline{Y})^2$.

    d) Calculate the unexplained variance.

    e) Calculate the explained variance.

    f) Calculate the coefficient of determination. Show that the square root of that value equals r.

11. A manager of a political campaign found a correlation of 0.70 between the number of people at a party and the loaves of bread consumed.

NUMBER OF PEOPLE	NUMBER OF LOAVES
$\overline{X} = 50$	$\overline{Y} = 5$
$s_x = 15$	$s_y = 1.2$

    a) For a party of 60 people, calculate the predicted number of loaves needed.

    b) For a party of 35, calculate the predicted number of loaves needed.

    c) What is the $s_{est\ y}$? Interpret your answer.

Multivariate
Data
Analysis

11

11.1 Introduction

11.2 The Multivariate Contingency Table

11.3 Partial Correlation

11.4 Multiple Regression Analysis

11.1 Introduction

In the previous chapters we have examined univariate and bivariate statistical techniques. Often, however, the complexity of social science issues requires that several variables be analyzed simultaneously. When this is the case, we utilize **multivariate statistical techniques** which allow us to determine the nature of the relationships between three or more variables.

When the relationship between marital status and employment status was considered in Chapter 8 (An Introduction to Contingency Tables), we concluded that a weak relationship existed between the two variables. That is, knowing a woman's marital status improved our ability to predict her employment status. In the process of establishing that a relationship existed, however, we ignored other variables that might have been useful in determining why some of the women in the sample chose to work and others did not. What if we had divided our sample into two groups: those women who need to work for financial reasons and those who do not? It is very likely that introducing a third variable, in this instance the financial necessity of working, into the analysis would allow us to more fully understand the employment status of the women in our sample. Indeed, it is possible that a major reason for women working is the need for additional income. A fourth variable, the presence or absence of young children in the home, might also help clarify the complex reasons that determine whether or not a woman chooses to work outside the home.

In Chapter 9 (Correlation) you learned to assess the direction and strength of the relationship between two interval-level variables. While it is valuable to know the correlation between two variables, you were later warned in Chapter 10 not to equate correlation with causation, as a third variable might be responsible for the observed bivariate correlation.

You also may have concluded after studying Chapter 10 (Regression and Prediction) that using a single independent variable to predict a dependent variable was far too simplistic an approach given the complexity of human attitudes and behavior. Age by itself, for example, can be used as an independent variable to predict a person's life satisfaction. However, if several other independent variables, such as an individual's economic and health status, were used jointly with the individual's age, we would expect to improve our prediction of life satisfaction.

All of these examples require the use of multivariate data analysis. The multivariate techniques presented in this chapter are extensions of what you have learned in Chapters 8, 9, and 10. They include the topics of multivariate contingency table analysis, partial correlation analysis, and multiple regression and correlation analysis. You may wish to briefly review those chapters prior to reading this chapter.

11.2 The Multivariate Contingency Table

Table 11.1 shows the relationship between high school seniors' plans to attend college and their parents' level of educational attainment. The table has been percentaged down, and comparisons made across the table reveal a positive relationship between parental educational attainment and seniors' plans to attend college (gamma = 0.443). Therefore, the more formal education the parent has completed, the more likely the student has plans to attend college following graduation from high school.

Before we can conclude that parental educational attainment is a key determinant of plans to attend college, we must consider the possibility that the relationship is due to a third variable. Perhaps the relationship between the initial two variables differs by sex. We can assess the effect of sex on college plans by analyzing the female and male respondents separately, as shown in Tables 11.1(b) and (c). Note that the distribution of cases in Table 11.1(a) would result if we combined Tables 11.1(b) and (c). For example, totaling the 347,000 cases in the upper right cell of Table 11.1(b) plus the 362,000 cases in the upper right cell of Table 11.1(c) would result in the 709,000 cases in the upper right cell of Table 11.1(a).

Table 11.1 College plans by educational attainment for head of family

(a) For females and males

	(X) EDUCATIONAL ATTAINMENT FOR HEAD OF FAMILY			
(Y) COLLEGE PLANS	Elementary	High school	College	Total
Plan to attend	31.6 (149)	42.4 (714)	71.4 (709)	(1572)
May attend	28.9 (136)	28.2 (475)	17.7 (176)	(787)
Will not attend	39.5 (186)	29.4 (495)	10.9 (108)	(789)
Total	100.0 (471)	100.0 (1684)	100.0 (993)	N = 3148

Gamma = +0.443

Numbers in thousands. Reported in Current Population Reports, "College Plans of High School Seniors: October 1975," Series P-20, No. 299, November 1976.

(b) For females

	FEMALES (Z) EDUCATIONAL ATTAINMENT FOR HEAD OF FAMILY (X)			
(Y) COLLEGE PLANS	Elementary	High school	College	Total
Plan to attend	34.7 (70)	45.7 (391)	75.6 (347)	(808)
May attend	26.7 (54)	24.8 (212)	15.7 (72)	(338)
Will not attend	38.6 (78)	29.5 (252)	8.7 (40)	(370)
Total	100.0 (202)	100.0 (855)	100.0 (459)	N = 1516

Gamma = +0.468

Continued on next page.

Table 11.1 College plans by educational attainment for head of family (cont.)

(c) For males

MALES (Z)
EDUCATIONAL ATTAINMENT FOR HEAD OF FAMILY (X)

(Y) COLLEGE PLANS	Elementary	High school	College	Total
Plan to attend	29.4 (79)	39.0 (323)	67.8 (362)	(764)
May attend	30.5 (82)	31.7 (263)	19.5 (104)	(449)
Will not attend	40.1 (108)	29.3 (243)	12.7 (68)	(419)
Total	100.0 (269)	100.0 (829)	100.0 (534)	N = 1632

Gamma = +0.425

The relationship between parental educational attainment and college plans of female seniors is presented in Table 11.1(b), and male seniors are included in Table 11.1(c). Tables 11.1(b) and (c) are referred to as **conditional tables** or **partial tables** because they summarize the relationship between two variables for subgroups, categories, or conditions of one or more other variables. The variable sex in this instance is referred to as a **control or test variable** and is designated by the letter Z. **Statistical elaboration** refers to the introduction of a control or test variable into an analysis to determine if a two-variable relationship remains the same or varies under the different categories or conditions of the control or test variable. Tables 11.1(b) and (c) are also referred to as first-order conditional tables. The **order** of a conditional table is determined by the number of control or test variables that have been introduced. A zero-order table [like Table 11.1(a)] includes no control variables; a first-order conditional table [like Table 11.1(b) or (c)] includes one control variable; and a second-order conditional table contains two control variables employed simultaneously, etc.

The effect of sex, our control variable, has been physically removed in Table 11.1(b) which contains only females and in Table 11.1(c) which contains only males. Thus both tables are homogeneous in terms of the test variable. Put another way, sex has been controlled or held constant. This was not the case in Table 11.1(a) which contained both females and males.

Controlling for the effects of sex has had a very minor influence on the bivariate relationship in Table 11.1(a). While the relationship between parental educational attainment and college plans is slightly stronger for females than males (gamma = 0.468 versus 0.425), the difference between the gammas is slight. Furthermore, a comparison of the percentages in Table 11.1(a) with those in Table 11.1(b) and (c), as well as a comparison of the partial tables, supports the finding that controlling for the effects of sex had little impact.

Perhaps at this point you have asked yourself why sex was chosen as a control variable. Several other variables might have proven equally as important or more important than sex. The choice is based on logic and the theoretical context of the variables being investigated. Another potential test variable that could have an effect on the relationship between parental educational attainment and college plans of high school seniors is the student's academic aptitude. Studies have shown that lower socioeconomic students with a high academic aptitude are more likely to attend college than lower socioeconomic students with a low academic aptitude. Aptitude seems to have an influence independent of family economic and educational background. If we controlled for aptitude using an interval measure of a standardized test, it would be necessary to establish "cutting points." Dividing the scores into low, medium, and high categories would result in three, first-order partial tables, one for each category.

Controlling for an ordinal- or interval-level variable reduces but does not eliminate the influence of a test variable. For example, family income could be trichotomized into the following three categories: (1) less than $10,000, (2) $10,000–$20,000, and (3) over $20,000; yet within each of the categories, considerable variation remains. Family income cannot be held constant in the sense that all of the cases in a partial table are homogeneous with respect to the test variable, as was the case when we controlled for sex and analyzed males and females separately.

The variation within categories of an ordinally or intervally measured test variable can be *reduced* by increasing the number of categories; however, there may not be a sufficient number of cases for all the cells that would result; and also, too many cells can make the data very difficult to interpret. A 3 × 3 bivariate table and a test variable with three categories would result in 3 first-order partial tables each containing 9 cells for a total of 27 cells (3 × 9 = 27).

11.3 Partial Correlation

11.3.1 Introduction

A study of black ministers in Nashville, Tennessee, was conducted, from which a model of their political involvement was developed.* Two demographic variables, the minister's age and years of formal education, were moderately correlated with civic participation (CP), a variable based on a combination of such activities as having worked in political campaigns, telephoned a state or local official about a community problem, contributed money to support a political campaign, etc. On the basis of the bivariate correlations (age with CP, $r = 0.30$ and education with CP, $r = 0.49$), the authors concluded that both age and education were important determinants of the minister's civic participation (the dependent variable).

Model A describes the expected relationships between the three variables. Education is shown as an **intervening variable** linking age and civic participation.

MODEL A   AGE ⟶ EDUCATION ⟶ CIVIC PARTICIPATION

Two other models were considered. Model B assumes that age and civic participation are **spuriously related** due to their relationship with education, and in model C age serves as an intervening variable.† Neither model B or C was considered plausible due to the unlikely causal ordering of education and age.

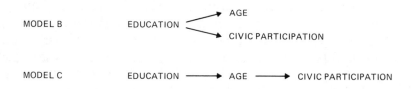

MODEL B   EDUCATION ⟨ AGE / CIVIC PARTICIPATION

MODEL C   EDUCATION ⟶ AGE ⟶ CIVIC PARTICIPATION

---

*   William M. Berenson, Kirk W. Elifson, and Tandy Tollerson III, Preachers in politics: a study of political activism among the black ministry, *Journal of Black Studies* 6, June 1976, pp. 373–383.

†   A **spurious relationship** is said to exist when the zero-order relationship between an independent and dependent variable disappears or becomes significantly weaker with the introduction of a control or test variable.

To test model A, we calculated a partial correlation coefficient before proceeding with a test of the overall model. A **partial correlation coefficient** is a measure of the linear relationship between two interval-level variables, controlling for one or more other variables.* In this instance, the partial correlation coefficient between age and civic participation, controlling for educational attainment, was -0.04, suggesting that education was an intervening variable causally linking age and civic participation.

11.3.2 The Logic of the Partial Correlation Coefficient

Technically, the partial correlation of variables X and Y controlling for variable Z is the correlation of the residuals of the regressions of X on Z and Y on Z. In other words, you can think of the control variable (Z) serving as an independent variable to predict X and Y in two separate regression analyses. Two sets of residuals are computed in the process.[†] The first set of residuals represents the variation in X not explained by Z, and the second set of residuals represents the variation in Y not explained by Z. The partial correlation coefficient between X and Y controlling for Z can be computed by correlating these two sets of residuals.

Partial correlation coefficients can be computed using the computational equation presented in the following section; however, a thorough understanding requires that one conceptual example be presented in which the logic of correlating two sets of residuals is developed. The partial correlation between age and civic participation, controlling for education, will serve as an illustration. Figure 11.1 is based on the scores presented in Table 11.2 of the five hypothetical male black ministers. Civic participation and age have been individually regressed on education, which serves as the control variable.

---

* Many social scientists routinely use ordinal-level variables that are evenly distributed and have five or more categories with interval statistical techniques. Violating the assumption of interval-level data given these restrictions does not have a marked impact on the outcome (Labovitz, 1970). As Blalock has noted, "... if one becomes too much of a purist with respect to measurement criteria, there may be an equally dangerous tendency to throw away information that is not completely satisfactory" (1979, p. 444). Henkel (1975) offers an opposing argument.

† A *residual* was defined in Chapter 10 as the size and direction of the error in prediction or the distance from an observed data coordinate to the regression line.

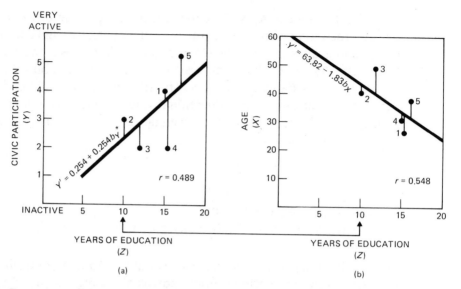

11.1 Regression of civic participation on years of education and regression of age on years of education.

Table 11.2 Raw scores, predicted scores, and residuals for five black ministers

BLACK MINISTER	AGE (X)	CIVIC PARTICIPATION (Y)	YEARS OF EDUCATION (Z)	Y' WITH Z INDEPENDENT	X' WITH Z INDEPENDENT	Y − Y'	X − X'
1	28	4	15	3.56	36.44	+0.44	−8.44
2	42	3	10	2.29	45.57	+0.71	−3.57
3	50	2	12	2.79	41.92	−0.79	+8.08
4	34	2	15	3.56	36.44	−1.56	−2.44
5	41	5	16	3.81	34.62	+1.19	+6.38

The *statistical control* procedure used in partial correlation analysis differs from the *physical control* procedure used with multivariate contingency tables in Section 11.2. Statistical control involves an adjustment process rather than placing the respondents into categories or subgroups as is the case when physical controls are used. Consider black minister 2 in Fig. 11.1(a) and (b). In Fig. 11.1(a) the coordinate for minister 2 lies above the regression line, indicating that he scored higher on civic participation than was predicted given his ten years of formal education. By the same token in Fig. 11.1(b), the coordinate for minister 2 lies below the regression line, indicating that he is younger than was predicted given his ten years of formal education. The residual for minister 2 in Fig. 11.1(a) indicates the extent to which his civic participation score is independent of, or unexplained by, his education score, and the residual shown in Fig. 11.1(b) for minister 2 indicates the extent to which his age is independent of, or unexplained by, his education score.

Note that both residuals are based on the same education score, hence, education has been controlled or held constant. This is also true of the four other ministers, and therefore the residuals or deviations from the regression line shown in Fig. 11.1(a) and (b) reflect the variation in civic participation and age that cannot be accounted for by education.

Table 11.2 includes the five ministers' ages, civic participation scores, and years of formal education, along with the predicted values for age (X') and civic participation (Y') when they were regressed independently on education (Z), the control variable. The actual values of the residuals shown in Fig. 11.1 are also provided. These residual values are used with Eq. (9.2) in Table 11.3 to compute the partial cor-

Table 11.3   Computational procedures for the partial correlation coefficient using residuals

BLACK MINISTER	REGRESSION RESIDUAL FOR CIVIC PARTICIPATION WITH EDUCATION INDEPENDENT $Y - Y'$	REGRESSION RESIDUAL FOR AGE WITH EDUCATION INDEPENDENT $X - X'$	$(Y - Y')^2$	$(X - X')^2$	$(X - X') \cdot (Y - Y')$
1	+0.44	−8.44	0.19	71.23	−3.71
2	+0.71	−3.57	0.50	12.74	−2.53
3	−0.79	+8.08	0.62	65.29	−6.38
4	−1.56	−2.44	2.43	5.95	+3.81
5	+1.19	+6.38	1.42	40.70	+7.59
$N = 5$	$\Sigma = -0.01*$	$\Sigma = 0.01*$	$\Sigma = 5.16$	$\Sigma = 195.91$	$\Sigma = -1.22$

$$r = \frac{N\Sigma XY - (\Sigma X)(\Sigma Y)}{\sqrt{[N\Sigma X^2 - (\Sigma X)^2][N\Sigma Y^2 - (\Sigma Y)^2]}} \tag{9.2}$$

$$= \frac{5(-1.22) - (0.01)(0.01)}{\sqrt{[5(195.91) - (0.01)^2][5(5.16) - (0.01)^2]}}$$

$$= \frac{6.10}{\sqrt{25,272.39}}$$

$$= -0.04$$

Note: When substituting into Eq. (9.2), $X - X' = X$ and $Y - Y' = Y$.

*Does not sum to zero due to rounding error.

relation coefficient between age and civic participation controlling for education.

11.3.3 The Computational Equation

Now that you fully understand the logic of partial correlation coefficients, computational Eq. (11.1) is the preferable means of computing first-order partial correlation coefficients.* Equation (11.1) requires that all of the relevant zero-order correlation coefficients be known, as shown in Table 11.4. It is not necessary to calculate the residuals to determine the partial correlation coefficient. Recall from Section 9.1 that the size of a correlation coefficient directly reflects the goodness of fit of the regression line to the coordinates and, therefore, provides us with an explicit description of the residuals.

Table 11.4   Zero-order correlation matrix

	X	Y	Z
(X) Age	1.00		
(Y) Civic participation	−0.30	1.00	
(Z) Education	−0.55	0.49	1.00

$$r_{XY.Z} = \frac{r_{XY} - r_{XZ}r_{YZ}}{\sqrt{(1 - r_{XZ}^2)(1 - r_{YZ}^2)}}$$ *(11.1)*

where

$$X = \text{the independent variable,}$$
$$Y = \text{the dependent variable, and}$$
$$Z = \text{the control variable.}$$

The partial correlation coefficient between X and Y controlling for Z is symbolized by $r_{XY.Z}$ (read "r sub XY dot Z"). The numerator of Eq. (11.1) provides for a measure of the variation that variables X and

---

* Computational equations for higher order partial correlation coefficients will not be discussed in this text. The **order of a partial correlation coefficient** indicates the number of control or test variables. The interested reader should consult Blalock, 1979.

Y share in common, less the product of the variation that X and Y share with Z, the control variable. The denominator involves taking the square root of the product of the variation in X and Y not explained by Z. Partial correlation coefficients can range from -1 to $+1$ and, when squared, are a measure of the percent of variation two variables have in common after controlling for the effects of a third (or third and fourth, etc.) variable. Squaring a partial correlation coefficient of -0.04 tells us that X and Y have less than 1% of their variance in common after Z has been controlled. Entering the correlation coefficients from Table 11.4 into Eq. (11.1):

$$
\begin{aligned}
r_{XY.Z} &= \frac{-0.30 - (-0.55)(0.49)}{\sqrt{(1 - (-0.55)^2)\,(1 - 0.49^2)}} \\
&= \frac{-0.03}{\sqrt{0.53}} \\
&= -0.04.
\end{aligned}
$$

This is the same value that was computed earlier. Comparison of the partial correlation coefficient (-0.04) with the zero-order correlation coefficient between age and civic participation ($r = -0.30$) suggests that education causally links age and civic participation.

Unfortunately, partial correlation coefficients cannot help us distinguish between a spurious relationship and one in which the control variable acts as an intervening variable as shown in the accompanying figure. Based on the logic developed in Section 11.3.1, the spurious interpretation appears to be far less plausible than an intervening variable interpretation.

SPURIOUS
MODEL

INTERVENING
MODEL

11.3.4 A Final Word

A partial correlation coefficient can also be defined as an average of the zero-order correlation between two variables within categories of the control variable. To the extent that the zero-order association between two variables is generally uniform across all categories of the control variable, a partial correlation coefficient can be a useful tool. However, when the association is not uniform within all categories of the control variable, the partial correlation coefficient is a misleading measure. A partial correlation coefficient of approximately zero could result, for instance, from two variables having a positive relationship below the median of the control variable and a negative relationship above the median. In the event that the relationship between X and Y is not uniform across all categories of the control variable, a multivariate contingency table should be used rather than partial correlation coefficients.

11.4 Multiple Regression Analysis
11.4.1 Introduction

Multiple regression analysis is a logical extension of simple regression, which was discussed in Chapter 10. The basic concepts are very similar. Differences stem from multiple regression's use of two or more independent variables as predictors of a dependent variable; simple regression involves only one independent and one dependent variable. Several predictor variables used jointly can potentially account for considerably more variation in a dependent variable than can a single predictor variable.

A multiple regression equation with two independent variables is written in the following manner:

$$Y' = a + b_1X_1 + b_2X_2. \qquad (11.2)$$

where

$Y' = $ the predicted value of the dependent variable,*
$a\ \ = $ the Y-intercept,
$b_1 = $ the slope associated with the first independent variable (X_1), and
$b_2 = $ the slope associated with the second independent variable (X_2).

---

* Some texts use a "hat" symbol (for example, $\hat{Y}$) to designate a predicted value.

The slopes are interpreted in the following manner: b_1 represents the units of change in Y given one unit increase in X_1, holding X_2 constant; and b_2 represents the units of change in Y given one unit increase in X_2, holding X_1 constant.

11.4.2 Visual Presentation of Multiple Regression Analysis

A visual presentation of multiple regression analysis with one dependent variable and two independent variables is shown in Fig. 11.2.

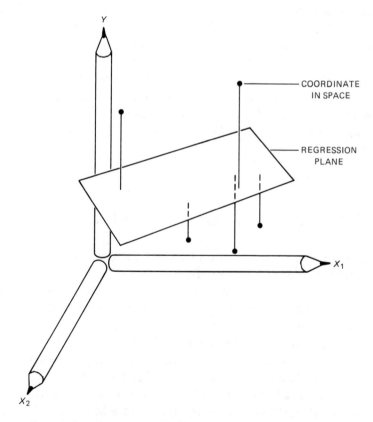

11.2 A three-dimensional portrayal of multiple regression analysis.

Visualize, with the aid of three pencils, some tape, a flat piece of cardboard, and Fig. 11.2, how the X_1-, X_2-, and Y-axes can be arranged to construct a three-dimensional model. Place two of the pencils on a flat surface such that the erasers touch, and the pencils form an angle similar to that shown between the X_1 and X_2 axes. Extend the third pencil, which represents the Y-axis, vertically from the intersection of the first two pencils. The pencils can be taped together in this position. To locate a hypothetical coordinate such as $X_1 = 1$, $X_2 = 2$, and $Y = 4$, begin at the intersection of the three axes, move one unit (say, 1/4 inch) out on the X_1 axis, and mark the distance. Move two units out on the X_2 axis and mark the distance. Now extend vertical lines from the one-unit value on the X_1 axis and the two-unit value on the X_2 axis in the direction that allows them to intersect. Finally, move up four units relative to the Y-axis from this point of intersection, and the coordinate $X_1 = 1$, $X_2 = 2$, and $Y = 4$ has been established. Once all of the coordinates have been located, you will have created a three-dimensional figure with a "cloud" of points in space formed by the coordinates.

A regression *line* of best fit relative to the coordinates in space would not be appropriate in this instance; however, a regression *plane* can be fitted to the points. A cardboard backing to a notepad can serve as the plane, as shown in Fig. 11.2. Those coordinates that lie above the plane are positive in value, and those that lie below the plane are negative in value. Two slopes are required to know how to tilt the plane relative to the X_1- and X_2-axes, and a is the point at which the plane crosses the Y-axis.

11.4.3 Multiple Regression: An Example

In the preceding discussion of partial correlation coefficients civic participation among black ministers was the dependent variable, age the independent variable, and education the control variable. It was determined that education acted as an intervening variable linking age and civic participation.

Multiple regression analysis allows us to use two or more interval-level or ratio-level variables simultaneously as predictors of an interval-level dependent variable. Equation (11.3) is based on a multiple regression analysis of the data in Table 11.2. Civic participation (Y) is the dependent variable, while age (X_1) and education (X_2) are the independent variables. The computation involved in multiple regression analysis is far too tedious and time-consuming for inclusion. While a computer was used to compute Eq. (11.3), the following example will allow you to understand the basic concepts of multiple regression.

$$Y' = 0.175 - 0.007X_1 + 0.242X_2. \qquad (11.3)$$

Equation (11.3) can be rewritten in words as:

$$\begin{array}{l} \text{a minister's} \\ \text{predicted} \\ \text{civic} \\ \text{participation} \\ \text{score} \end{array} = 0.175 - 0.007 \left(\begin{array}{c} \text{the} \\ \text{minister's} \\ \text{age} \end{array}\right) + 0.242 \left(\begin{array}{c} \text{the} \\ \text{minister's} \\ \text{education} \end{array}\right).$$

The value of a, which is 0.175, indicates that the regression plane intersects the Y-axis at 0.175. The **multiple regression coefficient*** associated with X_1 (age) indicates that Y' (civic participation) decreases 0.007 Y units with each one-year increase in X_1 (age), holding X_2 (education) constant. Similarly, the multiple regression coefficient associated with X_2 (education) indicates that Y' increases 0.242 Y units for each additional year of formal education holding X_1 (age) constant. As was true of simple regression coefficients, the sign associated with a multiple regression coefficient must always be the same as the sign of the zero-order correlation of the independent variable with the dependent variable.[†] For example, age is correlated -0.30 with civic participation (see Table 11.4), and the sign of the coefficient for age in Eq. (11.3) is also negative.

Given Eq. (11.3), we can predict the civic participation scores of the five ministers. From Table 11.2 we see that the first minister is 28 years old and has completed 15 years of formal education. Substituting into Eq. (11.3):

$$Y' = 0.175 - 0.007(28) + 0.242(15) = 3.61.$$

Therefore, the predicted civic participation score for the first minister is 3.61, whereas his actual score was 4 (see Table 11.2). The difference between the actual (Y_1) and predicted score (Y_1') yields the residual of

---

[*]   Multiple regression coefficients are also referred to as partial regression coefficients, unstandardized partial regression coefficients, and unstandardized multiple regression coefficients.

[†]   If the zero-order correlation coefficients between two or more variables in a multiple regression equation are close to 0.00, the signs may not be the same.

+0.39 for the first minister ($Y_1 - Y_1' = 4 - 3.61 = 0.39$), hence his coordinate lies 0.39 Y units above the regression plane. As was the case with simple regression, the five residuals (one for each minister) will sum to 0.

Equation (11.3) does not allow us to assess the *relative* importance of the independent variables in predicting civic participation, because the coefficients are based on different measurement units. Unlike r which is independent of the measurement units of the variables involved (Section 9.2.1), b is not independent of measurement units in either a simple or multiple regression context.

Computer programs for multiple regression analysis routinely calculate unstandardized regression coefficients based on the raw data (Equation 11.3) and the standardized regression coefficients based on standard scores.* The following standardized regression equation for the five ministers is mathematically equivalent to the unstandardized Eq. (11.3), if one converts the raw X_1, X_2, and Y scores to z scores and substitutes these standardized scores into the equation.

$$Y' = -0.043X_1 + 0.465X_2$$

Note that there is no Y-intercept in a standardized regression equation. The means of all the variables are 0 when they are standardized; therefore, because the regression plane must pass through the coordinate ($\overline{X}_1$, $\overline{X}_2$, $\overline{Y}$) or (0, 0, 0), the Y-intercept (a) always equals 0 in a standardized regression equation.

With standardized regression coefficients it is possible to assess the *relative* importance of age and education as predictors of civic participation, but these coefficients do not measure the *absolute* influence of each independent variable, which is possible with the unstandardized coefficients. The values (ignoring the signs) of the standardized regression coefficients are directly comparable, and, because they are expressed in standardized units, it is evident that education (X_2) is the strongest predictor. The standardized coefficient associated with X_1 is interpreted in the following manner: As X_1 increases one standard deviation unit, Y decreases 0.043 standard deviation unit when X_2 is held constant. By the same token, as X_2 increases one standard deviation unit, Y increases 0.465 standard deviation unit when X_1 is held con-

---

* The concept of standard scores was discussed in Section 7.2.

stant. Standardized regression coefficients are symbolized by the Greek letter β (beta).*

The relationship between the standardized and unstandardized coefficients is shown in Eq. (11.4):

$$\beta_{yx_1.x_2} = b_{yx_1.x_2} \left(\frac{s_{x_1}}{s_y}\right) \qquad (11.4)$$

where the subscript $yx_1.x_2$ indicates that Y is the dependent variable, X_1 is the independent variable for which the standardized regression coefficient is being computed, and the second independent variable (X_2) has been statistically controlled.

In words,

$$
\begin{array}{c}
\text{standardized} \\
\text{regression coefficient} \\
\text{for } X_1 \text{ controlling} \\
\text{for } X_2
\end{array}
=
\begin{array}{c}
\text{unstandardized} \\
\text{coefficient for} \\
X_1 \text{ controlling} \\
\text{for } X_2
\end{array}
\left(
\begin{array}{c}
\text{standard deviation} \\
\text{of } X_1 \\
\hline
\text{standard deviation} \\
\text{of } Y
\end{array}
\right).
$$

The unstandardized coefficient associated with X_1 in Eq. (11.3) is -0.007, the standard deviation of X_1 is 8.37, and the standard deviation of Y is 1.30. Therefore, substituting into Eq. (11.4):

$$\beta_{yx_1.x_2} = -0.007 \left(\frac{8.37}{1.30}\right)$$

$$= -0.045.^{\dagger}$$

Thus it is possible to convert an unstandardized equation into the standardized form if the standard deviations of the independent and dependent variables are known.

11.4.4 The Multiple Correlation Coefficient

The **multiple correlation coefficient** (**R**) is a measure of the linear relationship between a dependent variable and the combined effects of two or more independent variables. It is symbolized by a capital R to dif-

---

* Standardized regression coefficients are also referred to as *beta weights* and *standardized path coefficients*. These coefficients are based on sample data.

† Differs slightly from -0.043 computed earlier due to rounding error.

ferentiate it from *r,* the zero-order correlation coefficient which was discussed in Chapter 9. As was true of *r,* squaring R yields a far more interpretable measure. R^2, or the **coefficient of multiple determination,** indicates the proportion of total variation in a dependent variable that is explained jointly by two or more independent variables. For the black minister example, $R = 0.49$ and $R^2 = 0.24$; thus 24% of the variation in civic participation has been explained by age and education, and 76% (100% − 24% = 76%) of the variation has not been explained.*

Figure 11.3 presents the concept of the coefficient of multiple determination diagrammatically. The shaded area designated by letters *a, c,* and *d* in Figure 11.3 represents the total variation in *Y* explained jointly by X_1 and X_2. Area *c* represents the independent contribution of X_2 to an explanation of the variation in *Y* and area *d* the independent

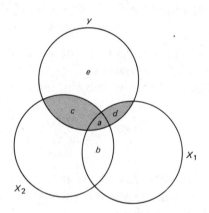

a = JOINT CONTRIBUTION OF X_1 AND X_2 TO AN EXPLANATION
      OF THE VARIATION IN *Y*.
b = VARIATION SHARED BY X_1 AND X_2 INDEPENDENT OF *Y*.
c = INDEPENDENT CONTRIBUTION OF X_2 TO AN EXPLANATION
      OF THE VARIATION IN *Y*.
d = INDEPENDENT CONTRIBUTION OF X_1 TO AN EXPLANATION
      OF THE VARIATION IN *Y*.
e = VARIATION IN *Y* NOT ACCOUNTED FOR BY X_1 AND X_2
      (EQUIVALENT TO $1 - R^2$).
a + *b* = VARIATION SHARED BY X_1 AND X_2.
a + *c* + *d* = TOTAL VARIATION IN *Y* ACCOUNTED FOR JOINTLY BY X_1 AND X_2
      (EQUIVALENT TO R^2).

11.3 A diagrammatic presentation of the coefficient of multiple determination.

---

* The unexplained variation is sometimes referred to as the *coefficient of nondetermination* and is symbolized by $1 - R^2$.

contribution of X_1. Area a represents the joint contribution of X_1 and X_2. The total area represented by areas a and b indicates the variation shared by X_1 and X_2. Area e indicates the variation in Y that has not been accounted for by the two independent variables, X_1 and X_2.

Unfortunately, there is no way to partition the variation in Y that was explained by X_1 and the variation that was explained by X_2 unless X_1 and X_2 are unrelated. In this example X_1 and X_2 are statistically associated as shown by their intersection (areas a and b), and it is impossible to know what proportion of the variation in Y, which is indicated by area a, is attributable to X_1 and what proportion is attributable to X_2.

Summary

In this chapter we have learned about multivariate statistical techniques including multivariate contingency tables, partial correlation analysis, multiple regression analysis, and multiple correlation analysis. All of these techniques involve examining the relationship between three or more variables simultaneously.

Multivariate contingency tables allow us to statistically elaborate the relationship between two variables by introducing a control variable.

Partial correlation analysis provides us with a measure of the linear relationship between two interval-level variables having statistically controlled for one or more other interval-level variables. The concept of statistical control was presented, as was the computational procedure. The partial correlation coefficient should not be used when the zero-order association between two variables is not uniform across all categories of the control variable.

Multiple regression analysis allows us to use two or more interval-level variables simultaneously as predictors of an interval-level dependent variable. The interpretation and relationship of unstandardized and standardized multiple regression coefficients was presented. The interpretation and use of multiple regression coefficients was also discussed.

Terms to Remember

Coefficient of multiple determination (R^2)   A measure of the proportion of total variation in a dependent variable that is explained jointly by two or more independent variables.

Conditional table   A contingency table that summarizes the relationship between two variables for categories or conditions of one or more other variables. Same as a partial table.

Control variable   A variable whose subgroups or categories are used in a conditional table to examine the relationship between two variables. Same as a test variable.

Intervening variable   A variable that causally links two other variables.

Multiple correlation coefficient (R)   A measure of the linear relationship between a dependent variable and the combined effects of two or more independent variables.

Multiple regression coefficient   A measure of the influence of an independent variable on a dependent variable when the effects of all other independent variables in the multiple regression equation have been held constant.

Multivariate statistical technique   A statistical technique used to determine the nature of the relationships between three or more variables.

Order of a conditional or partial table   The order of a table is the number of control or test variables that are included.

Order of a partial correlation coefficient   The order of a partial correlation coefficient indicates the number of control or test variables.

Partial table   A contingency table that summarizes the relationship between two variables for categories or conditions of one or more other variables. Same as a conditional table.

Partial correlation coefficient   A measure of the linear relationship between two interval-level variables controlling for one or more other interval-level variables.

Spurious relationship Exists when the zero-order relationship between an independent and dependent variable disappears or becomes significantly weaker with the introduction of a control variable.

Statistical elaboration The process of introducing a control variable to determine whether a bivariate relationship remains the same or varies under the different categories or conditions of the control variable.

Test variable A variable whose subgroups or categories are used in a conditional table to examine the relationship between two variables. Same as a control variable.

Exercises

1. Discuss the advantages of using multivariate statistical procedures rather than bivariate statistical procedures.

2. Differentiate between physical control as used with multivariate contingency tables and statistical control as used with partial correlation analysis.

3. Given the following bivariate contingency table, introduce a control variable (for example, education or another variable of your choice) and manipulate the data to show its influence on the relationship of age and attitude toward socialized medicine.

(Y) FAVOR SOCIALIZED MEDICINE?	(X) AGE		
	Young	Old	Total
Yes	34.7 (51)	63.6 (82)	(133)
No	65.3 (96)	36.4 (47)	(143)
Total	100.0 (147)	100.0 (129)	N = 276

4. The following table presents data obtained from a sample of college students. Based on the data, answer the following questions:

a) Identify the independent, dependent, and control variables that are included in this table.

b) Without doing any statistical operations beyond examining the percentages and numbers in the cells of the table, give an interpretation about the relationship between the variables in the table (what does this table tell us?) and describe the effect of the control variable.

c) Reconstruct or "collapse" this three-variable table into a two-variable table in a way that would allow you to examine the simple relationship between age and political orientation, and indicate whether it appears that young or old students are more conservative (again, it is not necessary to do any statistical operation other than simple examination of percentages and numbers of cases).

POLITICAL ORIENTATION	YOUNG LEVEL IN SCHOOL		OLD LEVEL IN SCHOOL	
	Undergraduate	*Graduate*	*Undergraduate*	*Graduate*
Liberal	75.0 (30)	25.0 (15)	60.0 (30)	50.0 (25)
Conservative	25.0 (10)	75.0 (45)	40.0 (20)	50.0 (25)
Total	100.0 (40)	100.0 (60)	100.0 (50)	100.0 (50) $N = 200$

5. The data in Table A show the relationship between the length of prison sentence (in months) received by defendants charged with violations of the Federal Drug Abuse Prevention and Control Act for 1977 and 1978. Table B shows the relationship between length of sentence and year by the type of substance involved. Percentage the tables and discuss the influence of the control variable.

Table A

SENTENCE IN MONTHS	YEAR		
	1977	*1978*	*Total*
1 – 12	1404	885	2289
13 – 35	811	623	1434
36 – 59	1143	956	2099
60+	1310	1141	2451
Total	4668	3605	$N = 8273$

Table B

SENTENCE IN MONTHS	MARIJUANA		NARCOTICS		CONTROLLED SUBSTANCES (PRESCRIBED DRUGS)	
	1977	1978	1977	1978	1977	1978
1 – 12	466	241	757	412	181	232
13 – 35	281	158	410	328	120	137
36 – 59	274	245	749	518	120	193
60 +	183	139	1022	739	105	263
Total	1204	783	2938	1997	526	825
	N = 1987		N = 4935		N = 1351	

Source: Administrative Office of the U.S. Courts, *Annual Report of the Director* as cited in *Statistical Abstract of the United States,* 1979, U.S. Department of Commerce, Bureau of the Census.

6. You are studying the relationship between church attendance, self-reported delinquent acts, and drug usage among a sample of high school students. Given the following correlation matrix, compute the first-order partial correlation coefficient between church attendance and reported delinquent acts controlling for drug usage. Interpret the results.

	1.	2.	3.
1. *Self reported delinquent acts*	1.00		
2. *Church attendance*	-0.32	1.00	
3. *Drug usage*	+0.41	-0.17	1.00

7. Explain why technically the partial correlation of variables X and Y controlling for variable Z is the correlation of the residuals of the regressions of X on Z and Y on Z.

8. Why should a partial correlation coefficient not be computed when the association between two variables is not uniform across all categories of the control variable?

9. If $r_{xy} = +0.80$, $r_{xz} = +0.61$, and $r_{yz} = +0.43$, compute $r_{xy.z}$.

10. Income (X_1) and seniority (X_2) have been used as predictors of job satisfaction (Y) for a sample of municipal employees. Given the following regression equation, interpret the Y-intercept and the regression coefficients associated with income and seniority.

$$Y' = 1.1 + 0.35X_1 + 0.12X_2$$

11. Is the equation in Exercise 10 in standardized or unstandardized form? How can you tell?

12. Given the scores for five of the employees included in the study summarized in Exercise 10, compute the predicted Y scores and the residuals for each of the five employees.

	X_1	X_2	Y
Employee 1	10	2	6
2	15	3	5
3	9	6	7
4	21	9	9
5	46	15	21

13. Assume that the following table contains the results of regression analyses of data from a national sample survey of 11th grade black and white students. The purpose of the analysis is to see what factors account for students' performance levels on math achievement tests (the dependent variable). The independent variables are: (1) the student's cumulative grade average, (2) the student's family's socioeconomic status (SES), (3) the student's reading ability test score, (4) the percentage of the student's class that is white, and (5) a measure of the teaching ability of the student's math teacher. Using the standardized regression coefficients and other statistics in the table, answer the following questions.

INDEPENDENT VARIABLES	WHITE STUDENTS	BLACK STUDENTS
1. Cumulative grade average	0.10	0.24
2. Family SES	0.42	0.49
3. Reading test score	0.39	0.05
4. % white in class	0.37	-0.19
5. Teacher's ability	0.17	0.34
Multiple correlation coefficient (R)	0.72	0.38
Coefficient of multiple determination (R^2)	0.52	0.14

a) For which group, blacks or whites, does the regression model using these five variables do a better job in predicting or explaining math achievement scores? How do you know this?

b) Of the variables listed, which one seems to be the most important contributing factor to the math test achievement score? Which are the least important factors in this analysis?

c) In multiple regression analysis, what is the meaning or interpretation of a standardized regression coefficient such as those in the table? How would you interpret the fact that for white students the standardized regression coefficient of teacher's ability is 0.17 and for black students it is 0.34. That is, what do these coefficients tell us?

Inferential Statistics Parametric Tests of Significance

12 Probability

13 Introduction to Statistical Inference

14 Statistical Inference and Continuous Variables

15 Statistical Inference with Two Independent
Samples and Correlated Variables

16 An Introduction to the Analysis of Variance

Probability

12

12.1 An Introduction to Probability

12.2 The Concept of Randomness

12.3 Approaches to Probability

12.4 Formal Properties of Probability

12.5 Probability and Continuous Variables

12.6 Probability and the normal-curve model

12.7 One- and Two-Tailed p Values

12.1 An Introduction to Probability

Probability is not as unfamiliar as many would think. Indeed, in everyday life you are constantly called upon to make probability judgments, although you may not recognize them as such. For example, suppose that, for various reasons, you are unprepared for today's class. You seriously consider not attending class. What are the factors that will influence your decision? Obviously, one consideration would be the likelihood that the instructor will discover you are not prepared. If the risk is high, you decide not to attend class; if low, then you will attend.

Let us look at this example in slightly different terms. There are two alternative possibilities:

Event *A:* Your lack of preparation *will* be detected.
Event *B:* Your lack of preparation *will not* be detected.

There is uncertainty in this situation because more than one alternative is possible. Your decision whether or not to attend class will depend upon the degree of assurance you associate with each of these alternatives. Thus if you are fairly certain that the first alternative will prevail, you will decide not to attend class.

Suppose that your instructor frequently calls upon students to participate in class discussion. In fact, you have noted that most of the students are called upon in any given class session. This is an example of a situation in which the first alternative is likely. Stated another way, the probability of event A is higher than the probability of event B. Thus you decide not to attend class. Although you have not used any formal probability laws in this example, you have actually made a judgment based upon an *intuitive* use of probability.

As social scientists we concern ourselves with such questions as the probability that a prisoner will commit another crime after being released from prison, or the probability that a particular candidate will win an election?

You may have noted that many of the questions raised in the exercises earlier in the book began with, "What is the likelihood that . . .?" These questions were in preparation for the formal discussion of probability occurring in the present and subsequent chapters. However, before discussing the elements of probability, it is desirable to understand one of the most important concepts in inferential statistics, that of randomness.

12.2 The Concept of Randomness

You will recall that, when discussing the role of inferential statistics in Chapter 1, we pointed out the fact that population parameters are rarely known or knowable. It is for this reason that we are usually forced to draw samples from a given population and estimate the parameters from the sample statistics. We want to select these samples in such a way that they are representative of the populations from which they are drawn. One way to achieve representativeness is to employ simple **random sampling:** *selecting samples in such a way that each sample of a given size has precisely the same probability of being selected* or, alternatively, *selecting the events in the sample such that each event is equally likely to be selected in a sample of a given size.*

Consider selecting samples of $N = 2$ from a population of five numbers: 0, 1, 2, 3, 4. If, for any reason, any number is more likely to be drawn than any other number, each sample would *not* have an equal likelihood of being drawn. For example, if for any reason the number 3 were twice as likely to be drawn as any other number, there would be a preponderance of samples containing the value 3. Such sampling procedures are referred to as being **biased.** When we are interested in learning the characteristics of the general population for a given variable, we should not select our sample from automobile registration lists or at random on a street corner in New York City. The dangers of generalizing to a population from such biased samples should be clear to you. Unless the condition of randomness is met, we may never know to what population we should generalize our results.

Moreover, the statistical tests presented in this text require **independent** random sampling. Two events are said to be independent if the selection of one has no effect upon the probability of selecting the other event. We can most readily grasp independence in terms of games of chance, assuming they are played honestly. Knowledge of the results of one toss of a coin, one throw of a die, one spin of the roulette wheel, or one selection of a card from a well-shuffled deck (assuming replacement of the card after each selection) will not aid us in our predictions of future outcomes.

It is beyond the scope of this text to delve deeply into sampling procedures, since that topic is a full course by itself. However, consider how we achieve randomness when drawing a sample from a population. Let us suppose we wish to select 25 persons randomly from a population totaling 300. One method to achieve randomness would be to place the names of each of the 300 persons in the population on a folded slip of paper. After shuffling the slips of paper and placing them in a container, we could reach into the container and draw 25 names.

An alternative method would be to use the table of random digits (Table M in Appendix C). Since the digits in this table have already been randomized by a computer, the effect of shuffling has been achieved. We assign each of our 300 persons comprising the population a number from 001 to 300. We may start with any row or column of digits in Table M. For example, if we start with the first row and choose consecutive sets of three digits, we obtain the following numbers: 100, 973, 253, 376, 520, 135 and so on. If any number over 300 or any repeated number appears, we disregard it, and we continue until we have chosen 25 "usable" numbers, which represent the 25 people that will constitute our sample.

12.3 Approaches to Probability

Probability may be regarded as a theory that is concerned with the possible outcomes of a study. It must be possible to list every outcome that can occur, and we must be able to state the expected relative frequencies of these outcomes. It is the method of assigning relative frequencies to each of the possible outcomes that distinguishes the classical from the empirical approach to probability theory.

12.3.1 Classical Approach to Probability

The theory of probability has always been closely associated with games of chance. For example, suppose we want to know the probability that a single card selected from a 52-card deck will be an ace of spades. There are 52 possible outcomes. We assume an ideal situation in which we expect that each outcome is equally likely to occur. Thus the probability of selecting an ace of spades is one out of 52 or 1/52. This kind of reasoning has led to the following classical definition of probability:

$$p(A) = \frac{\text{number of outcomes in which event } A \text{ occurs}}{\text{total number of outcomes}}. \qquad (12.1)$$

It should be noted that probability is defined as a proportion (p). The most important point in the classical definition of probability is the assumption of an *ideal* situation in which the composition of the population is known; that is, the total number of possible outcomes (N) is known. In this example, the total number of possible outcomes was 52 (all cards in the deck), and each outcome was assumed to have an equal likelihood of occurrence. Thus p (ace of spades) = 1/52; p (king of hearts) = 1/52, and so on.

12.3.2 Empirical Approach to Probability

Although it is usually easy to assign expected relative frequencies to the possible outcomes of games of chance, we cannot do this for most real life situations. In actual situations, expected relative frequencies are assigned on the basis of empirical findings. Thus we may not know the exact proportion of students in a university who have worked in a political campaign, but we may conduct a simple random sample of students in the university and estimate the proportion who have worked in a campaign. Once we have arrived at an estimate, we can use classical probability theory to answer questions such as: What is the probability that in a sample of ten students, drawn at random from the student body, three or more will have worked in a political campaign?

If, in a simple random sample of 100 students, we found that 30 had worked in a campaign, we could estimate that the proportion of experienced campaigners in the university was 0.30, using Eq. (12.1):

$$p(\text{campaigners}) = \frac{30}{100} = 0.30.$$

Thus the probability is 0.30 that a student in the university will have campaigned previously. (*Note:* This represents an *empirical* probability; that is, the expected relative frequency was assigned on the basis of empirical findings.)

Although we use an idealized model in our forthcoming discussion about the properties of probability, we can apply the same principles to many practical problems.

12.4 Formal Properties of Probability

12.4.1 Probabilities Vary Between 0 and 1.00

From the classical definition of probability, p is always between 0 and 1, inclusively. If an event is certain to occur, its probability is 1; if it is certain not to occur, its probability is 0. For example, the probability of drawing the ace of spades from an ordinary deck of 52 playing cards is 1/52. The probability of drawing a red ace of spades is zero because the ace of spades is never red. However, the probability of drawing a card with *some* marking on it is $p = 1$. Thus for any given event, say A, $0 \le p(A) \le 1.00$, in which the symbol $\le$ means "less than or equal to."

12.4.2 Expressing Probability

In addition to expressing probability as a proportion, several other ways are often used. It is sometimes convenient to express probability as a *percentage* or as the *number of chances* in 100.

To illustrate: If the probability of an event is 0.05, we expect this event to occur 5% of the time, or *the chances that this event will occur are 5 in 100*. This same probability may be expressed by saying that the odds are 95 to 5 *against* the event occurring, or 19 to 1 against it.

When expressing probability as the *odds against* the occurrence of an event, we use Eq. (12.2):

Odds against event A
= (total no. of outcomes — no. of times event A occurs)
to no. of times event A occurs.   *(12.2)*

Thus if $p(A) = 0.01$, the *odds against* the occurrence of event A are 99 to 1.

12.4.3 The Addition Rule

The **addition rule** can be stated as follows:

> If A and B are events, the probability of obtaining *either* of them is equal to the probability of A plus the probability of B minus the probability of their joint occurrence.

In symbolic form this reads

$$p(A \text{ or } B) = p(A) + p(B) - p(A \text{ and } B). \qquad (12.3)$$

where $p(A \text{ and } B)$ indicate the joint occurrence of A and B, that is, the probability that *both* will occur.

Let us consider an example with cards. Suppose that we draw one card from a standard deck of 52 cards. What is the probability that the card will be either an ace or a heart? The probability of drawing an ace is 4/52; the probability of drawing a heart is 13/52; and the probability of drawing the ace of hearts 1/52. Therefore, using Eq. (12.3)

$$p(A \text{ or } B) = \frac{4}{52} + \frac{13}{52} - \frac{1}{52} = \frac{16}{52} = \frac{4}{13}.$$

Now imagine that you have a population consisting of seven scores: 0, 1, 2, 3, 4, 5, 6. You write each of these numbers on slips of paper and place them in a hat. Then you draw a number, record it, place it back in the hat, and draw a second number. You add the second number to the first to obtain a sum. You continue drawing *pairs of scores* and obtaining their sums until you have obtained all possible combinations of these seven scores, taken two at a time. The following table shows all possible results from drawing samples of $N = 2$ from this population of seven scores. The values within each cell show the sum of the two scores.

		FIRST DRAW					
	0	*1*	*2*	*3*	*4*	*5*	*6*
0	0 + 0 = 0	1 + 0 = 1	2 + 0 = 2	3 + 0 = 3	4 + 0 = 4	5 + 0 = 5	6 + 0 = 6
1	0 + 1 = 1	1 + 1 = 2	2 + 1 = 3	3 + 1 = 4	4 + 1 = 5	5 + 1 = 6	6 + 1 = 7
2	0 + 2 = 2	1 + 2 = 3	2 + 2 = 4	3 + 2 = 5	4 + 2 = 6	5 + 2 = 7	6 + 2 = 8
3	0 + 3 = 3	1 + 3 = 4	2 + 3 = 5	3 + 3 = 6	4 + 3 = 7	5 + 3 = 8	6 + 3 = 9
4	0 + 4 = 4	1 + 4 = 5	2 + 4 = 6	3 + 4 = 7	4 + 4 = 8	5 + 4 = 9	6 + 4 = 10
5	0 + 5 = 5	1 + 5 = 6	2 + 5 = 7	3 + 5 = 8	4 + 5 = 9	5 + 5 = 10	6 + 5 = 11
6	0 + 6 = 6	1 + 6 = 7	2 + 6 = 8	3 + 6 = 9	4 + 6 = 10	5 + 6 = 11	6 + 6 = 12

Second draw (row labels)

If you constructed a table of all possible draws of $N = 2$ from this population and determined the *mean* of each of those two draws, you would obtain the frequency distribution of 49 possible means shown in Table 12.1.

By dividing each frequency by $N_{\overline{X}}$,* we obtain a probability distribution of means of sample size $N = 2$. Note that we are now in a

Table 12.1   Frequency and probability distribution of means of samples of size $N = 2$, drawn with replacement from a population of seven scores (0, 1, 2, 3, 4, 5, 6)

$\overline{X}$	f	$p(\overline{X})$
6.0	1	0.0204
5.5	2	0.0408
5.0	3	0.0612
4.5	4	0.0816
4.0	5	0.1020
3.5	6	0.1224
3.0	7	0.1429
2.5	6	0.1224
2.0	5	0.1020
1.5	4	0.0816
1.0	3	0.0612
0.5	2	0.0408
0.0	1	0.0204
	$N_{\overline{X}} = 49$*	$\Sigma p(\overline{X}) = 0.9997$†

*We use $N_{\overline{X}}$ to refer to the total number of means obtained in our sampling experiment and to distinguish it from N, which is the number of scores upon which each mean is based.

†$\Sigma p(\overline{X})$ should equal 1.000. The discrepancy of 0.0003 represents rounding error.

---

* $N_{\overline{X}}$ is equal to the total number of means. In this sampling experiment $N_{\overline{X}} = 49$.

position to answer such questions as: In a single draw of a mean when $N = 2$, what is the probability of obtaining a mean equal to 0, 2, or 5.5? The answers are, respectively, 0.0204, 0.1020, and 0.0408, as shown in Table 12.1.

In inferential statistics we often want to determine the probability of one of several different events. For example, we may be interested in answering such questions as: What is the probability of obtaining a mean of 4.5 to 5.0 *or* 5.0 or greater? We may suspect that the probability of either of these events can be determined by finding the probability of obtaining a mean of 4.5 or 5.0 (4/49 + 3/49) and adding this to the probability of obtaining a mean of 5.0 or greater (3/49 + 2/49 + 1/49). This would add up to 4/49 + 3/49 + 3/49 + 2/49 + 1/49 = 13/49. Note, however, that in arriving at this total we have counted one event (a mean of 5.0) twice. Obviously we should count this event in only one of the totals. We could add the probability of all the events together and subtract out the number that *share* both characteristics, since we previously added them twice. Thus our probability becomes:

$$p = \frac{4}{49} + \frac{3}{49} + \frac{3}{49} + \frac{2}{49} + \frac{1}{49} - \frac{3}{49} = \frac{10}{49} = 0.20.$$

You can now see that these calculations conform to the general formulation of addition rule (12.3).

Note that if the events A and B are **mutually exclusive** (that is, if both events *cannot* occur simultaneously), the last term disappears. Thus the addition rule with mutually exclusive events becomes

$$p(A \text{ or } B) = p(A) + p(B). \tag{12.4}$$

Furthermore, the addition rule for mutually exclusive events (12.4) can be rewritten to include more than two events. For example,

$$P(A \text{ or } B \text{ or } C) = p(A) + p(B) + p(C).$$

Consider a population comprised of 30% Catholics, 25% Methodists, 40% Baptists, and 5% Lutherans. What is the probability of getting a Catholic or a Methodist or a Baptist in a single draw?

$$\frac{30}{100} + \frac{25}{100} + \frac{40}{100} = \frac{95}{100} = 0.95.$$

Equation (12.4) can be extended to include any number of mutually exclusive events. Thus

$$p(A \text{ or } B \text{ or } \ldots Z) = p(A) + p(B) + \cdots + p(Z).$$

Let's look at another example, using the data in Table 12.1, involving mutually exclusive events. For example, if we draw a single sample of $N = 2$, what is the probability of obtaining a mean of 5 or greater? To answer this question, we add together the separate probabilities of the events that are included in the statement "5 or greater," *as long as the separate events are mutually exclusive* (that is, two or more cannot occur simultaneously). In the present example, as shown in Table 12.1, these events are means of 5.0, 5.5, and 6.0. Note that these events are mutually exclusive. If you obtain a mean of 5.5 on a single draw, it cannot have any other value. The probability of obtaining a mean equal to or greater than 5 becomes

$$p(\overline{X} = 5) + p(\overline{X} = 5.5) + p(\overline{X} = 6.0) = 0.0612$$
$$+ \ 0.0408 + 0.0204 = 0.1224.$$

Note that we can raise and answer such additional questions as: In a single sample of $N = 2$, what is the probability of obtaining a mean equal to or less than 2.0? Shown symbolically:

$$p(\overline{X} \leq 2.0) = p(\overline{X} = 2) + p(\overline{X} = 1.5) + p(\overline{X} = 1.0)$$
$$+ \ p(\overline{X} = 0.5) + p(\overline{X} = 0)$$
$$= 0.1020 + 0.0816 + 0.0612 + 0.0408 +$$
$$0.0204$$
$$= 0.3060.$$

We can also ask, "What is the probability of selecting a mean between 3.5 and 5.5, inclusive?" We simply add together the separate probabilities of the following events:

$$p(\overline{X} = 3.5) + p(\overline{X} = 4.0) + p(\overline{X} = 4.5) + p(\overline{X} =$$
$$5.0)$$
$$+ \ p(\overline{X} = 5.5)$$
$$= 0.1224 + 0.1020 + 0.0816 + 0.0612 +$$
$$0.0408$$
$$= 0.4080.$$

In Chapters 13 and 18 we will be dealing with problems based on *dichotomous,* yes/no, or *two-category* variables, in which the events in question are not only mutually exclusive, but are also **exhaustive.** For example, if women constitute 52% of the population in a given city and men constitute 48% of the population, what is the probability of drawing a woman and a man in one draw? Not only is it impossible to obtain both events simultaneously (that is, they are mutually exclusive) but there is *no possible outcome other than a woman or a man.* In the case of mutually exclusive and exhaustive events, we arrive at the very useful formulation:

$$p(A) + p(B) = 1.00. \qquad (12.5)$$

In treating dichotomous populations, it is common practice to use the two symbols P and Q to represent, respectively, the probability of the occurrence of an event and the probability of the nonoccurrence of an event. Thus if we are flipping a single coin, we can let P represent the probability of occurrence of a head and Q the probability of the nonoccurrence of a head (that is, the occurrence of a tail). These considerations lead to three useful formulations:

$$P + Q = 1.00, \qquad (12.6)$$
$$P = 1.00 - Q, \qquad (12.7)$$
$$Q = 1.00 - P, \qquad (12.8)$$

when the events are *mutually exclusive and exhaustive.*

12.4.4 The Multiplication Rule

In the preceding section, we were concerned with determining the probability of obtaining one event or another based upon a *single* draw (or trial) from a set of 49 means. In statistical inference, we are often faced with the problem of ascertaining the probability of the **joint or successive occurrence** of two or more events when *more* than one draw or trial is involved.

When sampling with replacement   Let us return to the sampling experiment related to Table 12.1. Recall that, in obtaining the sample means based upon $N = 2$, we selected a score from the hat, recorded it, returned it to the hat, and randomly selected a second score. Imagine that, on each draw, you obtain a score equal to zero. You ask, "What is the probability that, by chance, you would obtain a mean equal to zero on two successive draws from the population of seven scores?" Actually, we have previously obtained this answer when we constructed Table 12.1 (frequency and probability distribution of means of samples of size $N = 2$, drawn from a population of seven scores). We found a probability equal to 0.0204.

We may obtain the same result by an alternative method—the application of the **multiplication rule.** This rule states:

The probability of the simultaneous or successive occurrence of two events is the product of the separate probabilities of each event.

In symbolic form:

$$p(A \text{ and } B) = p(A)p(B). \tag{12.9}$$

Since there are seven scores, the probability that you will obtain a zero on the first draw is $p(A) = 1/7$.

The probability that you will obtain a zero on the second draw, $p(B)$, is also 1/7. Thus

$$p(A \text{ and } B) = \left(\frac{1}{7}\right)\left(\frac{1}{7}\right)$$

$$= \frac{1}{49} = 0.0204.$$

In this example, the occurrence of A is not dependent upon the occurrence of B, and vice versa. The events are said to be independent. It is only when events are independent that the simple multiplication rule shown in Eq. (12.9) applies. The reason that the events are independent is that we sampled *with replacement*. Recall that we returned each score to the hat after each draw. As long as we thoroughly mix the scores after each draw, the outcome of the first draw can have no effect upon the outcome of the second draw.

Before proceeding to the case where the events are dependent, let us look at two additional examples.

Example What is the probability of obtaining a zero on the first draw (event A) and a 6 on the second draw (event B)?

$$p(A) = \frac{1}{7}; \qquad p(B) = \frac{1}{7};$$

$$p(A \text{ and } B) = \left(\frac{1}{7}\right)\left(\frac{1}{7}\right)$$

$$= \frac{1}{49} = 0.0204.$$

Example What is the probability of obtaining a mean equal to 3 on two successive draws from the population of seven numbers? The answer to this question is not as straightforward because there are a number of different ways to obtain a mean of 3. The various ways in which a mean of 3 may be obtained are listed in the following table.

FIRST DRAW		SECOND DRAW	$p(A$ AND $B)$
0	and	6	0.0204
1	and	5	0.0204
2	and	4	0.0204
3	and	3	0.0204
4	and	2	0.0204
5	and	1	0.0204
6	and	0	0.0204

We may ask, "What is the probability of obtaining 0 and 6 *or* 1 and 5 *or* 2 and 4, and so on?" To answer this question, we must combine the addition and multiplication rules

$$
\begin{aligned}
p(\overline{X} = 3) =\ & p(0 \text{ and } 6) \text{ or } p(1 \text{ and } 5) \text{ or } p(2 \text{ and } 4) \text{ or} \\
& p(3 \text{ and } 3) \text{ or } p(4 \text{ and } 2) \text{ or } p(5 \text{ and } 1) \text{ or } p(6 \text{ and } 0) \\
=\ & p(0 \text{ and } 6) + p(1 \text{ and } 5) + p(2 \text{ and } 4) \\
& + p(3 \text{ and } 3) + p(4 \text{ and } 2) + p(5 \text{ and } 1) + p(6 \text{ and } 0) \\
=\ & \left(\frac{1}{7}\right)\left(\frac{1}{7}\right) + \left(\frac{1}{7}\right)\left(\frac{1}{7}\right) + \left(\frac{1}{7}\right)\left(\frac{1}{7}\right) + \left(\frac{1}{7}\right)\left(\frac{1}{7}\right) + \left(\frac{1}{7}\right)\left(\frac{1}{7}\right) + \left(\frac{1}{7}\right)\left(\frac{1}{7}\right) \\
& + \left(\frac{1}{7}\right)\left(\frac{1}{7}\right) \\
=\ & \frac{1}{49} + \frac{1}{49} + \frac{1}{49} + \frac{1}{49} + \frac{1}{49} + \frac{1}{49} + \frac{1}{49} \\
=\ & \frac{7}{49} = 0.1429.
\end{aligned}
$$

Note that this answer agrees with the probability of obtaining a mean equal to 3 shown in Table 12.1.

When sampling without replacement Let's contrast sampling with replacement with a technique known as sampling *without replacement*. Imagine that after selecting the first score we do *not* return it to the hat. It would then be impossible to obtain the same score in the second draw. Indeed, it would be impossible to obtain a mean equal to 6 or a mean equal to 0. Why?

When sampling without replacement, then, the results of the first draw influence the possible outcomes of the second draw. The events are said to be **dependent**. For related or dependent events, the multiplication rule becomes:

> Given two events A and B, the probability of obtaining both A and B jointly is the product of the probability of obtaining one of these events times the conditional probability of obtaining one event, given that the other event has occurred.

Stated symbolically,

$$p(A \text{ and } B) = p(A)p(B \mid A) = p(B)p(A \mid B). \qquad (12.10)$$

The symbols $p(B \mid A)$ and $p(A \mid B)$ are referred to as **conditional probabilities.** The symbol $p(B \mid A)$ means *the probability of B given that A has occurred*. The term *conditional probability* takes into account the possibility that the probability of B may depend on whether or not A has occurred, and conversely for $p(A \mid B)$. (*Note:* Either of the two events may be designated A or B since the symbols merely represent a convenient language for discussion, and there is no time order implied in the way they occur.)

Example   A playing card example can be used to demonstrate the multiplication rule for dependent events. What is the probability of drawing two queens in a row from a standard deck of playing cards? The probability of drawing one queen is 4/52; however, having removed one queen from the deck (and not replacing it) reduces the probability of drawing another queen on the second draw. The 51 remaining cards in the deck contain three queens (because one was removed on the first draw); therefore, the probability of drawing a queen on the second draw is 3/52. We now can determine the probability of drawing two queens in a row without replacement utilizing Eq. (12.10). The probability of obtaining a queen on the first draw is 4/52, and the probability of obtaining a second queen on the second draw having already removed one queen on the first draw is 3/51. Therefore, we can multiply these probabilities and determine the probability of drawing two successive queens.

$$\left(\frac{4}{52}\right)\left(\frac{3}{52}\right) = \frac{1}{221}.$$

Example A second example of the multiplication rule for related or dependent events involves a contingency table. Given the following population in which the relationship between attitude toward busing for integration purposes and race is summarized, what is the probability of drawing a nonwhite person who favors busing?

	RACE		
ATTITUDE TOWARD BUSING	Nonwhite	White	Total
Favor	200	300	500
Oppose	50	450	500
Total	250	750	1000

Event A is the probability of drawing a nonwhite person, and event B is the probability of drawing a person who favors busing. The probability of event A occurring is 0.25 because 250 of the 1000 persons are nonwhite, and the probability of event B occurring is 0.5 because 500 of the 1000 persons in the population favor busing. We also need to know the probability of B given A or A given B. The probability of A given B is 0.4 because 500 persons favor busing and 200 of these persons are nonwhite. The probability of B given A is 0.8 because 250 persons are nonwhite and, of these persons, 200 favor busing. Therefore, applying Eq. (12.10) we can solve the problem in one of two ways:

$$p(A \text{ and } B) = p(A)p(B \mid A) = (0.25)(0.8) = 0.2.$$

But we also know that

$$p(A \text{ and } B) = p(B)p(A \mid B).$$

Therefore,

$$p(A \text{ and } B) = (0.5)(0.4) = 0.2.$$

Our answers agree and can be confirmed by locating the "favor–nonwhite" cell in the contingency table. Two hundred of the 1000 persons in the population are nonwhite and favor busing; therefore, they constitute 20% or 0.2 of the population and we have confirmed our answer.

12.5 Probability and Continuous Variables

Up to this point we have considered probability in terms of the expected relative frequency of an event. In fact, as you will recall, probability was defined in terms of frequency and expressed as the following proportion [Eq. (12.1)]:

$$p(A) = \frac{\text{number of outcomes in which event } A \text{ occurs}}{\text{total number of outcomes}}.$$

However, this definition presents a problem when we are dealing with continuous variables such as age or income. As we pointed out in Section 3.7.1, it is generally advisable to represent frequency in terms of areas under a curve when we are dealing with continuous variables. Thus, for continuous variables, we may express probability as the following proportion:

$$p = \frac{\text{area under portions of a curve}}{\text{total area under the curve}}. \qquad (12.11)$$

Since the total area in a probability distribution is equal to 1.00, we define p as the proportion of total area under portions of a curve.

Chapters 13 through 17 use the standard normal curve as the probability model. Let us examine the probability-area relationship in terms of this model.

12.6 Probability and the Normal-Curve Model

In Section 7.4 we stated that the standard normal distribution has a mean of 0, a standard deviation of 1, and a total area that is equal to 1.00. We saw that when scores on a normally distributed variable are transformed into z scores, we are, in effect, expressing these scores in units of the standard normal curve. This permits us to express the difference between any two scores as proportions of total area under the curve. Thus we may establish probability values in terms of these proportions, as in Eq. (12.11). Let us look at several examples that illustrate the application of probability concepts to the normal-curve model.

12.6.1 Illustrative Problems*

For all problems, assume $\mu = 100$ and $\sigma = 16$.

Problem 12.1

What is the probability of selecting at *random,* from the general population, a person with an aptitude score of at least 132? The answer to this question is given by the proportion of area under the curve above a score of 132 (Fig. 12.1).

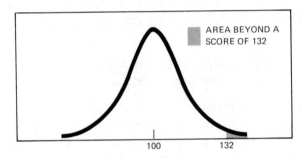

AREA BEYOND A
SCORE OF 132

100   132

12.1 Proportion of area above a score of 132 in a normal distribution with $\mu = 100$ and $\sigma = 16$.

First, we must find the z score corresponding to $X = 132$.

$$z = \frac{132 - 100}{16} = 2.00.$$

In column C of Table A in Appendix C, we find that 0.0228 of the area lies at or beyond a z of 2.00. Therefore, the probability of selecting, at random, a score of at least 132 is 0.0228.

Problem 12.2

What is the probability of selecting, at random, an individual with a score of at least 92?

---

*   See Section 7.6.

We are dealing with two mutually exclusive and exhaustive areas under the curve. The area under the curve above a score of 92 is P; the area below a score of 92 is Q. In solving our problem, we may therefore apply Eq. (12.7):

$$P = 1.00 - Q.$$

By expressing a score of 92 in terms of its corresponding z, we may obtain the proportion of area below $X = 92$ (that is Q) directly from column C (Table A).

The z score corresponding to $X = 92$ is

$$z = \frac{92 - 100}{16} = -0.50.$$

The proportion of area below a z of -0.50 is 0.3085. Therefore, the probability of selecting, at random, a score of at least 92 becomes

$$P = 1.00 - 0.3085 = 0.6915.$$

Figure 12.2 illustrates this relationship.

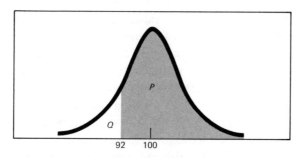

12.2 Proportion of area above (P) and below (Q) a score of 92 in a normal distribution with $\mu = 100$ and $\sigma = 16$.

Problem 12.3

Let us look at an example involving the multiplication law. Given that sampling with replacement is used, what is the probability of drawing, at random, three individuals with scores equaling or exceeding 124? For this problem, we will again assume that $\mu = 100$ and $\sigma = 16$.

The z score corresponding to $X = 124$ is

$$z = \frac{124 - 100}{16} = 1.5.$$

In column C (Table A), we find that 0.0668 of the area lies at or beyond $X = 124$. Therefore,

$$\begin{aligned} p(A, B, C) = p(A \text{ and } B \text{ and } C) &= p(A) \times p(B) \times p(C) \\ &= (0.0668)(0.0668)(0.0668) \\ &= 0.0003. \end{aligned}$$

12.7 One- and Two-Tailed p Values

In the first problem we posed the question: What is the probability of selecting a person with a score as high as 132? We answered the question by examining only one tail of the distribution, namely, scores as high as or higher than 132. For this reason, we refer to the probability value that we obtained as being a **one-tailed p value.**

In statistics and research, the following question is frequently asked: "What is the probability of obtaining a score (or statistic) this *deviant* from the mean? . . . or a score (or statistic) this *rare* . . . or a result this *unusual*?" Clearly, when the frequency distribution of scores is symmetrical, a score of 68 or lower is every bit as deviant from a mean of 100 as is a score of 132. That is, both are two standard deviation units away from the mean. When we express the probability value, taking into account both tails of the distribution, we refer to the p value as being **two tailed.** In symmetrical distributions, two-tailed p values may be obtained merely by doubling the one-tailed probability value. Thus in the preceding problem, the probability of selecting a person with a score as rare or unusual as 132 is $2 \times 0.0228 = 0.0456$.

We may illustrate the distinction between one- and two-tailed p values by referring to the sampling experiment we have used throughout the text. Figure 12.3 shows a probability histogram based upon Table 12.1. Recall that this distribution was obtained by selecting, with replacement, all possible samples of $N = 2$ from a population of seven scores.

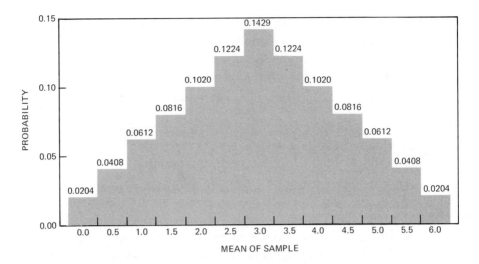

12.3 Probability histogram of means based upon selecting, with replacement, samples of $N = 2$ from a population of seven numbers.

If we ask "What is the probability of obtaining a mean equal to zero?" we need refer only to the left tail of the probability histogram in order to find the answer of 0.0204. However, if we ask "What is the probability of obtaining as deviant an outcome as a sample mean equal to zero?" we would have to look at both the left and the right tails. Since a mean of 6 is equally deviant from the mean of the distribution, we use the addition rule to obtain the two-tailed value

$$p(\overline{X} = 0 \text{ or } \overline{X} = 6)$$
$$= p(\overline{X} = 0) \text{ and } p(\overline{X} = 6)$$
$$= 0.0204 + 0.0204 = 0.0408.$$

Let's look at two additional examples.

Example What is the probability of obtaining a mean as low as 0.5 (that is, a mean of 0.5 or lower)?

We find the probability of obtaining a mean of 0.5 and a mean of 0.0 (the only mean lower than 0.5), and add these probabilities together to obtain the one-tailed p value:

$$p(\overline{X} \le 0.5) = p(\overline{X} = 0.05) + p(\overline{X} = 0.0)$$
$$= 0.0408 + 0.0204 = 0.0612.$$

Example   What is the probability of obtaining a mean as deviant from the distribution mean as a sample mean of 0.5?

The answer to this question calls for a two-tailed p value obtained by applying the addition rule to

$$p(\overline{X} = 0.5) + p(\overline{X} = 0.0) + p(\overline{X} = 5.5) + p(\overline{X} = 6.0).$$

However, since the distribution is symmetrical and means of 5.5 and 6.0 are equally as deviant as 0.5 and 0.0, respectively, we need only double the p value obtained in the preceding example:

$$p(0.5 \geq \overline{X} \geq 5.5) = 2(0.0612)$$
$$= 0.1224.$$

Incidentally, the left-hand member of this expression is read: *The probability of a mean equal to or less than* 0.5 *or equal to or greater than* 5.5.

The distinction between one- and two-tailed probability values takes on added significance as we progress into inferential statistics.

Summary

In this chapter we discussed: The importance of the concept of randomness in inferential statistics. Randomness refers to selecting the events in the sample such that each event is equally likely to be selected in a sample of a given size. Independent random sampling refers to the fact that the selection of one event has no effect upon the probability of selecting another event. Although the individual events are unpredictable, collections of random events take on characteristic and predictable forms. The normal curve was cited in this regard.

The theory of probability is concerned with the outcomes of studies. We can distinguish between probabilities established by assuming *idealized* relative frequencies and those established empirically by determining relative frequencies. Probability was defined as

$$p(A) = \frac{\text{number of outcomes in which } A \text{ occurs}}{\text{total number of outcomes}}.$$

The formal properties of probability were also discussed.

Probabilities vary between 0 and 1.00.

The addition rule is as follows: If A and B are two events, the probability of obtaining either of them is equal to the probability of A

plus the probability of B minus the probability of their joint occurrence. Thus

$$p(A \text{ or } B) = p(A) + p(B) - p(A \text{ and } B).$$

If events A and B are *mutually exclusive,* the addition rule becomes:

$$p(A \text{ or } B) = p(A) + p(B).$$

If the two events are *mutually exclusive and exhaustive,* we obtain:

$$p(A) + p(B) = 1.00.$$

Allowing P to represent the probability of occurrence and Q to represent the probability of nonoccurrence, we find that three additional useful formulations for mutually exclusive and exhaustive events are:

$$P + Q = 1.00, \quad P = 1.00 - Q, \quad Q = 1.00 - P.$$

The multiplication rule varies for independent and dependent events. *When events are independent* and when sampling with replacement, the selection on one trial is *independent* of the selection on another trial. Given two events, A and $B,$ the probability of obtaining both A and B in successive trials is the product of the probability of obtaining one of these events times the probability of obtaining the second of these events:

$$p(A \text{ and } B) = p(A)p(B).$$

When events are dependent and when sampling without replacement, the selection of one event affects the probability of selecting each remaining event, so they are *dependent.* Thus given two events, A and $B,$ the probability of obtaining both A and B jointly or successively is the product of the probability of obtaining one of the events times the conditional probability of obtaining one event, given that the other event has occurred. Symbolically:

$$p(A \text{ and } B) = p(A)p(B \mid A) = p(B)p(A \mid B).$$

We discussed the application of probability theory to continuously distributed variables. Probability is expressed in terms of the proportion of area under a curve. Hence

$$p = \frac{\text{area under portions of a curve}}{\text{total area under a curve}}.$$

We saw how we may use z scores and the standard normal curve to establish various probabilities for normally distributed variables.

Finally, we distinguished between one- and two-tailed probability values.

Terms to Remember

Addition rule   If A and B are mutually exclusive events, the probability of obtaining *either of them* is equal to the probability of A plus the probability of B. Symbolically,

$$p(A \text{ or } B) = p(A) + p(B).$$

Bias   In sampling, when selections favor certain events or certain collections of events.

Conditional probability   The probability of an event given that another event has occurred. Represented symbolically as $p(A \mid B)$, the probability of A given that B has occurred.

Dependence   The condition that exists when the occurrence of a given event affects the probability of the occurrence of another event.

Exhaustive   Two or more events are said to be exhaustive if they exhaust all possible outcomes. Symbolically,

$$p(A \text{ or } B \text{ or } \ldots) = 1.00.$$

Independence   The condition that exists when the occurrence of a given event will not affect the probability of the occurrence of another event. Symbolically,

$$p(A \mid B) = p(A) \quad \text{and} \quad p(B \mid A) = p(B).$$

Joint occurrence   The occurrence of two events simultaneously. Such events cannot be mutually exclusive.

Multiplication rule   Given two events A and B, the probability of obtaining both A and B jointly is the product of the probability of obtaining one of these events times the conditional probability of obtaining one event, given that the other event has occurred. Symbolically,

$$p(A \text{ and } B) = p(A)p(B \mid A) = p(B)p(A \mid B).$$

Mutually exclusive   Events A and B are said to be mutually exclusive if both cannot occur simultaneously. Symbolically, for mutually exclusive events,

$$p(A \text{ and } B) = 0.00.$$

One-tailed p values   Probability values obtained by examining only one tail of the distribution.

Probability   A theory concerned with possible outcomes of studies. Symbolically,

$$p(A) = \frac{\text{number of outcomes in which } A \text{ occurs}}{\text{total number of outcomes}}.$$

Random sampling   Samples are selected in such a way that each sample of a given size has precisely the same probability of being selected.

Two-tailed p values   Probability values that take into account both tails of the distribution.

Exercises

1. Imagine that we have a population of the following four scores: 0, 3, 6, and 9.

   a) Construct a probability distribution and histogram of all possible means when sampling with replacement, $N = 2$.

   b) Construct a probability histogram of all possible means when sampling with replacement, $N = 3$. (*Hint:* The table for finding the means follows. The values appearing in the cells represent the means of the three draws.)

First draw		0				3				6				9		
Second draw	0	3	6	9	0	3	6	9	0	3	6	9	0	3	6	9
Third draw 0	0	1	2	3	1	2	3	4	2	3	4	5	3	4	5	6
3	1	2	3	4	2	3	4	5	3	4	5	6	4	5	6	7
6	2	3	4	5	3	4	5	6	4	5	6	7	5	6	7	8
9	3	4	5	6	4	5	6	7	5	6	7	8	6	7	8	9

2. The original population of the four scores in Exercise 1 was rectangular (they all had the same associated frequency of 1). Compare the probability distributions in 1(a) and 1(b) and attempt to form a generalization about the form and the dispersion of the distribution of sample means as we increase the sample size.

3. Answer the following questions based upon the probability histograms obtained in Exercise 1.

   a) Drawing a single sample of $N = 2$, what is the probability of obtaining a mean equal to zero? Contrast this result with the probability of randomly selecting a mean equal to zero when $N = 3$.

   b) For each distribution, determine the probability of selecting a sample with a mean as rare or as unusual as 9.

   c) From each probability histogram, determine the probability of selecting a sample with a mean as low as 3.

   d) From each probability histogram, determine the probability of selecting a sample mean as deviant from the population mean as a mean of 3.

4. For the probability distribution of $N = 2$ [Exercise 1(a)] find

   a) $p(\overline{X} < 6)$ b) $p(\overline{X} \geq 7.5)$ c) $p(\overline{X} = 4.5)$

5. For the probability distribution of $N = 3$ [Exercise 1(b)] find

   a) $p(3 \leq \overline{X} \leq 6)$ b) $p(4 \leq \overline{X} \leq 5)$ c) $p(\overline{X} = 2 \text{ or } \overline{X} = 8)$

6. Let's now imagine a different type of sampling experiment. You have selected all possible samples of $N = 2$ from a population of scores and obtained the following means: 1, 2, 2, 3, 3, 3, 4, 4, 5. You now place paper slips in a hat with these means written on them. You select one mean, record it, and replace it in the hat. You select a second mean, *subtract* it from the first, and then replace it in the hat. The table for describing all possible *differences between means* of $N = 2$ follows.

		FIRST DRAW OF MEAN								
		1	2	2	3	3	3	4	4	5
	1	0	1	1	2	2	2	3	3	4
	2	-1	0	0	1	1	1	2	2	3
	2	-1	0	0	1	1	1	2	2	3
Second draw of mean	3	-2	-1	-1	0	0	0	1	1	2
	3	-2	-1	-1	0	0	0	1	1	2
	3	-2	-1	-1	0	0	0	1	1	2
	4	-3	-2	-2	-1	-1	-1	0	0	1
	4	-3	-2	-2	-1	-1	-1	0	0	1
	5	-4	-3	-3	-2	-2	-2	-1	-1	0

a) Construct a frequency distribution of differences between means.

b) Construct a probability distribution of differences between means.

c) Find the mean and the standard deviation of the differences between means.

7. Based upon the responses to Exercise 6, answer the following questions. Drawing two samples at random and with replacement from the population of means, and subtracting the second mean from the first, what is the probability that you will select

    a) A difference between means equal to zero?

    b) A difference between means equal to or less than 1 *or* equal to or greater than -1? (*Note:* -2, -3, -4 are all less than -1.)

    c) A difference between means equal to -4?

    d) A difference between means as rare or as deviant as -4?

    e) A difference between means equal to or greater than 3?

    f) A difference between means equal to or less than -3?

    g) A difference between means as rare or as unusual as -3?

    h) A difference between means equal to or less than 2 or equal to or greater than -2?

8. List all the possible outcomes of a coin that is tossed three times. Calculate the probability of

    a) 3 heads   **b)** 3 tails

    c) 2 heads and 1 tail   **d)** at least 2 heads

9. A card is drawn at random from a deck of 52 playing cards. What is the probability that

    a) It will be the ace of spades?

    b) It will be an ace?

    c) It will be an ace or a face card?

    d) It will be a spade or a face card?

10. Express the probabilities in Exercises 8 and 9 in terms of *odds against*.

11. In a single throw of two dice, what is the probability that

    a) A 7 will appear?

    b) A doublet (two of the same number) will appear?

    c) A doublet or an 8 will appear?

    d) An even number will appear?

12. On a slot machine (commonly referred to as a "one-armed bandit"), there are three reels with five different fruits plus a star on each reel. After inserting a coin and pulling the handle, the player sees that the three reels revolve independently several times before stopping. What is the probability that

a) Three lemons will appear?

b) Any three of a kind will appear?

c) Two lemons and a star will appear?

d) Two lemons and any other fruit will appear?

e) No star will appear?

13. Three cards are drawn at random (without replacement) from a deck of 52 cards. What is the probability that

a) All three will be hearts?

b) None of the three cards will be hearts?

c) All three will be face cards?

14. Calculate the probabilities in Exercise 13 if each card is replaced after it is drawn.

15. A well-known test of intelligence is constructed so as to have normally distributed scores with a mean of 100 and a standard deviation of 16.

a) What is the probability that someone picked at random will have an I.Q. of 122 or higher?

b) There are I.Q.'s so *high* that the probability is 0.05 that such I.Q.'s would occur in a random sample of people. Those I.Q.'s are beyond what value?

c) There are I.Q.'s so *extreme* that the probability is 0.05 that such I.Q.'s would occur in a random sample of people. Those I.Q.'s are beyond what values?

d) The next time you shop you will undoubtedly see someone who is a complete stranger to you. What is the probability that his I.Q. will be between 90 and 110?

e) What is the probability of selecting two people at random

i) with I.Q.'s of 122 or higher?

ii) with I.Q.'s between 90 and 110?

iii) one with an I.Q. of 122 or hgher, the other with an I.Q. between 90 and 110?

f) What is the probability that on leaving your class, the first student you meet will have an I.Q. below 120? Can you answer this question on the basis of the information provided previously? If not, why not?

16. Which of the following selection techniques will result in random samples? Explain your answers.

   a) *Population:* Viewers of a given television program. *Sampling technique:* On a given night, interviewing every fifth person in the studio audience.

   b) *Population:* A homemade pie. *Sampling technique:* A wedge selected from any portion of the pie.

   c) *Population:* All the students in a suburban high school. *Sampling technique:* Selecting one student sent to you by each homeroom teacher.

17. The proportion of people who are Baptist in a particular city is 0.20. What is the probability that

   a) A given individual, selected at random, will be a Baptist?

   b) Two out of two individuals will be Baptists?

   c) A given individual will *not* be a Baptist?

   d) Two out of two individuals will *not* be Baptists?

18. A bag contains six blue marbles, four red marbles, and two green marbles. If you select a single marble at random from the bag, what is the probability that it will be

   a) Red?   b) Blue?   c) Green?   d) White?

19. Selecting *without* replacement from the bag described in Exercise 18, what is the probability that

   a) Three out of three will be blue?

   b) Two out of two will be green?

   c) None out of four will be red?

20. Selecting *with* replacement from the bag described in Exercise 18, what is the probability that

   a) Three out of three will be blue?

   b) Two out of two will be green?

   c) None out of four will be red?

21. Forty percent of the students at a given college major in business administration. Seventy percent of these are male and thirty percent female. Sixty percent of the students in the school are male. What is the probability that

    a) One student selected at random will be a BA major?

    b) One person selected at random will be a female BA major?

    c) Two students selected at random will be BA majors?

    d) Two persons selected at random will be BA majors, one male, one female?

22. What is the probability that a score chosen at random from a normally distributed population with a mean of 66 and a standard deviation of 8 will be

    a) Greater than 70?

    b) Less than 60?

    c) Between 60 and 70?

    d) In the 70s?

    e) Either equal to or less than 54 or equal to or greater than 72?

    f) Either less than 52 or between 78 and 84?

    g) Either between 56 and 64 or between 80 and 86?

23. Given the following population, what is the probability that an upper-class person favors a tax cut? The probability that a lower-class person opposes a tax cut?

| ATTITUDE TOWARD | SOCIAL CLASS | | |
A TAX CUT	Lower	Upper	Total
Favor	75	25	100
Oppose	25	50	75
Total	100	75	N = 175

24. A political scientist is studying all 67 grass roots political organizations in a major American city. The following table shows the clarification of the political organizations by the party allegiance and socioeconomic composition of the members.

| PARTY | SOCIOECONOMIC STATUS | | |
	Low	Moderate	High
Republican	5	11	19
Democrat	10	12	6
Socialist	2	1	1

If one of the political organizations were chosen at random from this city, what is the probability that the members will have these characteristics?

    a) high income

    b) Socialist

    c) Socialist of high income

    d) not Republican

    e) not Democrat and low income

    f) not Republican and not Socialist

    g) not Republican and not moderate income

25. Given the information in Exercise 24, what is the conditional probability that if an organization is selected it will have the membership characteristics of

    a) moderate income, given that it is Republican

    b) Democrat, given that it is low income

    c) not Socialist, given high income

Introduction to Statistical Inference

13

13.1 Why Sample?

13.2 The Concept of Sampling Distributions

13.3 Binomial Distribution

13.4 Testing Statistical Hypotheses: Level of Significance

13.5 Testing Statistical Hypotheses:
Null Hypothesis and Alternative Hypothesis

13.6 Testing Statistical Hypotheses: The Two Types of Error

13.7 A Final Word of Caution

13.1 Why Sample?

You are the leader of a religious denomination, and for the purpose of planning recruitment you want to know what proportion of the adults in the United States claim church membership. How would you go about getting this information?

You are a sociologist, and you want to study the differences in child-rearing practices among parents of delinquent and nondelinquent children.

You are a market researcher, and you want to know what proportion of individuals prefer certain car colors and their various combinations.

You are a park attendant, and you want to determine whether the ice is sufficiently thick to permit safe skating.

You are a gambler, and you want to determine whether a set of dice is "biased."

What do each of these problems have in common? You are asking questions about the parameter of a population to which you want to generalize your answers, but you have no hope of ever studying the *entire* population. Earlier (Section 1.2) we defined a **population** as a *com-*

plete or theoretical set of individuals, objects, or measurements having some common observable characteristic. As has been noted, it is frequently impossible to study all the members of a given population because the population as defined either has an infinite number of members, or is so large that it defies exhaustive study. Moreover, when we refer to the *population* we are often dealing with a hypothetical entity. In some research situations the actual population may not exist (for example, the population of all babies regardless of whether or not they have been born) or certain elements may be very difficult to locate (for example, young black males are frequently undercounted in the census due to their high rate of geographical mobility).

Since populations can rarely be studied exhaustively, we must depend on **samples** as a basis for arriving at a hypothesis concerning various characteristics, or parameters, of the population. Note that our interest is not in descriptive statistics *per se,* but in making inferences from data. Thus if we ask 100 people how they intend to vote in a forthcoming election, our primary interest is not in knowing how these 100 people will vote, but in estimating how the members of the entire voting population will cast their ballots.

Almost all research involves the observation and the measurement of a limited number of individuals or events. These measurements are presumed to tell us something about the population. In order to understand how we are able to make inferences about a population from a sample, it is necessary to introduce the concept of sampling distributions.

13.2 The Concept of Sampling Distributions

In actual practice, inferences about the parameters of a population are made from statistics that are calculated from a sample of N observations drawn at random from this population. If we continued to draw samples of size N from this population, we should not be surprised if we found some differences among the values of the sample statistics obtained. Indeed, it is this observation that has led to the concept of sampling distributions. A **sampling distribution** is a theoretical probability distribution of the possible values of some sample statistic that would occur if we were to draw all possible samples of a fixed size from a given population.

Imagine drawing simple random samples (with replacement) of $N = 50$ from the population of students at a medium-sized university. You calculate the mean age for the first sample and find that it is 20.2. Suppose you continue to draw an infinite number of samples of $N = 50$ and calculate the sample mean for each of the samples drawn. By treating each of the sample means as a raw score, you could draw a frequency curve of the sample mean scores as shown in Fig. 13.1. The original dispersion of ages for the population has been superimposed for comparative purposes. Note that the variation of the sampling distribution of means is far smaller than that of the actual population distribution of ages, a topic that is addressed in the following chapter.

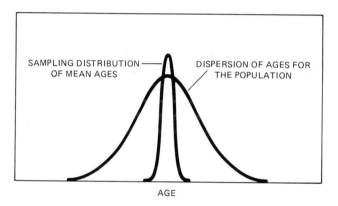

AGE

13.1 Hypothetical sampling distribution of mean ages and the original dispersion of ages for the population.

The sampling distribution is one of the most important concepts in inferential statistics. You are already familiar with several sampling distributions, although we have not previously named them as such. Recall the various sampling problems we have introduced throughout the earlier chapters in the text. In Chapter 12, we started with a population of seven scores, and selected, with replacement, samples of $N = 2$. We obtained all possible combinations of these scores, two at a time, and then found the mean of each of these samples. We then constructed a frequency distribution and probability distribution of means drawn from that population with a fixed sample size of $N = 2$.

Recall also that, in Exercise 1 at the end of Chapter 12, we constructed sampling distributions based upon drawing, with replacement, all possible samples of $N = 2$ and $N = 3$ from a population of four scores (0, 3, 6, 9). Table 13.1 shows these two sampling distributions, plus the sampling distribution of the mean when $N = 4$.

Table 13.1 Sampling distributions of means drawn from a population of four scores (0, 3, 6, 9; $\mu = 4.5$, $\sigma = 3.94$) and sample sizes $N = 2$, $N = 3$, and $N = 4$. Sampling with replacement.

N = 2		N = 3		N = 4	
$\overline{X}$	$p(\overline{X})$	$\overline{X}$	$p(\overline{X})$	$\overline{X}$	$p(\overline{X})$
9.0	0.0625	9.0	0.0156	9.0	0.0039
7.5	0.1250	8.0	0.0469	8.25	0.0156
6.0	0.1875	7.0	0.0938	7.50	0.0391
4.5	0.2500	6.0	0.1562	6.75	0.0781
3.0	0.1875	5.0	0.1875	6.00	0.1211
1.5	0.1250	4.0	0.1875	5.25	0.1562
0.0	0.0625	3.0	0.1562	4.50	0.1719
		2.0	0.0938	3.75	0.1562
		1.0	0.0469	3.00	0.1211
		0.0	0.0156	2.25	0.0781
				1.50	0.0391
				0.75	0.0156
				0.00	0.0039
$\overline{X} = 4.5$		$\overline{X} = 4.5$		$\overline{X} = 4.5$	
$s_{\overline{x}} = 2.37$		$s_{\overline{x}} = 1.94$		$s_{\overline{x}} = 1.68$	

Why is the concept of a sampling distribution so important? Once you are able to describe the sampling distribution of any statistic (be it mean, standard deviation, proportion), you are in a position to entertain and test a wide variety of different hypotheses. For example, you draw four numbers at random from some population. You obtain a mean equal to 6.00. You ask: "Is this mean an ordinary event or is it a rare event?" In the absence of a frame of reference, this question is meaningless. However, if we know the sampling distribution for this statistic, we would have the necessary frame of reference and the answer would be straightforward. If we were to tell you that the appropriate sampling distribution is given in Table 13.1 when $N = 4$, you would have no trouble answering the question. A mean of 6.00 would be drawn about 12% ($p = 0.1211$) of the time; and a mean of 6 or greater would occur almost 26% of the time ($p = 0.1211 + 0.0781 + 0.0391 + 0.0156 + 0.0039 = 0.2578$).

Whenever we estimate a population parameter from a sample, we shall ask questions such as: "How good an estimate do I have? Can I conclude that the population parameter is identical with the sample statistic? Or is there likely to be some error? If so, how much?" To answer each of these questions, we will compare our sample results with the "expected" results. The expected results are, in turn, given by the appropriate sampling distribution. But what does the sampling distribution of a particular statistic look like? How can we ever know the form of the distribution, and thus, what the expected results are? Since the inferences we will be making imply knowledge of the *form* of the sampling distribution, it is necessary to set up certain idealized *models*. The normal curve and the **binomial distribution** are two models whose mathematical properties are known. Consequently, these two distributions are frequently used as models to describe particular sampling distributions. Thus, for example, if we know that the sampling distribution of a particular statistic takes the form of a normal distribution, we may use the known properties of the normal distribution to make inferences and predictions about the statistic. The following sections should serve to clarify these important points.

13.3 Binomial Distribution

Let us say that you have a favorite coin that you use constantly in everyday life as a basis of "either-or" decision making. For example, you may ask, "Should I study tonight for the statistics quiz, or should I relax at one of the local movie houses? Heads, I study, tails, I don't." Over a period of time, you have sensed that the decision has more often gone against you than for you (in other words, you have to study more often than relax!). You begin to question the accuracy and the adequacy of the coin. Does the coin come up heads more often than tails? How might you find out?

One thing is clear. The true proportion of heads and tails characteristic of this coin can never be known. You could start tossing the coin this very minute and continue for a million years (granting a long life and a remarkably durable coin) and you would not exhaust the population of possible outcomes. In this instance, the true proportion of heads and tails is unknowable because the universe, or population, is unlimited.

The fact that the *true* value is unknowable does not prevent us from trying to estimate what it is. We have already pointed out that since populations can rarely be studied exhaustively, we must depend on samples to estimate the parameters.

Returning to our problem with the coin, we clearly see that in order to determine whether or not the coin is biased, we will have to obtain a sample of the behavior of that coin and arrive at some generalization concerning its possible bias. Let us define an *unbiased* coin as one in which the probability of heads is equal to the probability of tails. We may employ the symbol P to represent the probability of the occurrence of a head, and Q, the probability of the nonoccurrence of a head (that is, the occurrence of a tail). Since we are dealing with two mutually exclusive and exhaustive outcomes, if $P = Q = 1/2$, the coin is *unbiased*. Conversely, if $P \neq Q \neq 1/2$, the coin is *biased*.

How do we determine whether a particular coin is biased or unbiased? Suppose we conduct the following experiment. We toss the coin ten times and obtain nine heads and one tail. This may be viewed as a sample of the behavior of this coin. On the basis of this result, are we now justified in concluding that the coin is biased? Or is it reasonable to expect as many as nine heads from a coin that is unbiased? Before we answer these questions, it is necessary to look at the sampling distribution of *all* possible outcomes. Let us see how we might construct this sampling distribution, using a hypothetical coin.

13.3.1 Construction of Binomial Sampling Distributions by Enumeration

First, we must assume that this coin is unbiased and there are only two possible outcomes resulting from each toss of the coin: heads or tails. It will not stand on its side; it will not become lost; it cannot turn up both heads and tails at the same time.

If we toss this coin twice ($N = 2$), there are four possible ways the coin may fall: *HH, HT, TH,* and *TT.* The two middle ways may be thought of as the same outcome in that each represents one head and one tail. Thus tossing an unbiased coin twice results in the following theoretical frequency distribution:

OUTCOME	NUMBER OF WAYS FOR SPECIFIED OUTCOME TO OCCUR
2H	1
1H, 1T	2
2T	1
	4

Note that in N tosses of a coin, there are $N + 1$ different possible outcomes and 2^N different ways to obtain these $N + 1$ outcomes. Thus when $N = 2$, there are four ways of obtaining the three different outcomes.

We may calculate the probability associated with each outcome by dividing the number of ways each outcome may occur by 2^N. For example, using Eq. (12.1), the probability of obtaining one head and one tail in two tosses of an unbiased coin is

$$p(1H,1T) = \frac{2}{4} = 0.50.$$

We have seen how we can enumerate all the possible outcomes of tossing a hypothetical coin when $N = 2$, and construct corresponding frequency and probability distributions. This probability distribution represents the sampling distribution of outcomes when $N = 2$. Similarly, we may enumerate all the possible outcomes for any number of tosses of our hypothetical coin ($N = 3$, $N = 4$, etc.) and then construct the corresponding frequency and probability distributions.

Let us illustrate the construction of a probability distribution when $N = 5$. First, all possible outcomes are enumerated, as in Table 13.2. When $N = 5$, there are 32 ways (2^5) of obtaining the six different outcomes ($5 + 1$). By placing the six different outcomes along the baseline and representing their frequency of occurrence along the

Table 13.2   All possible outcomes obtained by tossing an unbiased coin five times ($N = 5$)

	NUMBER OF HEADS				
0H	1H	2H	3H	4H	5H
TTTTT	HTTTT	HHTTT	HHHTT	HHHHT	HHHHH
	THTTT	HTHTT	HHTHT	HHHTH	
	TTHTT	HTTHT	HTHHT	HHTHH	
	TTTHT	HTTTH	THHHT	HTHHH	
	TTTTH	THHTT	HHTTH	THHHH	
		THTHT	HTHTH		
		THTTH	THHTH		
		TTHHT	HTTHH		
		TTHTH	THTHH		
		TTTHH	TTHHH		

ordinate, we have constructed the theoretical frequency distribution for the various outcomes when $N = 5$ (see Fig. 13.2).

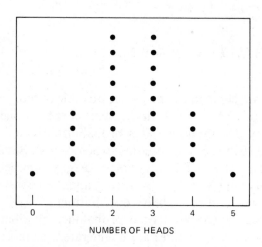

0   1   2   3   4   5

NUMBER OF HEADS

13.2 Theoretical distribution of various numbers of heads obtained by tossing an unbiased coin five times ($N = 5$).

It is now possible to calculate the probability associated with each outcome. For example, using Eq. (12.1), the probability of obtaining four heads and one tail in five tosses of an unbiased coin is

$$p(4H, 1T) = \frac{5}{32} = 0.156.$$

The complete probability distribution when $N = 5$ is shown in Fig. 13.3.

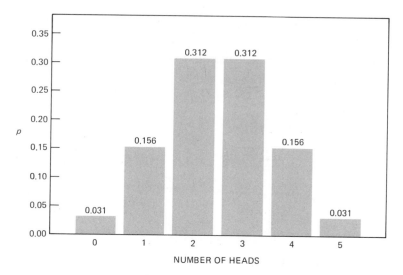

13.3 Histogram of theoretical probability distribution of various numbers of heads obtained by tossing an unbiased coin five times ($N = 5$).

13.3.2 Construction of Binomial Sampling Distributions Using the Binomial Expansion

Thus far, in order to calculate the probability associated with each outcome, we have had to enumerate all the possible ways in which various outcomes occur (see Table 13.2). As N increases, the process of listing the outcomes becomes exceedingly laborious since the number of ways for the outcomes to occur (2^N) doubles with each additional toss of the coin.*

The simplest method for demonstrating the binomial expansion is to remember the equality $(P + Q)^N = 1.00$, in which N represents the number of trials, such as tosses of a coin.

---

* In addition, when $P \neq Q \neq \frac{1}{2}$, it is not possible to obtain the probabilities associated with the various outcomes by simple enumeration. It is necessary to use the binomial expansion to obtain these probabilities.

When you toss a coin once and P represents the probability of a head, and Q represents the probability of the nonoccurrence of a head (that is, a tail), their joint probability is:

$$(P + Q)^N = (P + Q)^1 = \frac{1}{2} + \frac{1}{2} = 1.00.$$

When you toss the coin twice:

$$(P + Q)^N = (P + Q)^2 = P^2 + 2PQ + Q^2$$

$$= \frac{1}{4} + 2\left(\frac{1}{4}\right) + \frac{1}{4}$$

$$= \frac{1}{4} + \frac{1}{2} + \frac{1}{4} = 1.00.$$

The progression, then, proceeds as follows:

1. One toss: $(P + Q)^1$.
2. Two tosses: $(P + Q)^2 = (P + Q)(P + Q)$; that is,

$$\begin{array}{r} P + Q \\ \times \quad P + Q \\ \hline P^2 + PQ \\ + PQ + Q^2 \\ \hline P^2 + 2PQ + Q^2. \end{array}$$

3. Three tosses: $(P + Q)^3 = (P + Q)^2(P + Q)^1 = [P^2 + 2PQ + Q^2] \cdot [P + Q]$; that is,

$$\begin{array}{r} P^2 + 2PQ + Q^2 \\ \times \quad P + Q \\ \hline P^3 + 2P^2Q + PQ^2 \\ + P^2Q + 2PQ^2 + Q^3 \\ \hline P^3 + 3P^2Q + 3PQ^2 + Q^3. \end{array}$$

4. Four tosses: $(P + Q)^4 = (P + Q)^3(P + Q)^1 = [P^3 + 3P^2Q + 3PQ^2 + Q^3] [P + Q]$; that is,

$$\begin{array}{r} P^3 + 3P^2Q + 3PQ^2 + Q^3 \\ \times \quad P + Q \\ \hline P^4 + 3P^3Q + 3P^2Q^2 + PQ^3 \\ + P^3Q + 3P^2Q^2 + 3PQ^3 + Q^4 \\ \hline P^4 + 4P^3Q + 6P^2Q^2 + 4PQ^3 + Q^4. \end{array}$$

If you are interested in constructing the sampling distribution of the binomial when $N = 4$, you stop at this point and substitute the appropriate value of P and Q. In the coin-tossing experiment, $P = Q = 1/2$. Therefore,

$$(P + Q)^4 = P^4 + 4P^3Q + 6P^2Q^2 + 4PQ^3 + Q^4 = 1.00;$$

$$(P + Q)^4 = \left(\frac{1}{2}\right)^4 + 4\left(\frac{1}{2}\right)^3\left(\frac{1}{2}\right) + 6\left(\frac{1}{2}\right)^2\left(\frac{1}{2}\right)^2 + 4\left(\frac{1}{2}\right)\left(\frac{1}{2}\right)^3 + \left(\frac{1}{2}\right)^4$$

$$= \left(\frac{1}{16}\right) + \left(\frac{4}{16}\right) + \left(\frac{6}{16}\right) + \left(\frac{4}{16}\right) + \left(\frac{1}{16}\right)$$

$$= 0.0625 + 0.2500 + 0.3750 + 0.2500 + 0.0625 = 1.00.$$

The last line represents the sampling distribution for the binomial when $N = 4$ and $P = Q = 1/2$.

Equation (13.1) presents the binomial expansion in its general form:

$$(P + Q)^N = P^N + \frac{N}{1} P^{N-1}Q + \frac{N(N-1)}{(1)(2)}P^{N-2}Q^2$$

$$+ \frac{N(N-1)(N-2)}{(1)(2)(3)} P^{N-3}Q^3 + \cdots + Q^N.$$

(13.1)

There are $N + 1$ terms to the right of this equation, each representing a different possible outcome. The first term on the right-hand side of the equation (P^N) provides the probability of all events in the P category; the second term provides the probability of all events in the P category, except one; and, finally, the last term (Q^N) is the probability of all events in the Q category.

13.3.3 A Final Comment

By now you have probably noticed the similarity of shape between the binomial sampling distribution and the normal distribution. In fact, if P and Q are approximately equal to $1/2$ and N is large, the shape of the binomial distribution becomes very similar to that of the normal distribution. Therefore, the normal curve may be used as an approximation to the binomial distribution as the number of cases increases. The value of the normal curve serving as an approximation of the binomial distribution is particularly attractive given the extent to which the binomial distribution becomes extremely cumbersome as N increases.

Finally, it should be noted that the binomial distribution has a status parallel to that of the normal distribution. Whereas the normal distribution is a model for population or sampling distributions with continuous variables, the binomial distribution serves the same purpose for discrete, random variables such as the outcome of a true-false test and for many small-group research applications.

13.4 Testing Statistical Hypotheses: Level of Significance

At this point, you may wonder what happened to the experiment we were about to perform to determine whether our "decision-making" coin was biased. Having learned to calculate probability values, let us now address ourselves to the experiment. We are going to toss the coin a given number of times and determine whether or not the outcome is within certain expected limits. For example, if we toss our coin ten times and obtain five heads and five tails, would we begin to suspect our coin of being biased? Of course not, since this outcome is exactly a 50–50 split, and is in agreement with the hypothesis that the coin is not biased. What if we obtained six heads and four tails? Again, this is not an unusual outcome. In fact, if we expand the binomial, we can answer the question, "Given a theoretically perfect coin, how often would we expect an outcome at least this much different from a 50–50 split?" Reference to Fig. 13.4, which represents the theoretical probability distribution of various numbers of heads when $N = 10$, reveals that departures from a 50–50 split are quite common. Indeed, whenever we obtain either six or more heads, or four or fewer heads, we are departing from a 50–50 split. Such departures will occur fully 75.4% $(1 - 0.246 = 0.754 \times 100 = 75.4\%)$ of the time when we toss a perfect coin in a series of trials with ten tosses per trial.

What if we obtained nine heads and one tail? Clearly, we begin to suspect the honesty of the coin. Why? At what point do we change from attitudes accepting the honesty of the coin to attitudes rejecting its honesty? This question takes us to the crux of the problem of inferential statistics. We have seen that the more unusual or rare the event, the more prone we are to look for nonchance explanations of the event. When we obtained six heads in ten tosses of our coin, we felt no necessity to find an explanation for its departure from a 50–50 split, other than to state that such a departure would occur frequently "by

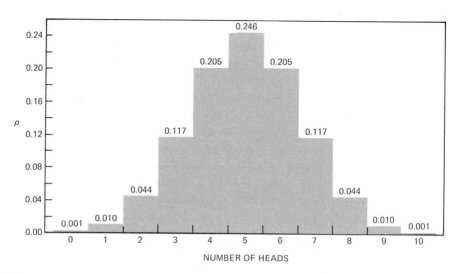

13.4 Bar graph of binomial sampling distribution of various numbers of heads obtained by tossing an unbiased coin ten times ($N = 10$), or by tossing ten coins one time.

chance." However, when we obtained nine heads, we had an uncomfortable feeling concerning the honesty of the coin. Nine heads out of ten tosses is such a rare occurrence that we begin to suspect that the explanation may be found in terms of the characteristics of the coin rather than in the so-called "laws of chance." The critical question is, "Where do we draw the line that determines what inferences we make about the coin?"

The answer to this question reveals the basic nature of science: its probabilistic rather than its absolutistic orientation. In the social sciences, most researchers have adopted one of the following two cutoff points as the basis for *inferring the operation of nonchance factors.*

1. When the event or one more deviant would occur 5% of the time or less, *by chance,* some researchers are willing to assert that the results are due to nonchance factors. This cutoff point is known variously as the 0.05 **significance level,** or the 5.00% significance level.

2. When the event or one more deviant would occur 1% of the time or less, *by chance,* other researchers are willing to assert that the results are due to nonchance factors. This cutoff point is known as the 0.01 **significance level,** or the 1.00% significance level.

The level of significance set by the researchers for inferring the operation of nonchance factors is known as the **alpha (α) level.** Thus, when using the 0.05 level of significance, $\alpha = 0.05$; when using the 0.01 level of significance, $\alpha = 0.01$.

In order to determine whether the results were due to nonchance factors in the present coin experiment, we need to calculate the probability of obtaining an event as rare as nine heads out of ten tosses. In determining the rarity of an event, we must consider the fact that the rare event can occur in both directions (for example, nine tails and one head) and that it includes more extreme events. In other words, the probability of an event as rare as nine heads out of ten tosses or as rare as one or zero heads out of ten tosses as well is equal to

$$p(9 \text{ heads}) + p(10 \text{ heads}) + p(1 \text{ head}) + p(0 \text{ heads}).$$

Since this distribution is symmetrical,

$$p(9 \text{ heads}) = p(1 \text{ head}) \quad \text{and} \quad p(10 \text{ heads}) = p(0 \text{ heads}).$$

Thus

$$p(9 \text{ heads}) + p(10 \text{ heads}) + p(1 \text{ head}) + p(0 \text{ heads})$$
$$= 2[p(9 \text{ heads}) + p(10 \text{ heads})].$$

These p values may be obtained from Fig. 13.4 as follows:

$$p(9 \text{ heads}) = 0.010, \quad \text{and} \quad p(10 \text{ heads}) = 0.001.$$

Therefore the **two-tailed probability** of an event as rare as nine heads out of ten tosses or as rare as one or zero heads out of ten tosses is $2(0.010 + 0.001) = 0.022$ or 2.2%.

Applying the 0.05 significance level ($\alpha = 0.05$), we would conclude that the coin was biased (that is, the results were due to nonchance factors). However, if we employed the 0.01 significance level ($\alpha = 0.01$), we would not be able to assert that these results were due to nonchance factors.

It should be noted and strongly emphasized that you do not conduct a study, analyze the results, arrive at a probability value, and then decide upon the α-level. The α-level must be specified *prior* to the study as part of the overall strategy of designing the study.

13.5 Testing Statistical Hypotheses: Null Hypothesis and Alternative Hypothesis

At this point, many students become disillusioned by the arbitrary nature of decision making in science. Let us examine the logic of statistical inference a bit further and see if we can resolve some of the doubts. Prior to the beginning of any study, the researcher sets up two mutually exclusive hypotheses:

1. The **null hypothesis** (H_0), which specifies hypothesized values for one or more of the *population parameters*. For example, we could state a null hypothesis in which we expect there to be no difference between the incomes of male and female professors in the following manner:

$$H_0: \mu \text{ males } = \mu \text{ females.}$$

2. The **alternative hypothesis** (H_1), which asserts that the *population parameter* is some value other than the one hypothesized. An alternative hypothesis to the null hypothesis just stated could be:

$$H_1: \mu \text{ males } \neq \mu \text{ females,}$$

where $\neq$ means "does not equal."

In the present coin experiment these two hypotheses read as follows:

$$H_0: \text{ the coin is unbiased; that is,}$$

$$P = Q = \frac{1}{2},$$

where

$$P = \text{ probability of heads;}$$

$$Q = \text{ probability of tails;}$$

$$H_1: \text{ the coin is biased; that is,}$$

$$P \neq Q \neq \frac{1}{2}.$$

The alternative hypothesis may be either **directional** or **nondirectional**. When H_1 asserts *only* that the population parameter is *different from* the one hypothesized, it is referred to as a nondirectional or two-tailed hypothesis, for example,

$$P \neq Q \neq \frac{1}{2}.$$

Very frequently in the social sciences hypotheses are **directional** or *one-tailed*. In this instance, in addition to asserting that the population parameter is different from the one hypothesized, we assert the *direction* of that difference, for example,

$$P > Q \text{ or } P < Q.$$

In evaluating the outcome of a study, **one-tailed probability** values should be used whenever our alternative hypothesis is directional. When the alternative hypothesis is directional, so is the null hypothesis. For example, if the alternative hypothesis is that μ males $> \mu$ females, the null hypothesis is μ males $\leq \mu$ females. Conversely, if H_1 is μ males $< \mu$ females, H_0 reads: μ males $\geq \mu$ females.

13.5.1 The Notion of Indirect Proof

Careful analysis of the logic of statistical inference reveals that the null hypothesis can never be proved. For example, if we had obtained exactly five heads on ten tosses of a coin, would this prove that the coin was unbiased? The answer is a categorical "No!" A bias, if it existed, might be of such a small magnitude that we failed to detect it in ten trials. But what if we tossed the coin one hundred times and obtained fifty heads? Wouldn't this prove something? Again, the same considerations apply. No matter how many times we tossed the coin, we could never exhaust the population of possible outcomes. We can make the assertion, however, that *no basis exists for rejecting* the hypothesis that the coin is biased.

How, then, can we prove the alternative hypothesis that the coin is biased? Again, we cannot prove the alternative hypothesis directly. Think, for the moment, of the logic involved in the following problem.

Draw two lines on a paper and determine whether they are of different lengths. You compare them and say, "Well, certainly they are not equal. Therefore they must be of different lengths." By rejecting equality (in this case, the null hypothesis) you assert that there is a difference.

Statistical logic operates in exactly the same way. We cannot prove the null hypothesis, nor can we directly prove the alternative hypothesis. However, if we can *reject* the null hypothesis, we can assert its alternative, namely, that the population parameter is some value other than the one hypothesized.* Applied to the coin problem, if we can reject the null hypothesis that $P = Q = 1/2$, we can assert the alternative, namely, that $P \neq Q \neq 1/2$. Note that the support of the alternative hypothesis is always *indirect*. We have supported it by rejecting the null hypothesis. On the other hand, since the alternative hypothesis can neither be proved nor disproved directly, we can *never prove the null hypothesis* by rejecting the alternative hypothesis. The strongest statement we are entitled to make in this respect is that we *failed to reject the null hypothesis.*

What, then, are the conditions for rejecting the null hypothesis? Simply this: when using the 0.05 level of significance, you reject the null hypothesis when a given result occurs, by chance, 5% of the time or less. When using the 0.01 level of significance, you reject the null hypothesis when a given result occurs, by chance, 1% of the time or less. Under these circumstances, of course, you *affirm* the alternative hypothesis.

In other words, one rejects the null hypothesis when the results occur, by chance, 5% of the time or less (or 1% of the time or less), *assuming that the null hypothesis is the true distribution.* That is, one assumes that the null hypothesis is true, calculates the probability on the basis of this assumption, and if the probability is small, one rejects the assumption.

13.6   Testing Statistical Hypotheses: The Two Types of Error

You may now ask, "But aren't we taking a chance that we will be wrong in rejecting the null hypothesis? Is it not possible that we have, in fact, obtained a statistically rare occurrence by chance?"

The answer to this question must be a simple and humble "Yes." This is precisely what we mean when we say that science is probabilistic. If there is any absolute statement that scientists are entitled to make, it is that we can never assert with complete confidence that our findings or propositions are true. There are countless examples in science in which

---

*   The null hypothesis can be thought of as the hypothesis to be nullified by a statistical test.

an apparently firmly established conclusion has had to be modified in the light of further evidence.

In the coin experiment, even if all the tosses had resulted in heads, it is possible that the coin was not, in fact, biased. By chance, once in every 1024 experiments, "on the average," the coin will turn up heads 10 out of 10 times. When we use the 0.05 level of significance, approximately 5% of the time we will be wrong when we reject the null hypothesis and assert its alternative.

These are some of the basic facts of the reality of inductive reasoning to which the student must adjust. The student of behavior who insists upon absolute certainty before speaking on an issue is a student who has been mute throughout life, and who will remain so (probably). These considerations have led statisticians to formulate two types of errors that may be made in statistical inference.

13.6.1 Type I Error (Type α Error)

In a **type I error,** we reject the null hypothesis when it is actually true. The probability of making a type I error is α. We have already pointed out that if we set our rejection point at the 0.05 level of significance, we will mistakenly reject H_0 approximately 5% of the time. It would seem, then, that in order to avoid this type of error we should set the rejection level as low as possible. For example, if we were to set $\alpha = 0.001$, we would risk a type I error only about one time in every thousand. It should be noted that the 0.05 level is rather routinely used in the social and behavioral sciences unless there is a particular reason to be extremely conservative about making a type I error. For example, suppose we were comparing a totally new teaching method to the technique currently in use. Suppose also that the null hypothesis were really true, that is, there were *no* difference between the two methods. If a type I error were made and the null hypothesis falsely rejected, this could conceivably lead to an extremely costly and time-consuming changeover to a method that was in fact no better than the one being used. In situations such as these we might want to set a more conservative level of significance (for example $\alpha = 0.01$). To familiarize you with the use of both α-levels, we have arbitrarily employed the $\alpha = 0.01$ and $\alpha = 0.05$ levels in examples presented throughout the text. However, the lower we set α the greater is the likelihood that we will make a type II error.

13.6.2 Type II Error (Type β Error)

In a **type II** error, we fail to reject the null hypothesis when it is actually false. Beta (β) is the probability of making a type II error. This type of error is far more common than a type I error. For example, if we apply the 0.01 level of significance as the basis of rejecting the null hypothesis and then conduct a study in which the result we obtained would have occurred by chance only 2% of the time, we cannot reject the null hypothesis.

It is clear, then, that the lower we set the rejection level, the less the likelihood of a type I error and the greater the likelihood of a type II error. Conversely, the higher we set the rejection level, the greater the likelihood of a type I error and the smaller the likelihood of a type II error.

The fact that the rejection level is set as low as it is attests to the conservatism of scientists, that is, the greater willingness on the part of the scientist to make an error in the direction of *failing* to claim a result than to make an error in the direction of *claiming* a result when he or she is wrong.

Table 13.3 summarizes the type of error made as a function of the true status of the null hypothesis and the decision we have made. We should note that type I and type II errors are sampling errors and refer to samples drawn from hypothetical populations.

Table 13.3 The type of error made as a function of the true status of H_0 and the statistical decision we have made. To illustrate, if H_0 is true (column 1) and we have rejected H_0 (row 2), we have made a type I error. If H_0 is false (column 2) and we have rejected H_0, we have made a correct decision.

		TRUE STATUS OF H_0	
		H_0 *True*	H_0 *False*
Decision	Accept H_0	Correct $1 - \alpha$	Type II error β
	Reject H_0	Type I error α	Correct $1 - \beta$

Let's look at a few examples, in which for illustrative purposes we supply the following information about the underlying population:

H_0, α-level, obtained p, statistical decision made, and the true status of H_0. Let's ascertain what type of error, if any, has been made.

1. H_0: $\mu_1 = \mu_2$, $\alpha = 0.05$, two-tailed test. Obtained $p = 0.03$, two-tailed value. Statistical decision: H_0 is false. Actual status of H_0: True.
   Error: Type I—rejecting a true H_0.

2. H_0: $\mu_1 = \mu_2$, $\alpha = 0.05$, two-tailed test. Obtained $p = 0.04$, two-tailed value. Statistical decision: H_0 is false. Actual status of H_0: False.
   Error: No error has been made. A correct conclusion was drawn since H_0 is false and the statistical decision was that H_0 is false.

3. H_0: $\mu_1 = \mu_2$, $\alpha = 0.01$, two-tailed test. Obtained $p = 0.10$, two-tailed value. Statistical decision: fail to reject H_0. Actual status of H_0: False.
   Error: Type II—failing to reject a false H_0.

4. H_0: $\mu_1 = \mu_2$, $\alpha = 0.01$, two-tailed test. Obtained $p = 0.006$, two-tailed value. Statistical decision: Reject H_0. Actual status of H_0: False.
   Error: No error has been made since the statistical decision has been to reject H_0 when H_0 is actually false.

You may now ask, "In actual practice, how can we tell when we are making a type I or a type II error?" The answer is simple: We can't! If we examine once again the logic of statistical inference, we shall see why. We have already stated that with rare exceptions we cannot or will not know the true parameters of a population. Without this knowledge, how can we know whether our sample statistics have approximated or have failed to approximate the true value? How can we know whether or not we have mistakenly rejected a null hypothesis? If we did know a population value, we could know whether or not we made an error. Under these circumstances, however, the need for sampling statistics is eliminated. We collect samples and draw inferences from samples only because our population values are unknowable for one reason or another. When they become known, the need for statistical inference is lost.

Is there no way, then, to know which surveys or experiments reporting significant results are accurate and which are not? The answer is a conditional "Yes." If we were to repeat the survey and obtain similar results, we would have increased confidence that we were not making a type I error. For example, if we tossed our coin in a second series of ten trials and obtained nine heads, we would feel far more confident that our coin was biased. Parenthetically, replication of studies is one of the weaker areas in social science research. The general attitude is that a study is not much good unless it is "different" and is therefore making a novel contribution. Replicating studies, when they are performed, frequently go unpublished. In consequence we may feel assured that in studies using the 0.05 significance level, approximately 1 out of every 20 that reject the null hypothesis is making a type I error.*

13.7 A Final Word of Caution

There is a very fine line between a significant and a nonsignificant finding. For example, if you tested a hypothesis at the 0.05 significance level and found that the probability of the results occurring by chance was 0.06, you would not reject the null hypothesis. If a friend tested the same hypothesis at the 0.05 level using another sample from the population for which you tested your hypothesis and found the probability of the results occurring by chance to be less than 0.05 (sometimes written as $P < 0.05$), he or she would reject the null hypothesis.

Consider another example in which you hypothesize that the mean age of two groups of people does not differ. The mean age of group one is 36 and the mean age of group two is 37. You test the hypothesis at the 0.01 level of significance and conclude they differ significantly. What you have found is that a mean difference of one year is statistically significant but also possibly trivial. Significance must be interpreted within meaningful context. Had you found a mean difference of one year between two groups of preschoolers, we would be far less likely to call the difference trivial. In fact, in the latter instance, the difference probably would be considered both practically and statistically significant because of the difference one year can have on the maturational process of a preschooler.

Finally, as you study the material in the following chapters, be aware that large samples are more likely to result in a statistically significant finding than are small samples due to the influence of the sample size (N) when testing hypotheses.

---

* The proportion is probably even higher, since our methods of accepting research reports for publication are heavily weighted in terms of the statistical significance of the results. Thus if four identical studies were conducted independently, and only one obtained results that permitted rejection of the null hypothesis, *this* one would most likely be published. There is virtually no way for the general scientific public to know about the three studies that *failed* to reject the null hypothesis.

Summary

We have seen that one of the basic problems of inferential statistics involves estimating population parameters from sample statistics.

In inferential statistics we are frequently called upon to compare our *obtained* values with *expected* values. The expected values are given by the appropriate sampling distribution, which is a theoretical probability distribution of the possible values of a sample statistic.

We have seen how to use sampling distributions to interpret sample statistics.

We have seen that there are two mutually exclusive and exhaustive statistical hypotheses in every study: the null hypothesis (H_0) and the alternative hypothesis (H_1).

If the outcome of a study is rare (here *rare* is defined as some arbitrary but accepted probability value), we reject the null hypothesis and assert its alternative. If the event is not rare (that is, the probability value is greater than what we have agreed upon as being significant), we fail to reject the null hypothesis. However, in no event are we permitted to claim that we have *proved H_0*.

The researcher is faced with two types of errors in establishing a cutoff probability value that he or she will accept as significant:

Type I: rejecting the null hypothesis when it is true.

Type II: failing to reject ("accepting") the null hypothesis when it is false.

The basic conservatism of scientists causes them to establish a low level of significance, resulting in a greater incidence of type II errors than of type I errors. Without replication of studies we have no basis for knowing when a type I error has been made, and even with replication we cannot claim knowledge of absolute truth.

Finally, and perhaps most important, we have seen that scientific knowledge is probabilistic and not absolute.

Terms to Remember

Alpha (α) level   The level of significance set by the researcher for inferring the operation of nonchance factors.

Alternative hypothesis (H_1)   A statement specifying that the population parameter is some value other than the one specified under the null hypothesis.

Binomial distribution   A model with known mathematical properties used to describe the distribution of discrete random variables.

Directional hypothesis   An alternative hypothesis that states the direction in which the population parameter differs from the one specified under H_0.

Nondirectional hypothesis   An alternative hypothesis (H_1) that states only that the population parameter is *different* from the one specified under H_0.

Null hypothesis (H_0)   A statement that specifies hypothesized values for one or more of the population parameters. Commonly, although not necessarily, involves the hypothesis of "no difference."

One-tailed probability value   Probability value obtained by examining only one tail of the distribution.

Population   A complete set of individuals, objects, or measurements having some common observable characteristic.

Sample   A subset of a population or universe.

Sampling distribution   A theoretical probability distribution of a statistic that would result from drawing all possible samples of a given size from some population.

Significance level   A probability value that is considered so rare in the sampling distribution specified under the null hypothesis that one is willing to assert the operation of nonchance factors. Common significance levels are 0.05, 0.01, and 0.001.

Two-tailed probability value   Probability value that takes into account both tails of the distribution.

Type I error (type α error)   The rejection of H_0 when it is actually true. The probability of a type I error is given by the α- level.

Type II error (type β error)   The probability of accepting H_0 when it is actually false. The probability of a type II error is given by β.

Exercises

1. Explain, in your own words, the nature of drawing inferences in social science. Be sure to specify the types of risks that are taken and the ways in which the researcher attempts to keep these risks within specifiable limits.

2. Give examples of studies in which

   a) A type I error would be considered more serious than a type II error.

   b) A type II error would be considered more serious than a type I error.

3. After completing a study, Nelson W. concluded, "I have proved that no difference exists between the two groups." Criticize his conclusion according to the logic of drawing inferences in science.

4. Explain what is meant by the following statement: "It can be said that the purpose of any study is to provide the occasion for rejecting the null hypothesis."

5. In a ten-item true-false examination,

   a) What is the probability that Alice, an unprepared student, will obtain all correct answers by chance?

   b) If eight correct answers constitute a passing grade, what is the probability that she will pass?

   c) What are the odds against her passing?

6. Identify H_0 and H_1 in the following:

   a) The population mean age is 31.

   b) The proportion of Democrats in Watanabe County is not equal to 0.50.

   c) The population mean age is not equal to 31.

   d) The proportion of Democrats in Watanabe County is equal to 0.50.

7. Suppose that you are a welfare eligibility employee responsible for approving clients for welfare. What type of error would you be making if:

   a) The hypothesis that a client is qualified for welfare is erroneously accepted?

   b) The hypothesis that a client is qualified for welfare is erroneously rejected?

   c) The hypothesis that a client is qualified for welfare is correctly accepted?

   d) The hypothesis that a client is qualified for welfare is correctly rejected?

8. A public opinion analyst recommends a political campaign strategy for a candidate on the basis of hypotheses she has formulated about future trends in the candidate's popularity. What type of error is she making if she makes the following predictions under the given conditions?

   a) H_0: The candidate's popularity will remain stable.
      Fact: It goes up precipitously.

   b) H_0: The candidate's popularity will remain stable.
      Fact: It falls abruptly.

   c) H_0: The candidate's popularity will remain stable.
      Fact: It shows only minor fluctuation about a central value.

9. An investigator sets $\alpha = 0.01$ for rejection of H_0. He conducts a study in which he obtains a p value of 0.02 and fails to reject H_0. *Discuss:* Is it more likely that he is accepting a true or a false H_0?

10. *Comment:* A student has collected a mass of data to test 100 different null hypotheses. On completion of the analysis he finds that 5 of the 100 comparisons yield p values ≤ 0.05. He concludes: "Using $\alpha = 0.05$, I have found a true difference in five of the comparisons."

11. Does the null hypothesis in a one-tailed test differ from the null hypothesis in a two-tailed test? Give an example.

12. Does the alternative hypothesis in a one-tailed test differ from the alternative hypothesis in a two-tailed test? Give an example.

13. In rejecting the null hypothesis for a one-tailed test, do all deviations count equally? Explain.

14. Discuss the similarities and differences between the normal curve and the binomial curve.

15. Suppose you want to test the hypothesis that there is not an equal number of male and female executives in a given large company. The appropriate null hypothesis would be:

  a) There are more female than male executives.

  b) The numbers of male and female executives are equal.

  c) There are more male than female executives.

16. Assume that there are exactly the same number of males and females in the population of employees qualified for executive work. What is the probability of each sex becoming an executive, if the executives are selected for the jobs solely on the basis of qualification? If you sampled 30 executives, what would be the expected ratio of males to females?

17. To test the hypotheses in the previous two exercises, you randomly sampled the sex of 12 executives and found that 10 were male, and 2 were female. What is the probability of this distribution occurring by chance?

18. In Exercise 17, if you found six female and six male executives, could you state with absolute certainty that the number of male and female executives in that company is equal? Explain your answer.

In Exercises 19 through 23, H_0, α, obtained p, and true status of H_0 are given. State whether or not an error in statistical decision has been made. If so, state the type of error.

19. H_0: $P = Q$, $\alpha = 0.01$, one-tailed test. Obtained $p = 0.008$, one-tailed value (in predicted direction). Actual status of H_0: True.

20. H_0: $P = Q$, $\alpha = 0.05$, two-tailed test. Obtained $p = 0.08$, two-tailed value. Actual status of H_0: True.

21. H_0: $P = Q$, $\alpha = 0.05$, two-tailed test. Obtained $p = 0.06$, two-tailed value. Actual status of H_0: False.

22. H_0: $P = Q$, $\alpha = 0.05$, two-tailed test. Obtained $p = 0.03$, two-tailed value. Actual status of H_0: False.

23. H_0: $P = Q$, $\alpha = 0.01$, two-tailed test. Obtained $p = 0.005$, two-tailed value. Actual status of H_0: False.

24. If $\alpha = 0.05$, what is the probability of making a type I error?

25. Construct a binomial sampling distribution when $N = 7$, $P = Q = 1/2$, and answer the following questions: What is the probability that:

  a) Six or more will be in the P category?

  b) Four or fewer will be in the P category?

c) One or fewer *or* six or more will be in the P category?

d) Two or more will be in the P category?

Statistical Inference and Continuous Variables

14

14.1 Introduction

14.2 Sampling Distribution of the Mean

14.3 Testing Statistical Hypotheses: Parameters Known

14.4 Estimation of Parameters: Point Estimation

14.5 Testing Statistical Hypotheses with Unknown Parameters: Student's t

14.6 Estimation of Parameters: Interval Estimation

14.7 Confidence Intervals and Confidence Limits

14.8 Test of Significance for Pearson's r: One-Sample Case

14.9 Test of Significance for Goodman's and Kruskal's Gamma

14.1 Introduction

You were first introduced to the concepts of *sample, population,* and *statistical inference* in Chapter 1. In the process of studying these terms you learned that a statistic is a number that describes a characteristic of a sample and that a parameter describes a characteristic of a population. Furthermore, you learned that in many instances (but not always) the parameter is unknown and must be estimated from information about a sample drawn from the population. Statistical inference involves a number of procedures that allow us to estimate parameters from sample statistics.

Recall the example in Chapter 1 in which a poll was conducted for a U.S. Senator to estimate the percentage of a state's registered voters who approve of nuclear energy as an energy source. A simple random sample was drawn from the list of the state's registered voters, and the percentage favoring nuclear energy was computed. How confident are we that the statistic we computed from the sample of voters provides a good estimate of the population parameter? Or, how do we test the hypothesis that a sample of prisoners drawn from a prison is representative of the entire prison population from which the sample was drawn? If we find that the Pearson correlation coefficient between educational attainment and income is $+0.65$ for those persons attending a school's ten-year reunion, can we be sure that if we contacted the population of all former graduates of the school the correlation would be +0.65? The statistical inference procedures to answer these and other questions are the topic of this chapter. Before we can consider these issues, however, two new terms, the sampling distribution of the mean and the standard error of the mean must be presented. In Chapter 13, we illustrated the use of a sampling distribution for a discrete two-category nominal variable (the binomial distribution) and for all possible means when drawing samples of a fixed *N* from a population of four scores. Table 13.1 showed the frequency and probability distributions of means when all possible samples of a given size were selected from the population of four scores. Figure 14.1 shows probability histograms, with superimposed curves obtained by connecting the midpoints of each bar.

Before proceeding with the discussion of sampling distributions for interval or ratio-scaled variables, examine Table 13.1 and Fig. 14.1 carefully. See if you can answer the following questions.

1. How does the mean of each sampling distribution of means compare with the mean of the population from which the samples were drawn?

2. How does the variability or dispersion of the sample means change as we increase the sample size upon which each sampling distribution is based?

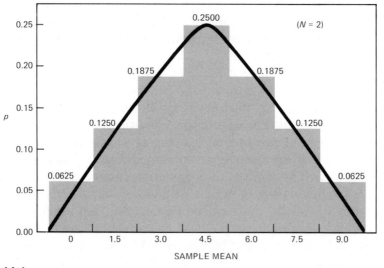

14.1

(a)

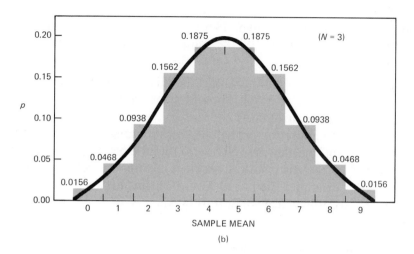

(b)

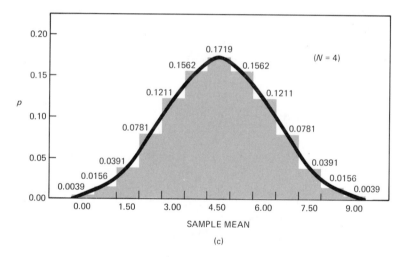

(c)

14.1 Probability histograms based upon sampling distributions of means drawn, with replacement, from a population of four scores and sample sizes $N = 2$, $N = 3$, and $N = 4$.

Now compare your answers with ours:

1. The mean of the population of four scores is 4.5. The mean of each sampling distribution of means is 4.5. Thus the mean of a sampling distribution of means is the same as the population mean from which

the sample means were drawn. This statement is true for all sizes of N. In other words, the mean of the sampling distribution does not vary with the sample size.

2. As you increase the sample size, the dispersion of sample means becomes less. A greater proportion of means are close to the population mean, and extreme deviations are rarer as N becomes larger. To verify these statements, note the probability of obtaining a mean as rare as 0 or 9 at different sample sizes. Note also that the proportion of means in the middle of the distribution becomes greater as sample size is increased. For example, the proportion of means between and including 3 and 6 is 0.6250 when $N = 2$, 0.6874 when $N = 3$, and 0.7265 when $N = 4$.

Finally, the standard deviation of the sample means—which we'll call the **standard error of the mean ($s_{\bar{x}}$)** from this point forward—shows that the dispersion of sample means decreases as sample size is increased.*

Figure 14.2 is a line drawing showing the decreasing magnitude of $s_{\bar{x}}$ as the sample size increases. Shown are the population standard

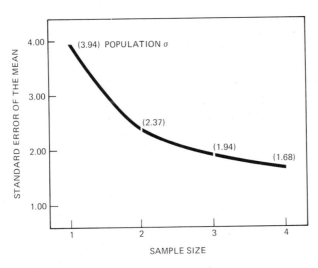

14.2 Magnitude of the standard error of the mean as a function of sample size.

---

* The relationship between the sample size and the standard error of the mean is specified by the **law of large numbers**; that is, as the sample size increases, the standard error of the mean decreases.

deviation of four scores and the standard error of the mean for the sampling distributions when $N = 2$, $N = 3$, and $N = 4$. Note that the decrease is not linear.

14.2   Sampling Distribution of the Mean

Let us imagine that we are conducting a sampling experiment in which we randomly draw (with replacement)* a sample of two scores from a population in which $\mu = 5.00$ and $\sigma = 0.99$ (see Table 14.1). For example, we might draw scores of 3 and 6. We calculate the sample mean and find $\overline{X} = 4.5$. Now suppose we continue to draw samples of $N = 2$ (for example, we might draw scores of 2, 8; 3, 7; 4, 5; 5, 6; etc.) until we obtain an indefinitely large number of samples. If we calculate the sample mean for each sample drawn and treat each of these sample means as a raw score, we may set up a frequency distribution of these sample means.

Table 14.1   An approximately normally distributed population with $\mu = 5.00$ and $\sigma = 0.99$

X	f	$p(X)$
2	4	0.004
3	54	0.054
4	242	0.242
5	400	0.400
6	242	0.242
7	54	0.054
8	4	0.004
	$N = 1000$	$\Sigma p(X) = 1.000$

Let us repeat these procedures with increasingly larger sample sizes, for example, $N = 5$, $N = 15$. We now have three frequency distributions of sample means based on three different sample sizes.

Intuitively, what might we expect these distributions to look like? Since we are selecting at random from the population, we would expect

---

*   If the population is infinite or extremely large, the difference between sampling with or without replacement is negligible.

the mean of the distribution of sample means to approximate the mean of the population.

How might the dispersion of these sample means compare with the variability in the original distribution of scores? In the original distribution, when $N = 1$ the probability of obtaining an extreme score such as 8 is 4/1000 or 0.004 (see Table 14.1). The probability of obtaining a sample *mean* equal to 8 when $N = 2$ (for example, drawing scores of 8, 8) is equal to 0.004 × 0.004 or 0.000016 [Eq. (12.9)]. In other words, the probability of selecting a sample with an extreme *mean* is less than the probability of selecting a single score that is equally extreme. What if we increased our sample size to $N = 4$? The probability of obtaining results this extreme ($\overline{X} = 8$) is exceedingly small $(0.004)^4 = 0.0000000003$. Generalizing, the probability of drawing extreme values of the sample mean is less as N increases. Since the standard deviation is a direct function of the number of extreme scores (see Chapter 6), it follows that a distribution containing proportionately fewer extreme scores will have a lower standard deviation. Therefore, if we treat each of the sample means as a raw score and then calculate the standard deviation ($\sigma_{\overline{X}}$, referred to as the **standard error of the mean***), it is clear that, as the sample size increases, the variability of the sample means decreases. The standard error of the mean is interpreted in the same way that the standard deviation is interpreted with respect to the normal curve. Looking at the frequency curve in Fig. 14.3 that is based on $N = 2$ we see that $\sigma_{\overline{X}} = 0.70$; therefore, we know that approximately 68% of the area under the curve lies within ±0.70 units of the population mean ($\mu = 5$). Therefore, approximately 68% of the area lies between 4.30 (5 − 0.70) and 5.70 (5 + 0.70).

If the sampling experiment that was based on samples of $N = 2$, 5, and 15 was actually conducted, the frequency curves of sample means would be obtained as in Fig. 14.3. There are three important lessons that may be learned from a careful examination of Fig. 14.3.

1. The distribution of sample means, drawn from a normally distributed population, is bell-shaped or "normal."

2. The mean of the sample means ($\mu_{\overline{X}}$) is equal to the mean of the population (μ) from which these samples were drawn.

---

* This notation represents the standard deviation of a sampling distribution of means. This is purely a theoretical notation since, with an infinite number of sample means, it is not possible to assign a specific value to the number of sample means involved.

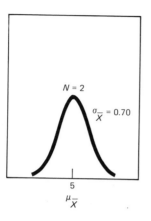 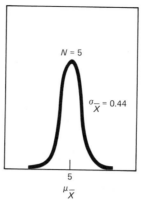 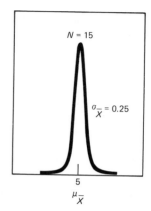

14.3 Frequency curves of sample means drawn from a population in which $\mu = 5.00$ and $\sigma = 0.99$.

3. The distribution of sample means becomes more and more compact as we increase the size of the sample. For example, in Fig. 14.3 $\sigma_{\bar{X}} = 0.70$ for the curve based on $N = 2$, $\sigma_{\bar{X}} = 0.44$ for the curve based on $N = 5$, and $\sigma_{\bar{X}} = 0.25$ for the curve based on $N = 15$. This is an extremely important point in statistical inference, about which we soon shall have a great deal more to say.

If we base our estimate of the population mean on a *single* sample drawn from the population, our approximation to the parameter is likely to be closer as we increase the size of the sample. In other words, if it is true that the dispersion of sample means decreases with increasing sample size, it also follows that the mean of any single sample is more likely to be closer to the mean of the population as the sample size increases.

These three observations illustrate a rather startling theorem that is of fundamental importance in inferential statistics, that is, the **central limit theorem,** which states:

If random samples of a fixed N are drawn from *any* population (regardless of the form of the population distribution), as N becomes larger, the distribution of sample means approaches normality, with the overall mean approaching μ, the variance of the sample means $\sigma_{\bar{X}}^2$ being equal to σ^2/N, and a standard error $\sigma_{\bar{X}}$ of $\sigma/\sqrt{N}$.

Stated symbolically,

$$\sigma_{\bar{X}}^2 = \frac{\sigma^2}{N}, \qquad (14.1)$$

and

$$\sigma_{\bar{X}} = \frac{\sigma}{\sqrt{N}}. \qquad (14.2)$$

In essence, the central limit theorem allows us to understand that if the underlying population distribution is skewed, the distribution of sample means will approach the form of a normal curve as the sample size increases. The effect of the central limit theorem is shown in Fig. 14.4.

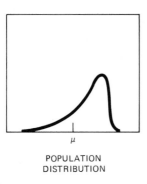
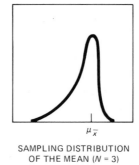
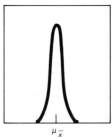

μ	$\mu_{\bar{X}}$	$\mu_{\bar{X}}$
POPULATION DISTRIBUTION	SAMPLING DISTRIBUTION OF THE MEAN ($N = 3$)	SAMPLING DISTRIBUTION OF THE MEAN ($N = 30$)

14.4 Effect of the central limit theorem.

14.3 Testing Statistical Hypotheses: Parameters Known

14.3.1 Finding the Probability That a Sample Mean Will Fall within a Certain Range

Let us briefly examine some of the implications of the relationships we have just discussed.

When μ and σ are *known* for a given population, it is possible to describe the form of the distribution of sample means when N is large (regardless of the form of the original distribution). It will be a normal distribution with a mean ($\mu_{\bar{X}}$) equal to μ and a standard error ($\sigma_{\bar{X}}$) equal to $\sigma/\sqrt{N}$. It now becomes possible to determine probability values in terms of areas under the normal curve. Thus we may use the known relationships of the normal probability curve to determine the probabilities associated with any sample mean (of a given N) randomly drawn from this population.

We have already seen (Section 7.3) that any normally distributed variable may be transformed into the normally distributed z scale. We have also seen (Section 12.6) that we may establish probability values in terms of the relationships between z scores and areas under the normal curve. That is, for any given raw score value (X) with a certain proportion of area beyond it, there is a corresponding value of z with the same proportion of area beyond it. Similarly, for any given value of a sample mean ($\overline{X}$) with a certain proportion of area beyond it, there is a corresponding value of z with the same proportion of area beyond it. Thus assuming that the form of the distribution of sample means is normal, we may establish probability values in terms of the relationships between z scores and areas under the normal curve.

To illustrate: Suppose we have a population of Baptist churches with a mean membership size of 250 ($\mu = 250$) and a standard deviation of 50 ($\sigma = 50$), from which we randomly draw 100 churches ($N = 100$). What is the probability that the sample mean ($\overline{X}$) will be equal to or greater than 255 members? Thus $H_0: \mu = \mu_0 = 250$ and $H_1: \mu \geq 255$. Figure 14.5 illustrates the area under the normal curve, which represents the probability that $\mu \geq 255$.

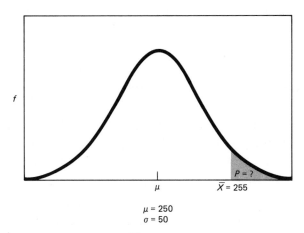

$\mu = 250$
$\sigma = 50$

14.5 The probability of $\overline{X} \geq 255$ if $\mu = 250$ and $\sigma = 50$.

The value of z corresponding to $\overline{X} = 255$ is obtained as follows:

$$z = \frac{\overline{X} - \mu_0}{\sigma_{\overline{X}}}, \qquad (14.3)$$

where μ_0 = value of the population mean under H_0,

$$\sigma_{\overline{X}} = \frac{\sigma}{\sqrt{N}} = \frac{50}{\sqrt{100}} = 5.00, \quad \text{and} \quad z = \frac{255 - 250}{5.00} = 1.00.$$

Looking up a z of 1.00 in column C (Table A of Appendix C), we find that 0.1587 of the sample means fall at or above $\overline{X} = 255$. Therefore there are approximately 16 chances in 100 of obtaining a sample mean equal to or greater than 255 from this population when $N = 100$. Note that $\sigma_{\overline{X}} = 5.00$ is considerably less than $\sigma = 50$. Why?

14.3.2 Testing Hypotheses about the Sample Mean

Let us extend this logic to a situation in which we do not know from what population a sample is drawn. We suspect that it may have been selected from the preceding population of churches with $\mu = 250$ and $\sigma = 50$, but we are not certain. We wish to test the hypothesis that our sample mean was indeed selected from this population. Let us imagine that we had obtained $\overline{X} = 263$ for $N = 100$. Is it reasonable to assume that this sample was drawn from the preceding population of churches?

Setting up this problem in formal statistical terms involves the following six steps, which are common to all hypothesis testing situations.

1. *Null hypothesis* (H_0): The mean membership size of the population (μ) from which the sample of churches was drawn equals 250, that is, $\mu = \mu_0 = 250$.
2. *Alternative hypothesis* (H_1): The mean of the population from which the sample was drawn does *not* equal 250; $\mu \neq \mu_0$. Note that H_1 is non-directional; consequently, a two-tailed test of significance will be used.
3. *Statistical test:* The z statistic is used since σ is known.
4. *Significance level:* $\alpha = 0.01$. If the difference between the sample mean and the specified population mean is so extreme that its associated probability of occurrence under H_0 is equal to or less than 0.01, we will reject H_0.

5. *Sampling distribution:* The normal probability curve.

6. *Critical region for rejection of H_0:* $|z| \geq |z_{0.01}| = 2.58.$* A **critical region** is that portion of area under the curve that includes those values of a statistic that lead to rejection of the null hypothesis.

The critical region is chosen to correspond with the selected level of significance. Thus for $\alpha = 0.01$, two-tailed test, the critical region is bounded by those values of $z_{0.01}$ that mark off a total of 1% of the area. Referring again to column C, Table A in Appendix C, we find that the area beyond a z of 2.58 is approximately 0.005. We double 0.005 to account for both tails of the distribution. Figure 14.6 depicts the critical region for rejection of H_0 when $\alpha = 0.01$, two-tailed test.

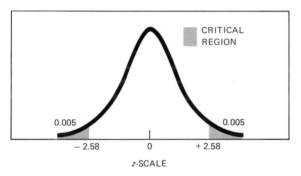

14.6 Critical region for rejection of H_0 when $\alpha = 0.01$, two-tailed test.

Therefore, in order to reject H_0 at the 0.01 level of significance, the absolute value of the obtained z must be equal to or greater than $|z_{0.01}|$ or 2.58. Similarly, if we are to be allowed to reject H_0 at the 0.05 level of significance, the absolute value of the obtained z must be equal to or greater than $|z_{0.05}|$ or 1.96.

---

* Since $z_{0.01} = \pm 2.58$ $|z| \geq |z_{0.01}|$ is equivalent to stating $z \geq 2.58$ or $z \leq -2.58$. Recall that the symbol $|\ |$ indicates the absolute value of the number, that is, without regard to sign.

In the present example, the value of z corresponding to $\overline{X} = 263$ is

$$z = \frac{\overline{X} - \mu_0}{\sigma_{\overline{X}}} = \frac{263 - 250}{5.00} = 2.60.$$

Decision   Since the obtained z falls within the critical region (that is, $2.60 >$ $z_{0.01}$), we may reject H_0 at the 0.01 level of significance. Thus we conclude that the sample in question ($\overline{X} = 263$) was not drawn from the population in which $\mu = 250$ and $\sigma = 50$. Having established α at the 0.01 level of significance, we know that the probability of rejecting the null hypothesis when it is actually true (and therefore making a type I error) is 1 out of 100, a rare event that is very unlikely to occur by chance.

14.4 Estimation of Parameters: Point Estimation

So far, we have been concerned with testing hypotheses when the population parameters are known. However, we have taken some pains in this book to point out that population values are rarely known, particularly when the population is extremely large. Every ten years, when the federal government undertakes that massive data collection effort called the census, we come close to knowing the parameters on the various questions. But knowledge of parameters is not the usual case. However, the fact that we do not know the population values does not prevent us from using the logic developed in Section 14.3.

Whenever we make inferences about population parameters from sample data, we compare our sample results with the expected results given by the appropriate sampling distribution. A hypothetical sampling distribution of sample means is associated with any sample mean. This distribution has a mean, $\mu_{\overline{X}}$, and a standard deviation, $\sigma_{\overline{X}}$. So far, in order to obtain the values of $\mu_{\overline{X}}$ and $\sigma_{\overline{X}}$, we have required a knowledge of μ and σ. In the absence of knowledge concerning the exact values of the parameters, we are forced to estimate μ and σ from the statistics calculated from sample data. Since in actual practice we rarely select more than one sample, our estimates are generally based on the statistics calculated from a single sample. All such estimates of population parameters involving the use of single sample values (for example, mean age of a sample) are known as **point estimates.**

14.4.1 Unbiased Estimate of the Population Variance

Throughout this book, we have turned to a sampling experiment whenever we wanted to illustrate a concept of fundamental importance in statistical analysis. Let us take a look at the denominator of the variance equation, and show that $N - 1$ in the denominator provides an unbiased estimate of the population variance, whereas N in the denominator underestimates the population variance.

Let us imagine the following sampling experiment. You place the following population of four scores in a hat: 1, 2, 3, 4. The mean of this population is 2.5 and the variance is 1.25. You select, with replacement, all possible samples of $N = 2$ and calculate the variance of each sample, using N and $N - 1$ in the denominator. Just as we previously placed the mean of each sample in the cell corresponding to both draws, we now place the *variance* of each sample in the appropriate cell.

First, let's do this using N in the denominator when calculating each sample variance.

Variance of each sample when using N in the denominator ($N = 2$)

		FIRST DRAW			
		1	*2*	*3*	*4*
Second draw	*1*	0.00	0.25	1.00	2.25
	2	0.25	0.00	0.25	1.00
	3	1.00	0.25	0.00	0.25
	4	2.25	1.00	0.25	0.00

Let us now construct a frequency distribution of these variances and calculate the mean variance.

s^2	f	fs^2
2.25	2	4.50
1.00	4	4.00
0.25	6	1.50
0.00	4	0.00
	$\Sigma f = 16$	$\Sigma fs^2 = 10.00$

The mean variance is found to be $10/16 = 0.625$. Recall that the variance of the population is 1.25. In this sampling experiment, using N in the denominator of the variance equation, the mean variance of all possible samples of $N = 2$ underestimates the population variance. Generalizing, sample variances that use N in the denominator provide a biased estimate of the population variance.

Now let us repeat the same procedures, using $N - 1$ to calculate the variance of each sample. The frequency distribution and mean of the sample variances are shown in the last table.

Variance of each sample when using $N - 1$ in the denominator ($N = 2$)

		FIRST DRAW			
		1	2	3	4
Second draw	1	0.00	0.50	2.00	4.50
	2	0.50	0.00	0.50	2.00
	3	2.00	0.50	0.00	0.50
	4	4.50	2.00	0.50	0.00

s^2	f	fs^2
4.50	2	9.00
2.00	4	8.00
0.50	6	3.00
0.00	4	0.00
	$\Sigma f = 16$	$\Sigma fs^2 = 20.00$

Now the mean variance is $20/16 = 1.25$. Note that this is identical to the original variance of the population. Thus using $N - 1$ in the denominator provides an unbiased estimate of the population variance.

14.4.2 Estimating $\sigma_{\overline{X}}$ from Sample Data

You will recall that we previously defined the variance of a sample as

$$s^2 = \frac{\Sigma(X - \overline{X})^2}{N}$$

in Eq. (6.3). We obtained the standard deviation, s, by finding the square root of this value. These definitions are perfectly appropriate as long as we are interested only in *describing* the variability of a sample. However, when our interest shifts to *estimating* the population variance from a sample value, we find that the preceding definition is inadequate since $\Sigma(X - \overline{X})^2/N$ tends, on the average, to *underestimate* the population variance. In other words, it provides a *biased estimate* of the population variance, whereas an unbiased estimate is required.

An **unbiased estimate** is an estimate that equals, on the average, the value of the parameter. That is, when we make the statement that a statistic is an unbiased estimate of a parameter, we are saying that the mean of the distribution of an extremely large number of sample statistics, drawn from a given population, tends to center upon the corresponding value of the parameter. It has been demonstrated in Section 14.4.1 that an unbiased estimate of the population variance may be obtained by dividing the sum of squares by $N - 1$. We shall use the symbol $\hat{s}^2$ to represent a sample variance providing an *unbiased estimate of the population variance,* and s to represent a sample standard deviation based on the unbiased variance estimate. Thus

$$\text{unbiased estimate of } \sigma^2 = \hat{s}^2 = \frac{\Sigma(X - \overline{X})^2}{N - 1}, \qquad (14.4)$$

and

$$\text{estimated } \sigma = \hat{s} = \sqrt{\hat{s}^2}. \qquad (14.5)$$

We are now able to estimate $\sigma_{\overline{X}}^2$ and $\sigma_{\overline{X}}$ from sample data. We will use the symbols $s_{\overline{X}}^2$ and $s_{\overline{X}}$ to refer to the estimated variance and standard error of the mean, respectively. Since we do not know σ^2, we accept the unbiased variance estimate ($\hat{s}^2$) as the best estimate we have of the population variance. Thus the equation for determining the variance of the mean from sample data is

$$\text{estimated } \sigma_{\overline{X}}^2 = s_{\overline{X}}^2 = \frac{\hat{s}^2}{N}. \qquad (14.6)$$

We estimate the standard error of the mean by finding the square root of this value:

$$\text{estimated } \sigma_{\overline{X}} = s_{\overline{X}} = \sqrt{\frac{\hat{s}^2}{N}} = \frac{\hat{s}}{\sqrt{N}}. \qquad (14.7)$$

If the sample variance (not the unbiased estimate) is used, we may estimate $\sigma_{\overline{X}}$ as:

$$\text{estimated } \sigma_{\overline{X}} = s_{\overline{X}} = \frac{s}{\sqrt{N-1}} = \sqrt{\frac{\sum(X - \overline{X})^2}{N(N-1)}} . \qquad (14.8)$$

Equation (14.8) is the one most frequently used in the social sciences to estimate the standard error of the mean. We will follow this practice.

Before proceeding further, let us review some of the symbols we have been discussing. Table 14.2 shows the various symbols for the means, variances, and standard deviations depending upon whether we are dealing with population parameters, unbiased population estimators, or sample statistics.

Table 14.2   Review of symbols

	POPULATION PARAMETERS (THEORETICAL)	PARAMETERS OF SAMPLING DISTRIBUTION OF MEAN (THEORETICAL)	UNBIASED POPULATION ESTIMATORS FOR SAMPLING DISTRIBUTION OF MEAN (EMPIRICAL)	SAMPLE STATISTICS (EMPIRICAL)
Means	μ, μ_0	$\mu_{\overline{X}}$	$\overline{X}$	$\overline{X}$
Variances	σ^2	$\sigma_{\overline{X}}^2$	$\hat{s}^2, s_{\overline{X}}^2$	s^2
Standard Deviations	σ	$\sigma_{\overline{X}}$	$\hat{s}, s_{\overline{X}}$	s

14.5 Testing Statistical Hypotheses with Unknown Parameters: Student's t

We previously pointed out that when the parameters of a population are known, it is possible to describe the form of the sampling distribution of sample means. It will be a normal distribution with $\sigma_{\overline{X}}$ equal to $\sigma/\sqrt{N}$. By employing the relationship between the z scale and the normal distribution, we were able to test hypotheses using

$$z = \frac{(\overline{X} - \mu_0)}{\sigma_{\overline{X}}}$$

as a test statistic. When σ is not known, we are forced to estimate its value from sample data. Consequently, estimated $\sigma_{\overline{X}}$ (that is, $s_{\overline{X}}$) must be based on the estimated σ (that is, s), that is,

$$s_{\overline{X}} = \frac{\hat{s}}{\sqrt{N}}.$$

Now, if substituting $\hat{s}$ for σ provided a reasonably good approximation to the sampling distribution of means, we could continue to use z as our test statistic, and the normal curve as the model for our sampling distribution. As a matter of fact, however, this is not the case. At the turn of the century, a statistician by the name of William Gosset, who published under the pseudonym of Student, noted that the approximation of $\hat{s}$ to σ is poor, particularly for small samples. This failure of approximation is due to the fact that, with small samples, $\hat{s}$ will tend to underestimate σ more than one-half of the time. Consequently, the statistic

$$\frac{\overline{X} - \mu_0}{\hat{s}/\sqrt{N}}$$

will tend to be spread out more than the normal distribution.

Gosset's major contribution to statistics consisted of his description of a distribution, or rather, a family of distributions, that permits the testing of hypotheses with samples drawn from normally distributed populations, when σ is not known. These distributions are referred to variously as the *t*-**distributions** or **Student's** *t*. The ratio used in the testing of hypotheses is known as the *t*-ratio:

$$t = \frac{\overline{X} - \mu_0}{s_{\overline{X}}}, \tag{14.9}$$

where μ_0 is the value of the population mean under H_0.

The t statistic is similar in many respects to the previously discussed z statistic [see Eq. (14.3)]. Both statistics are expressed as the deviation of a sample mean from a population mean (known or hypothesized) in terms of the standard error of the mean. By reference to the appropriate sampling distribution, we may express this deviation in terms of probability. When the z statistic is used, the standard normal curve is the appropriate sampling distribution. For the t statistic there is a family of distributions that varies as a function of **degrees of freedom (df).**

The term *degrees of freedom* refers to the number of values that are free to vary after we have placed certain restrictions on our data. To illustrate, let us imagine that we have four numbers: 18, 23, 27, 32. The sum is 100 and the mean is $\overline{X} = 100/4 = 25$. Recall that if we subtract the mean from each score, we should obtain a set of four deviations that add up to zero. Thus

$$(18 - 25) + (23 - 25) + (27 - 25) + (32 - 25) =$$
$$(- 7) + (- 2) + 2 + 7 = 0.$$

Note also that the four deviations are not independent. Once we have imposed the restriction that the deviations are taken from the mean, the values of only three deviations are free to vary. As soon as three deviations are known, the fourth is completely determined. Stated another way, the values of only three deviations are free to vary. For example, if we know three deviations to be $- 7$, $- 2$, and 7, we may calculate the unknown deviation by use of the equality:

$$(X_1 - \overline{X}) + (X_2 - \overline{X}) + (X_3 - \overline{X}) + (X_4 - \overline{X}) = 0.$$

Therefore

$$(X_4 - \overline{X}) = 0 - [(X_1 - \overline{X}) + (X_2 - \overline{X}) + (X_3 - \overline{X})].$$

In the present example,

$$(X_4 - \overline{X}) = 0 - (- 7) + (- 2) + 2$$
$$= 0 + 7 - 2 + 2 = 7.$$

To generalize: For any given sample on which we have placed a single restriction, the number of degrees of freedom is $N - 1$. In the preceding example, $N = 4$; therefore degrees of freedom are $4 - 1 = 3$.

Note that when $s/\sqrt{N - 1}$ [Eq. (14.8)] is used to obtain $s_{\overline{X}}$, the quantity under the square root sign ($N - 1$) is the degrees of freedom.

We noted above that the use of Student's t depends on the assumption that the underlying population is normally distributed. This requirement stems from a unique property of normal distributions. *Given that observations are independent and random, the sample means and sample variances are independent only when the population is normally distributed.* As we previously pointed out, two scores or statistics are independent only when the values of one do not depend on the values of the other and vice versa. Tests of significance of means which are discussed in the following sections demand that the means and variances be independent of one another. They cannot vary together in some systematic way—for example, with the variances becoming larger as the means become larger. If they do vary in a systematic way, the underlying population cannot be normal and tests of significance based on the assumption of normality may be invalid. It is for this reason that the assumption of normality underlies the use of **Student's t-ratio.**

One final note: Student's *t*-ratio is referred to as a **robust test,** meaning that statistical inferences are likely to be valid even when there are fairly large departures from normality in the population distribution. This robustness is another consequence of the central limit theorem. If we have serious doubts concerning the normality of the population distribution, it is wise to increase the N in each sample.

14.5.1 Characteristics of *t*-Distributions

Let us compare the characteristics of the *t*-distributions with the already familiar standard normal curve. First, both distributions are symmetrical about a mean of zero. Therefore the proportion of area beyond a particular positive *t*-value is equal to the proportion of area below the corresponding negative *t*.

Second, the *t*-distributions are more spread out than the normal curve. Consequently, the proportion of area beyond a specific value of *t* is *greater* than the proportion of area beyond the corresponding value of *z*. However, as sample size increases and, therefore, the greater the degrees of freedom, the more the *t*-distributions resemble the standard normal curve. In order that you may see the contrast between the *t*-distributions and the normal curve we have reproduced three curves in Fig. 14.7: the sampling distributions of *t* when degrees of freedom = 3, degrees of freedom = 10, and the normal curve.

Inspection of Fig. 14.7 permits several interesting observations. We have already seen that with the standard normal curve, $|z| \geq 1.96$ defines the region of rejection at the 0.05 level of significance. However, when degrees of freedom = 3, a $|t| \geq 1.96$ includes approximately 15% of the total area. Consequently, if we were to use the normal curve for testing hypotheses when N is small (therefore degrees of freedom is small) and σ is unknown, we would be in serious danger of making a type I error, that is, rejecting H_0 when it is true. A much larger

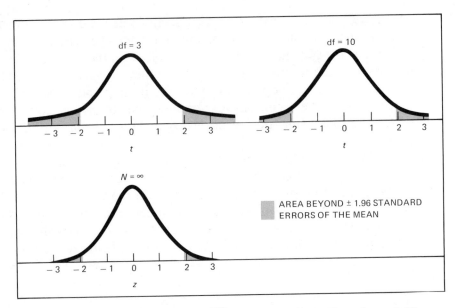

14.7 Sampling distributions of $t = (\overline{X} - \mu_0)/s_{\overline{x}}$ when df = 3, and 10, compared to the standard normal curve.

value of t is required to mark off the bounds of the critical region of rejection. Indeed, when degrees of freedom = 3, the absolute value of the obtained t must be equal to or greater than 3.18 to reject H_0 at the 0.05 level of significance (two-tailed test). However, as degrees of freedom increases, the differences in the proportions of area under the normal curve and the Student t-distributions become negligible.

In contrast to our use of the normal curve, the tabled values for t (Table C in Appendix C) are **critical values,** that is, *those values that bound the critical rejection regions corresponding to varying levels of significance.* Thus in using the table for the distributions of t, we locate the appropriate number of degrees of freedom in the left-hand column and then find the column corresponding to the chosen α. The tabled values represent the t-ratio required for significance. If the absolute value of our obtained t-ratio equals or exceeds this tabled value, we may reject H_0.

14.5.2 Illustrative Problem: Student's t

Let us now examine an example involving a small sample. A group of 17 male prisoners in federal, maximum security penitentiaries was randomly selected from the federal prison population to participate in a new rehabilitation program. The 17 prisoners had a mean score on a "dangerousness to society" index (low score = not dangerous, high score = extremely dangerous) of 84 and a standard deviation of 16. Therefore, we know:

$$\overline{X} = 84, \qquad s = 16, \qquad N = 17.$$

Can we assume that the 17 prisoners who were randomly chosen to participate in the rehabilitation program are representative of a prison population in which $\mu = 78$?

Let us set up this problem in more formal statistical terms.

1. *Null hypothesis (H_0):* The mean of the population from which the sample was drawn equals 78 ($\mu = \mu_0 = 78$).

2. *Alternative hypothesis (H_1):* The mean of the population from which the sample was drawn does not equal 78 ($\mu \neq \mu_0$).

3. *Statistical test:* The Student t-ratio is chosen because we are dealing with a normally distributed variable in which σ is unknown.

4. *Significance level:* $\alpha = 0.05$.

5. *Sampling distribution:* The sampling distribution is the Student t-distribution with df $= 16$. (*Note:* df $= N - 1$.)

6. *Critical region:* $|t_{0.05}| \geq 2.12$. Since H_1 is nondirectional, the critical region consists of all values of $t \geq 2.12$ and $t \leq -2.12$. In the present example, the value of t corresponding to $\overline{X} = 84$ is

$$t = \frac{\overline{X} - \mu_0}{s_{\overline{X}}} = \frac{84 - 78}{16/\sqrt{16}} = 1.50.$$

Decision   Since the obtained t does not fall within the critical region (that is $1.50 < t_{0.05}$), we fail to reject H_0 (see Fig. 14.8). In other words, we have statistical justification to believe that the sample of 17 prisoners is (in terms of their dangerousness to society) representative of the federal prison population; therefore, the rehabilitation program can proceed with the sample of 17 prisoners. Had our decision been that the sample was not representative, a new sample would have to be drawn and tested for representativeness before continuing with the program. Why? We wish to increase the likelihood that if the program is successful, it was not successful due to a unique sample. Once the program's success is established with a representative sample of prisoners, it should be effective for the prison population.

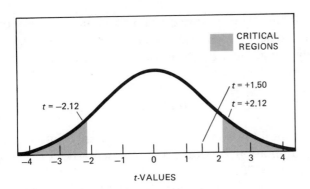

14.8 Sampling distribution of *t* and critical regions for a two-tailed test, $t_{0.05}$, df $= 16$.

14.6 Estimation of Parameters: Interval Estimation

We have repeatedly pointed out that one of the basic problems in inferential statistics is the estimation of the parameters of a population from statistics calculated from a sample. This problem, in turn, involves two subproblems: (1) *point estimation* and (2) *interval estimation*.

In Section 14.4 you learned that when we estimate parameters using single sample values, these estimates are known as point estimates. A single sample value drawn from a population provides an estimate of the population parameter. But how good an estimate is it? If a population mean were known to be 100, would a sample mean of 60 constitute a good estimate? How about a sample mean of 130, 105, 98, or 99.4? Under what conditions do we consider an estimate good? Since we know that the population parameters are virtually never known and that we generally use samples to estimate these parameters, is there any way to determine the amount of error we are likely to make? The answer to this question is "No." It is possible, however, not only to estimate the population parameter (**point estimation**), but also to state a range of values that we are confident includes or encompasses the population parameter (**interval estimation**). Moreover, we may express our confidence in terms of probabilities, and as a result social scientists generally prefer interval estimates over point estimates.

Let us say that we are to estimate the weight of a man based on physical inspection. Let us assume that we are unable to place him on a scale, and that we cannot ask him his weight. This problem is similar to many we have faced throughout this text. We cannot know the population value (the man's true weight) and hence we are forced to estimate it. Let us say that we have the impression that he weighs about 200 pounds. If we are asked, "How confident are you that he weighs *exactly* 200 pounds?," we would probably reply, "I doubt that he weighs exactly 200 pounds. If he does, you can credit me with a fantastically lucky guess. However, I feel reasonably confident that he weighs between 190 and 210 pounds." In doing this, we have stated the interval within which we feel confident that the true weight falls. After a moment's reflection, we might hedge slightly, "Well, he is almost certainly between 180 and 220 pounds. In any event, I feel perfectly confident that his weight is included in the interval between 170 and 230 pounds." Note that the greater the size of the interval, the greater is our feeling of certainty that the true value is encompassed between these limits. Note also that in stating these confidence limits, we are, in effect, making two statements: (1) We are stating the limits that include the man's weight, and (2) we are rejecting the possibility that his weight is not included in these limits. Thus if someone asks, "Is it conceivable that our subject weighs as much as 240 pounds or as little as 160 pounds?", our reply would be negative.

14.7 Confidence Intervals and Confidence Limits

In the preceding example, we were, in a sense, concerning ourselves with the problem of estimating a **confidence interval.** A confidence interval for μ specifies a range of values bounded by two endpoints (**confidence limits**) that include μ a certain percent of the time.

Let us consider a problem in which we apply the concepts of confidence interval and confidence limits. First, we will use the terms in a context of decision making concerning a hypothesis and, second, within an estimation context.

A local health systems agency is trying to decide on the feasibility of establishing a health care clinic in a particular neighborhood. In part, the decision will depend on estimates of the health care needs of the neighborhood. A limited budget does not allow assessing the needs of all the residents in the neighborhood. Consequently, we must be content with selecting a simple random sample from the neighborhood and basing our estimates on the sample. The medical team screens a simple random sample of 26 residents; each resident is assigned an overall health care needs score. We obtain the following results:

$$\overline{X} = 108, \qquad s = 15, \qquad N = 26.$$

Our best estimate of the population mean (that is, the mean health care needs score within the neighborhood) is 108. However, even though our sample statistics provide our best estimates of population values, we recognize that such estimates are subject to error. As with the weight problem, we would be very lucky if the mean of the neighborhood were actually 108. On the other hand, if we have used truly random selection procedures, we have a right to believe that our sample value is fairly close to the population mean. The critical question becomes: Between what confidence limits do we consider as likely, hypotheses concerning the value of the population mean (μ) in the neighborhood?

We have seen that the mean of the sampling distribution of sample means ($\mu_{\overline{X}}$) is equal to the mean of the population. We have also seen that since, for any given N, we may determine how far sample means are likely to deviate from any given or hypothesized value of μ, we may determine the likelihood that a particular $\overline{X}$ could have been drawn from a population with a mean of μ_0 where μ_0 represents the value of the population mean under H_0. Now, since we do not know the value of the population mean, we are free to hypothesize *any* value we desire.

It should be clear that we could entertain an unlimited number of hypotheses concerning the population mean and subsequently reject them, or fail to reject them, on the basis of the size of the t-ratios. This problem requires that we use a t-ratio due to the small sample size and unknown population parameter. Use of z values to estimate confidence intervals will be presented in the following example.

In the present problem, let us select a number of hypothetical population means. We will use the 0.05 level of significance (two-tailed test) and test the hypothesis that $\mu_0 = 98$. The value of t corresponding to $\overline{X} = 108$ is 3.333.

$$t = \frac{\overline{X} - \mu_0}{s_{\overline{X}}} = \frac{108 - 98}{15/\sqrt{25}} = \frac{10}{3} = 3.333.$$

In Table C (Appendix C), we find that $t_{0.05}$ for 25 df is 2.060. Since our obtained t is greater than this critical value, we reject the hypothesis that $\mu_0 = 98$. In other words, it is unlikely that $\overline{X} = 108$ was drawn from a population with a mean of 98.

Our next hypothesis is that $\mu_0 = 100$, which gives a t of

$$t = \frac{108 - 100}{15/\sqrt{25}} = 2.667.$$

Since $2.667 > t_{0.05}$ (or 2.060), we may reject the hypothesis that the population mean is 100.

If we hypothesize $\mu_0 = 102$, the resulting t-ratio of 2.000 is less than $t_{0.05}$. Consequently, we may consider the hypothesis that $\mu_0 = 102$ tenable. Similarly, if we obtained the appropriate t-ratios, we would find that the hypothesis $\mu_0 = 114$ is tenable, whereas hypotheses of values greater than 114 are untenable. Thus $\overline{X} = 108$ was probably drawn from a population whose mean falls in the interval 102–114 (note that these limits, 102 and 114, represent approximate limits, that is, the closest *integers*). The hypothesis that $\overline{X} = 108$ was drawn from a population with $\mu < 102$ or $\mu > 114$ may be rejected at the 0.05 level of significance. The interval within which the population mean probably lies is called the *confidence interval*. We refer to the limits of this interval as the *confidence limits*. Since we have been using $\alpha = 0.05$, we call it the *95% confidence interval*. Similarly, if we used $\alpha = 0.01$, we could obtain the *99% confidence interval*. Now that we have considered the hypothesis testing approach to confidence intervals, we will calculate the exact limits of the 95% confidence interval directly. It is not necessary to perform all of these calculations to establish the confidence limits.

To determine the upper limit for the 95% confidence interval:

$$\text{upper limit } \mu_0 = \overline{X} + t_{0.05}(s_{\overline{X}}) \qquad (14.10)$$

We know that $\overline{X} = 108$ and that $s_{\overline{X}} = 15/\sqrt{25}* = 3$. Table C shows that $t_{0.05}$ with 25 degrees of freedom equals 2.060. Multiplying the critical value of t times $s_{\overline{X}}$ (the estimated standard error) provides us with the number of units we must move to the right of the mean to encompass one-half of the 95% confidence interval. The resulting value is the upper limit of the 95% confidence interval.

Applying Eq. (14.10), we find that the upper 95% confidence limit in the preceding problem is

$$\text{upper limit } \mu_0 = 108 + (2.060)(3.0)$$
$$= 108 + 6.18 = 114.18.$$

---

* Applying Eq. (14.8).

To compute the lower limit:

$$\text{lower limit } \mu_0 = \overline{X} - t_{0.05}(s_{\overline{X}}). \qquad \textbf{(14.11)}$$

Similarly, using Eq. (14.11), we find that the lower confidence limit is

$$\text{lower limit } \mu_0 = 108 - (2.060)(3.0)$$
$$= 108 - 6.18 = 101.82.$$

Note that to compute the lower confidence limit we *subtracted* the same value (6.18) from the sample mean of 108 that we *added* to the sample mean when determining the upper limit. Hence the sample mean is equidistant from the upper and lower limits.

You should also note that Eqs. (14.10) and (14.11) are derived algebraically from Eq. (14.9):

$$t_{0.05} = \frac{\overline{X} - \mu_0}{s_{\overline{X}}}, \quad \text{therefore } \mu_0 = \overline{X} + t_{0.05}(s_{\overline{X}}).$$

Having established the lower and the upper limits as 101.82 and 114.18, respectively, we may now conclude: On the basis of our obtained mean and standard deviation, which were computed from scores drawn from a population in which the true mean is unknown, we assert that the population mean probably falls within the interval that we have established.

If we wish to compute the 99% confidence interval, it is necessary to substitute $t_{0.01}$ in the preceding equation. To compute the upper limit of the 99% confidence interval:

$$108 + (2.485)(3.0) = 108 + 7.46 = 115.46.$$

To compute the lower limit:

$$108 - (2.485)(3.0) = 108 - 7.46 = 100.54.$$

Notice that the 99% confidence interval is wider than the 95% confidence interval by comparing the confidence limits in the two examples.

Some words of caution in interpreting the confidence interval. In establishing the interval within which we believe the population mean falls, we have *not* established any probability that our obtained mean is correct. In other words, we cannot claim that the chances are 95 in 100 (or 99 in 100) that the population mean is 108. Our statements are valid only with respect to the interval and not with respect to any particular value of the sample mean. In addition, since the population mean is a fixed value and does not have a distribution, our probability statements never refer to μ. The probability we assert is about the interval, that is, the probability that the interval contains μ.

Finally, when we have established the 95% confidence interval of the mean, we are not stating that the probability is 0.95 that the particular interval we have calculated contains the population mean. It should be clear that, if we were to select repeated samples from a population, both the sample means and the sample standard deviations would differ from sample to sample. Consequently, our estimates of the confidence interval would also vary from sample to sample. When we have established the 95% confidence interval of the mean, we are stating that, if repeated samples of a given size are drawn from the population, 95% of the interval estimates will include the population mean. Figure 14.9 provides an illustration.

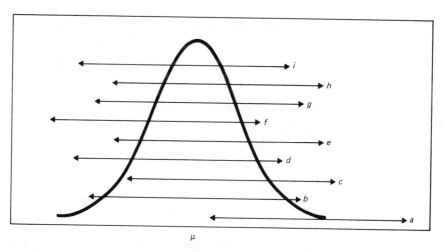

14.9 Confidence intervals based on repeated samples from a population. Interval a does not include μ.

14.7.1   Confidence Intervals and Limits for Large Samples

In many instances the sample size exceeds 50 and the z distribution may be used to compute confidence intervals for a population mean. Assume we have conducted a random sample of 600 Atlanta area adolescents, and their family income averaged $15,000 with a standard deviation of $2,500. We wish to compute the 95% confidence interval for the mean family income.

$$\overline{X} = \$15,000, \quad s = \$2500, \quad N = 600.$$

We can compute $s_{\overline{X}}$ in the following manner using Eq. (14.8):

$$s_{\overline{X}} = \frac{s}{\sqrt{N-1}} = \frac{2500}{\sqrt{599}} = \frac{2500}{24.47} = 102.17.$$

To compute the upper limit:

$$\text{upper limit } \mu_0 = \overline{X} + z(s_{\overline{X}}). \qquad (14.12)$$

To compute the lower limit:

$$\text{lower limit } \mu_0 = \overline{X} - z(s_{\overline{X}}). \qquad (14.13)$$

In Table A (Appendix C), we find that the z value which includes 47.5% of the area (95/2 = 47.5) to the right (or left) of the mean is 1.96. Substituting this value into Eqs. (14.12) and (14.13), we can now determine the upper and lower limits of the 95% confidence interval for μ, the unknown population parameter.

$$\text{upper limit} = 15,000 + 1.96(102.17) = \$15,200.25,$$
$$\text{lower limit} = 15,000 - 1.96(102.17) = \$14,799.75.$$

14.7.2 Confidence Intervals and Limits for Percentages

Confidence intervals and limits can also be computed for percentages and proportions. Suppose we drew a random sample of 200 from the list of registered voters in Wisconsin we discussed in Chapter 1 and in the introduction to this chapter in an effort to estimate the percent favoring nuclear power as an energy source. What if 56% favored nuclear energy? Hypothetically, it is possible that, if we had drawn four more samples from the list of registered voters (with replacement), the following percentages of voters would favor nuclear power as an energy source: 52%, 55%, 58%, and 57%.

Our concern is that of establishing a confidence interval around the 56% value determined by the actual survey. While it *appears* that a majority of voters prefer nuclear energy, it is possible that this is not the case. To establish the confidence interval we must first determine the **standard error of the proportion** [see Eq. (14.14)], which is an estimate of the standard deviation of the sampling distribution of proportions. Converting 56% to a proportion allows us to use Eq. (14.14) and determine the standard error of the proportion.

$$\sigma_P = \sqrt{\frac{P(1-P)}{N}} \qquad (14.14)$$

where

P = the sample proportion and
N = sample size.

Therefore

$$\sigma_P = \sqrt{\frac{0.56(0.44)}{200}} = \sqrt{\frac{0.2464}{200}} = 0.035.$$

The 95% confidence interval may now be computed by determining the upper and lower confidence limits of the proportion.

$$
\begin{aligned}
\text{Upper limit} &= P + z(\sigma_P) \\
&= 0.56 + 1.96(0.035) \\
&= 0.56 + 0.07 \\
&= 0.63.
\end{aligned}
\qquad (14.15)
$$

$$
\begin{aligned}
\text{Lower limit} &= P - z(\sigma_P) \\
&= 0.56 - 1.96(0.035) \\
&= 0.56 - 0.07 \\
&= 0.49.
\end{aligned}
\qquad (14.16)
$$

We now know the lower confidence limit to be 0.49 or 49% and the upper confidence limit to be 0.63 or 63%. The confidence interval therefore includes the percentages between 49 and 63%. We have *not* established that the percentage of all voters favoring nuclear energy is 56% or that the chances are 95 out of 100 that 56% favor nuclear energy. What we can say is that if we drew a large number of samples ($N = 200$) from the population, 95% of the interval estimates would include the actual population percentage. Had we increased the sample size, the standard error of the proportion would decrease and, therefore, the 95% confidence interval would be narrower. Examine Eq. (14.14) and demonstrate this to yourself.

14.8 Test of Significance for Pearson's *r*: One-Sample Case

In Chapter 9 we discussed the calculation of two statistics—r_s and Pearson's *r*—commonly used to describe the extent of the relationship between two variables. It will be recalled that the coefficient of correlation varies between ± 1.00, with $r = 0.00$ indicating the absence of a relationship. It is easy to overlook the fact that correlation coefficients based on sample data are only estimates of the corresponding population parameter and, as such, will distribute themselves about the population value. Thus it is quite possible that a sample drawn from a population in which the true correlation is zero may yield a high positive or negative correlation by *chance*. The null hypothesis most often investigated in the one-sample case is that the *population correlation coefficient* (ρ, pronounced "rho") is zero.

It is clear that a test of significance is called for. However, the test is complicated by the fact that the sampling distribution of ρ is usually nonnormal, particularly as ρ approaches the limiting values of ± 1.00. Consider the case in which ρ equals $+0.80$. You may determine that sample correlation coefficients drawn from this population will distribute themselves around $+0.80$ and can take on any value from -1.00 to $+1.00$. Remember, however, that there is a definite restriction in the range of values that sample statistics greater than $+0.80$ can assume, whereas there is no similar restriction for values less than $+0.80$. The result is a negatively skewed sampling distribution. The departure from normality will usually be less as the number of paired scores in the sample increases.

When the population correlation from which the sample is drawn is equal to zero, the sampling distribution is more likely to be normal. These relationships are demonstrated in Fig. 14.10, which illustrates the sampling distribution of the correlation coefficient when $\rho = -0.80, 0$, and $+0.80$.

14.8.1 Testing H_0: $\rho = 0$

When testing the null hypothesis that the population correlation coefficient is zero, the following *t*-test should be used:

$$t = \frac{r\sqrt{N - 2}}{\sqrt{1 - r^2}}. \qquad (14.17)$$

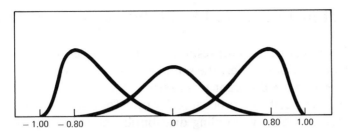

14.10 Illustrative sampling distributions of correlation
coefficients when $\rho = -0.80$ and when $\rho = +0.80$.

The number of degrees of freedom is equal to $N - 2$.

Let us look at an illustrative example. A team of social psychologists has developed a scale that purports to measure social isolation. The scores made on the scale by 15 respondents are correlated with their scores on an index revealing the degree of prejudice felt toward minority groups. They obtain a Pearson's *r* of 0.60. May they conclude that the obtained correlation is not likely to have been drawn from a population in which the true correlation is zero?

Let us set up this problem in formal statistical terms.

1. *Null hypothesis (H₀):* The population correlation coefficient from which this sample was drawn equals 0.00 ($\rho = 0.00$).

2. *Alternative hypothesis (H₁):* The population correlation coefficient from which the sample was drawn does *not* equal 0.00 ($\rho \neq 0.00$).

3. *Statistical test:* The *t*-test, with $N - 2$ degrees of freedom.

4. *Significance level:* $\alpha = 0.05$, two-tailed test.

5. *Sampling distribution:* The Student *t*-distribution with df = 13.

6. *Critical region:* $|t_{0.05}| \geq 2.160$. Since H_1 is nondirectional, the critical region consists of all values of $t \geq 2.160$ and $t \leq -2.160$.

In the present example,

$$t = \frac{0.60\sqrt{13}}{0.80} = 2.70.$$

Decision: Since the obtained $t > t_{0.05}$, it falls within the critical region for rejecting H_0. Thus it may be concluded that the sample was drawn from a population in which $\rho \neq 0.00$. Having rejected the null hypothesis also allows us to conclude that the regression coefficient also differs from 0.

14.8.2 Testing Other Null Hypotheses Concerning ρ

The t-test should not be used for testing hypotheses other than $\rho = 0.00$ since, as we pointed out earlier, the sampling distribution is not normal for values of ρ that are different from zero.

Fisher has described a procedure for transforming sample r's to a statistic z_r, which yields a sampling distribution more closely approximating the normal curve, even for samples employing small N's.

The test statistic is the normal deviate, z, in which

$$z = \frac{z_r - Z_r}{\sqrt{1/(N-3)}}, \qquad (14.18)$$

where

z_r = the transformed value of the sample r (see Table E), and
Z_r = the transformed value of the population correlation coefficient specified under H_0.

Using the same example, let us test H_0 that the population ρ from which the sample was drawn is 0.25. Recall that $r = 0.60$ and $N = 15$.

Obtaining z_r is greatly simplified by Table E, Appendix C, which shows the value of z_r, corresponding to each value of r, in steps of 0.01, between 0.00 and 0.99. Thus, referring to Table E, we see that an r of 0.60, for example, has a corresponding z_r of 0.693.

Similarly, ρ of 0.25 has a corresponding Z_r of 0.256. Substituting these values in the equation for z, we find:

$$z = \frac{0.693 - 0.256}{\sqrt{1/(15-3)}} = \frac{0.437}{0.289} = 1.51.$$

Decision: At the 0.05 level, two-tailed test, $|z_{0.05}| \geq 1.96$ is required for significance. Since $z = 1.51$ is less than the critical values, we cannot reject H_0. It is quite possible that the sample was drawn from a population in which the true correlation is 0.25.

14.8.3 Test of Significance of r_s: One-Sample Case

Table F in Appendix C, presents the critical values of r_s, one- and two-tailed tests, for selected values of N from 5 to 30. In Chapter 9, we demonstrated the calculation of r_s from data consisting of the social class backgrounds of 15 married couples. A correlation of 0.64 was computed. Since $N = 15$ is not listed in Table F, it is necessary to interpolate, using the critical values of $N = 14$ and 16. The critical value at the 0.05 level, two-tailed test, for $N = 14$ is 0.544; at $N = 16$ it is 0.506. By using linear interpolation, we may roughly approximate the critical value corresponding to $N = 15$.

With Table F, linear interpolation merely involves adding together the boundary values of r_s and dividing by 2. Thus

$$r_{s(0.05)} = \frac{0.544 + 0.506}{2} = 0.525.$$

Since our obtained r_s of 0.64 exceeds the critical value at the 0.05 level, we may conclude that the population value of the Spearman correlation coefficient from which the sample was drawn is greater than 0.00.

14.9 Test of Significance for Goodman's and Kruskal's Gamma

Gamma was discussed in Chapter 8 using an example in which the relationship between political liberalism and year in college was explored. Gamma was found to equal $+0.465$. Overall, there were 1167 concordant pairs and 426 discordant pairs. Let us test the hypothesis that in the population from which the sample was drawn, γ (the Greek letter symbolizing the parameter) is greater than 0. Therefore, our null hypothesis is $H_0: \gamma \leq 0$ and we will test it at the 0.01 level of significance with a one-tailed test using the following equation, which allows us to convert G to a z score:

$$z = (G - \gamma) \sqrt{\frac{C - D}{N(1 - G^2)}}$$

where

$$
\begin{aligned}
G &= 0.465, \\
C &= 1167, \\
D &= 426, \\
N &= 80, \text{ and} \\
\gamma &= 0 \text{ because } H_0: \gamma \leq 0.
\end{aligned}
$$

Therefore

$$z = (0.465 - 0) \sqrt{\frac{1167 - 426}{80(1 - 0.216)}}$$

$$= (0.465) \sqrt{\frac{741}{62.72}}$$

$$= (0.465)(3.44)$$

$$= +1.60.$$

The computed z score of 1.60 does not fall into the critical region that begins at $z = 2.33$ (see Table A in Appendix C); therefore, we must accept the null hypothesis that the association between political liberalism and year in college in the population from which the 80 students was drawn equals 0.

Summary

We have seen that if we take a number of samples from a given population, then:

1. The distribution of sample means tends to be normal.
2. The mean of these sample means ($\mu_{\bar{x}}$) is equal to the mean of the population (μ).
3. The standard error of the mean ($\sigma_{\bar{x}}$) is equal to $\sigma/\sqrt{N}$. As N increases, the variability decreases.

We used these relationships in the testing of hypotheses (for example, $\mu = \mu_0$), when the standard deviation of a population was known, using the z statistic and the standard normal curve.

When σ is not known, we demonstrated the use of sample statistics to estimate these parameters. We used these estimates of the parameters to test hypotheses, using the Student t-ratio and the corresponding sampling distributions. We compared these t-distributions, which vary as a function of degrees of freedom (df), with the standard normal curve.

We used the t-ratio and the normal distribution as a basis for establishing confidence intervals for means and proportions.

Finally, we demonstrated the test of significance for Pearson's r, Spearman's r_s, one-sample case, and for gamma.

Terms to Remember

Central limit theorem If random samples of a fixed N are drawn from *any* population (regardless of the form of the population distribution), as N becomes larger, the distribution of sample means approaches normality, with the overall mean approaching μ, the variance of the sample means $\sigma_{\bar{X}}^2$ being equal to σ^2/N, and a standard error $\sigma_{\bar{X}}$ of $\sigma/\sqrt{N}$.

Confidence interval A confidence interval for a parameter specifies a range of values bounded by two endpoints called confidence limits. Common confidence intervals are 95% and 99%.

Confidence limits The two endpoints of a confidence interval.

Critical region That portion of the area under a curve that includes those values of a statistic that lead to rejection of the null hypothesis.

Critical values of t Those values that bound the critical rejection regions corresponding to varying levels of significance.

Degrees of freedom (df) The number of values that are free to vary after we have placed certain restrictions upon our data.

Interval estimation The determination of an interval within which the population parameter is presumed to fall. (Contrast *Point estimation.*)

Law of large numbers As the sample size increases, the standard error of the mean decreases.

Point estimation An estimate of a population parameter that involves a single sample value, selected by the criterion of "best estimate." (Contrast *Interval estimation.*)

Robust test A statistical test from which statistical inferences are likely to be valid even when there are fairly large departures from normality in the population distribution.

Standard error of the mean A theoretical standard deviation of sample means, of a given sample size, drawn from some specified population. When based upon a known population standard deviation, $\sigma_{\bar{X}} = \sigma/\sqrt{N}$; when estimated from a single sample, $s_{\bar{X}} = s/\sqrt{N-1}$.

Standard error of the proportion   An estimate of the standard deviation of the sampling distribution of proportions.

t-distributions   Theoretical symmetrical sampling distributions with a mean of zero and a standard deviation that becomes smaller as degrees of freedom (df) increase. Used in relation to the Student t-ratio.

t-ratio   A test statistic for determining the significance of a difference between means (two-sample case) or for testing the hypothesis that a given sample mean was drawn from a population with the mean specified under the null hypothesis (one-sample case). Used when population standard deviation (or standard deviations) is not known.

Unbiased estimate of a parameter   An estimate that equals, on the average, the value of the parameter.

Exercises

1. Describe what happens to the distribution of sample means when you
    a) Increase the size of each sample.
    b) Increase the number of samples.
2. Explain why the standard deviation of a sample will usually underestimate the standard deviation of a population. Give an example.
3. Given that $\overline{X} = 24$ and $s = 4$ for $N = 15$, use the t-distribution to find
    a) The 95% confidence limits for μ.
    b) The 99% confidence limits for μ.
4. Given that $\overline{X} = 24$ and $s = 4$ for $N = 121$, use the t-distribution to find
    a) The 95% confidence limits for μ.
    b) The 99% confidence limits for μ.
    Compare the results with Exercise 3.
5. An instructor gives his class an examination which, as he knows from years of experience, yields $\mu = 78$ and $\sigma = 7$. His present class of 22 obtains a mean of 82. Is he correct in assuming that this is a superior class? Use $\alpha = 0.01$, two-tailed test.
6. An instructor gives his class an examination which, as he knows from years of experience, yields $\mu = 78$. His present class of 22 obtains $\overline{X} =$

82 and $s = 7$. Is he correct in assuming that this is a superior class? Use $\alpha = 0.01$, two-tailed test.

7. Explain the difference between Exercises 5 and 6. What test statistic is used in each case, and why? Why is the decision different?

   Generalize: What is the effect of knowing σ upon the likelihood of a type II error?

8. The Superintendent of Zody school district claims that the children in her district are brighter, on the average, than the general population of students. In order to determine the aptitude of school children in the district, a study was conducted. The results were as follows.

TEST SCORES
105
109
115
112
124
115
103
110
125
99

   The mean of the general population of school children is 106. Set this up in formal statistical terms (that is, H_0, H_1, etc.) and draw the appropriate conclusions. Use a one-tailed test, $\alpha = 0.05$.

9. For a particular population with $\mu = 28.5$ and $\sigma = 5.5$, what is the probability that, in a sample of 100, $\overline{X}$ will be:

   a) Equal to or less than 30.0?   **b)** Equal to or less than 28.0?

   c) Equal to or more than 29.5?   **d)** Between 28.0 and 29.0?

10. Given that $\overline{X} = 40$ for $N = 24$ from a population in which $\sigma = 8$, find

    a) The 95% confidence limits for μ.

    b) The 99% confidence limits for μ.

11. Overton University claims that, because of its superior facilities and close faculty supervision, its students complete the Ph.D. program earlier than usual. They base this assertion on the fact that the national mean age for completion is 32.11, whereas the mean age of their 26 Ph.D.'s is 29.61 with $s = 6.00$. Test the validity of their assumption.

12. Using the data in the preceding exercise, find the interval within which you are confident that the true population mean (average age for

Ph.D.'s at Overton University) probably falls, using the 95% confidence interval.

13. A sociologist asserts that the average length of courtship is longer before a second marriage than before a first. He bases this assertion on the fact that the average for first marriages is 265 days, whereas the average for second marriages (for example, his 626 respondents) is 268.5 days, with $s = 50$. Test the validity of his assumption.

14. Using the data in the preceding exercise, find the interval within which you are confident that the true population mean (average courtship days for a second marriage) probably falls, using the 99% confidence interval.

15. Random samples of size 2 are selected from the following finite population of scores: 1, 3, 5, 7, 9, and 11.

    a) Calculate the mean and standard deviation of the population.

    b) Construct a histogram showing the sampling distribution of means when $N = 2$. Sample *without* replacement.

    c) Construct a histogram showing the means of all possible samples that can be drawn. Sample *with* replacement.

16. Given (b) in Exercise 15, answer the following: Selecting a sample with $N = 2$, what is the probability that:

    a) A mean as high as 10 will be obtained?

    b) A mean as low as 2 will be obtained?

    c) A mean as deviant as 8 will be obtained?

    d) A mean as low as 1 will be obtained?

17. Given (c) in Exercise 15, answer the following: Selecting a sample with $N = 2$, what is the probability that:

    a) A mean as high as 10 will be obtained?

    b) A mean as low as 2 will be obtained?

    c) A mean as deviant as 8 will be obtained?

    d) A mean as low as 1 will be obtained?

18. A census field supervisor ranked 25 of her workers on the length of their employment with the Census Bureau and their rate of interview completion. She correlated these ranks and obtained an $r_s = -0.397$. Assuming $\alpha = 0.05$, two-tailed test, what do you conclude?

19. As a requirement for admission to Blue Chip University, a candidate must take a standardized entrance examination. The correlation between performance on this examination and college grades is 0.43.

a) The director of admissions claims that a better way to predict college success is by using high school averages. To test her claim, she randomly selects 52 students and correlates their college grades with their high school averages. She obtains $r = 0.54$. What do you conclude?

b) The director's assistant constructs a test which he claims is better for predicting college success than the one currently used. He randomly selects 67 students and correlates their grade point averages with performance on his test. The obtained $r = 0.61$. What do you conclude?

20. In 1978 there were 5,206,000 families in the United States that were headed by a woman with no husband present. Of these families, 42.8% were headed by a divorced woman. If the standard error of the percentage is 0.8%, what is the 68% confidence interval? What is the 95% confidence interval? (*Source:* U.S. Bureau of the Census, *Divorce, Child Custody and Child Support,* Current Population Reports, Series P-23, No. 84, June 1979.

21. You have conducted a survey of 420 students, and 62% have indicated that they are satisfied with the quality of the instruction they are receiving. Compute the standard error of the proportion and the 95% confidence interval. What are the confidence limits? Interpret your answer.

22. What are the statistics used to describe the distribution of a sample? The distribution of a sample statistic?

23. Is s^2 an unbiased estimate of σ^2? Why?

24. Is s^2 an unbiased estimate of σ^2? Why?

25. What is a confidence interval? A confidence limit? What is their relationship?

26. Give an example to show the effect of the α level on the precision of a confidence interval.

27. How do the t-distributions differ from the normal distribution? Are they ever the same?

28. In a study of marital success and failure, Bentler and Newcomb (1978) gave a personality questionnaire to 162 newly married couples. Four years later 77 couples from the original sample were located; of these, 53 were still married, while 24 had separated or divorced. Among the still-married group it was found that the correlation between the husband's and wife's score on an attractiveness scale was 0.59, and that the correlation between their scores on a generosity scale was 0.23. Among the divorced group the correlation between the ex-husband's and ex-wife's score on the attractiveness scale was 0.07, and the correlation between their scores on the generosity scale was 0.13.

a) Using $\alpha = 0.05$, set up and test the null hypothesis that the correlation between the married couples' scores on attractiveness is equal to zero. Also set up and test the null hypothesis that the correlation between the divorced couples' scores on attractiveness is equal to zero.

b) Using $\alpha = 0.05$, set up and test the null hypothesis that the correlation between the married couples' scores on generosity is equal to zero. Also set up and test the null hypothesis that the correlation between the divorced couples' scores on generosity is equal to zero.

29. Compute gamma for Exercise 1 in Chapter 8 and determine if it is significant at the 0.01 level assuming a one-tailed test.

30. Compute r_s for Exercise 17 in Chapter 9 and determine if it is significant at the 0.05 level assuming a one-tailed test.

31. Assume that the mean number of religious conversions a year per Methodist church is 50. A group of 16 Methodist ministers who attended a special seminar for increasing conversions show the following number of conversions in their churches during the year following the seminar:

NUMBER OF CONVERSIONS	
55	85
50	60
45	50
75	65
80	60
75	50
80	80
80	50

Calculate the value of t. Can you conclude that the seminar produced ministers with superior abilities to attract converts?

32. A Los Angeles criminologist found a r_s of 0.15 between the number of times 20 male prostitutes had been arrested and the number of years they had worked as prostitutes. Can she conclude that this correlation is significant?

33. Mr. Smith stated that his training program for selling life insurance enables a company to sell more life insurance than the "average" com-

pany. The mean amount of life insurance sold by all salesmen per month is $100,000. A sample of ten people who have been through the training program show monthly selling rates (in thousands) of:

SELLING RATES	
$100	90
120	130
130	135
120	140
125	110

If you were a supervisor of insurance salesmen, would you adopt Mr. Smith's training program? Calculate the value of t.

Statistical Inference with Two Independent Samples and Correlated Samples

15

15.1 Sampling Distribution of the Difference between Means

15.2 Estimation of $\sigma_{\bar{X}_1 - \bar{X}_2}$ from Sample Data

15.3 Testing Statistical Hypotheses: Student's t

15.4 The t-Ratio and Homogeneity of Variance

15.5 Statistical Inference with Correlated Samples

15.1 Sampling Distribution of the Difference between Means

Does the recidivism rate of juvenile offenders who are provided with "father figures" differ from those without "father figures"? Does life satisfaction differ for those under 60 and those over 60 years of age? Is there a difference in the achievement orientation of girls and boys?

Each of these problems involves the comparison of at least two samples or groups that were drawn independently. In the typical survey research study the assumption of sample independence is not a problem because the *overall* sample is randomly drawn. From within the overall sample we may wish to compare two subgroups or samples such as respondents over 60 and respondents under 60 years of age. The subgroup of respondents over 60 and the subgroup of respondents under 60 (or any other comparison groups included in the survey) were necessarily drawn independently because the overall sample was randomly drawn. That is, the random selection of respondents under 60 was independent of the random selection of respondents over 60.

Thus far we have restricted our examination of hypothesis testing to one-sample cases. However, most social science research involves the comparison of two or more samples or subgroups with respect to a specific variable (for example, age differences, life satisfaction differences) to determine whether or not these samples might have reason-

371

ably been drawn from the same population. If the means of two samples differ, must we conclude that these samples were drawn from two different populations?

Recall our previous discussion on the sampling distribution of sample means (Section 14.2). We saw that some variability in the sample statistics is to be expected, even when these samples are drawn from the same population. We were able to describe this variability in terms of the sampling distribution of sample means. In the two-sample case, imagine drawing *pairs of samples,* finding the difference between the means of each pair, and obtaining a distribution of these differences. The resulting distribution would be the *sampling distribution of the difference between means.*

To illustrate, imagine that from the population described in Table 14.1, in which $\mu = 5.00$ and $\sigma = 0.99$, we randomly draw (with replacement) two samples at a time. For illustrative purposes, let us draw two cases for the first sample (that is, $N_1 = 2$), and three cases for the second sample (that is, $N_2 = 3$). For example, we might draw scores of 5, 6 for our first sample and scores of 4, 4, 7 for our second sample. Thus, since $\overline{X}_1 = 5.5$ and $\overline{X}_2 = 5.0$, $\overline{X}_1 - \overline{X}_2 = 0.5$. Now suppose we continue to draw samples of $N_1 = 2$ and $N_2 = 3$ until we otain an indefinitely large number of pairs of samples. If we calculate the differences between these pairs of sample means and treat each of these differences as a raw score, we may set up a frequency distribution of these differences. What might we expect this distribution to look like? Since we are selecting pairs of samples at random from the *same* population, we would expect a normal distribution with a mean of zero.

Going one step further, we may describe the distribution of the difference between pairs of sample means, even when these samples are *not* drawn from the same population. It will be a normal distribution with a mean ($\mu_{\overline{X}_1 - \overline{X}_2}$) equal to $\mu_1 - \mu_2$ and a standard deviation ($\sigma_{\overline{X}_1 - \overline{X}_2}$, referred to as the **standard error of the difference between means**) equal to $\sqrt{\sigma_{\overline{X}_1}^2 + \sigma_{\overline{X}_2}^2}$.*

Thus the sample distribution of the statistic

$$z = \frac{(\overline{X}_1 - \overline{X}_2) - (\mu_1 - \mu_2)}{\sigma_{\overline{X}_1 - \overline{X}_2}}$$

is normal and therefore permits us to use the standard normal curve in the testing of hypotheses.

---

* The standard error of the difference between means is interpreted in the same manner as the standard deviation and the standard error of the mean.

15.2 Estimation of $\sigma_{\bar{X}_1 - \bar{X}_2}$ from Sample Data

The statistic z is used only when the population standard deviations are known or the sample size exceeds 50. Since it is rare that the population standard deviations are known we must estimate the standard error. Thus we are interested in the estimated $\sigma_{\bar{X}_1 - \bar{X}_2}$.

Historically, the estimated standard error of the difference between means was defined as follows:

$$s_{\bar{X}_1 - \bar{X}_2} = \sqrt{s_{\bar{X}_1}^2 + s_{\bar{X}_2}^2}.$$

However, this equation provides a biased estimate whenever N_1 is not equal to N_2.

If we randomly select a sample of N_1 observations from a population with unknown variance and a second random sample of N_2 observations also from a population with unknown variance, the following is an unbiased estimate of the standard error of the difference. This estimate, which assumes that the two samples are drawn from a population with the same variance, pools the sum of squares and degrees of freedom of the two samples to obtain a pooled estimate of the standard error of the difference. Hence

$$s_{\bar{X}_1 - \bar{X}_2} = \sqrt{\left(\frac{SS_1 + SS_2}{N_1 + N_2 - 2}\right)\left(\frac{1}{N_1} + \frac{1}{N_2}\right)}. \qquad (15.1)$$

However, if $N_1 = N_2 = N$, Eq. (15.1) is simplified to

$$s_{\bar{X}_1 - \bar{X}_2} = \sqrt{\frac{SS_1 + SS_2}{N(N - 1)}}, \qquad (15.2)$$

where

$$SS_1 = \sum X_1^2 - \frac{(\sum X_1)^2}{N_1},$$

and

$$SS_2 = \sum X_2^2 - \frac{(\sum X_2)^2}{N_2}.$$

15.3 Testing Statistical Hypotheses: Student's *t*

The statistic used in the testing of hypotheses, when population standard deviations are not known, is the familiar *t*-ratio

$$t = \frac{(\overline{X}_1 - \overline{X}_2) - (\mu_1 - \mu_2)}{s_{\overline{X}_1 - \overline{X}_2}}, \qquad (15.3)$$

in which $(\mu_1 - \mu_2)$ is the expected value as stated in the null hypothesis.

The most common null hypothesis tested is that both samples come from the same population of means, that is, $(\mu_1 - \mu_2) = 0$. However, there are times when the null hypothesis may specify a difference between population means. As an example, let us imagine that we know (or have a reasonable estimate of) the difference in the number of months that white and nonwhite felons convicted of armed robbery are sentenced by a judge. Imagine that we have reason to believe that the judge is prejudiced. In this case the null hypothesis would specify the known differences in sentences. Thus if the sentences of nonwhites are known to be five months longer than the sentences of whites on the average, the null hypothesis would read:

$$H_0: \mu_{\text{nonwhite}} - \mu_{\text{white}} = 5 \text{ months.}$$

You will recall that Table C, Appendix C, provides the critical values of t required for significance at various levels of α. Since the degrees of freedom for each sample is $N_1 - 1$ and $N_2 - 1$, the total df in the two-sample case is $N_1 + N_2 - 2$.

15.3.1 Illustrative Problem: Student's t

In Chapter 1 we described an experiment in which we sought to assess the effectiveness of two teaching techniques: the use of videotapes versus the traditional lecture method. Nine students in group 1 (experimental group) viewed a political scientist's videotaped lectures, and ten students in group 2 (control group) were lectured in person by the professor. We wish to determine which technique proved most effective as measured by the final examination scores (low score = poor performance). Let us set up the problem in formal statistical terms. The results of the experiment are shown in Table 15.1.

1. *Null hypothesis (H₀):* There is no difference between the population means of the students taught by videotape (experimental group) and the students taught by the traditional lecture method (control group), or the mean of the experimental group will exceed that of the control group; that is, $\mu_1 \geq \mu_2$.*

---

* Recall from Section 13.5 that whenever the alternative hypothesis is directional, the null hypothesis is also stated in directional terms.

Table 15.1 Scores of two groups on the final examination

GROUP 1 EXPERIMENTAL		GROUP 2 CONTROL	
X_1	X_1^2	X_2	X_2^2
12	144	21	441
14	196	18	324
10	100	14	196
8	64	20	400
16	256	11	121
5	25	19	361
3	9	8	64
9	81	12	144
11	121	13	169
		15	225
$\Sigma 88$	996	$\Sigma 151$	2445
$N_1 = 9$		$N_2 = 10$	
$\overline{X}_1 = 9.778$		$\overline{X}_2 = 15.100$	

2. *Alternative hypothesis (H₁):* There is a difference between the population means of the two groups and the mean of group 2 (control group) will exceed the mean of group 1 (experimental group), that is, $\mu_1 < \mu_2$. Thus we have a one-tailed test.

3. *Statistical test:* Since we are comparing two sample means presumed to be drawn from normally distributed populations with equal variances, the Student's *t*-ratio two-sample case is appropriate.

4. *Significance level:* $\alpha = 0.025$.

5. *Sampling distribution:* The sampling distribution is the Student's *t*-distribution with df $= N_1 + N_2 - 2$, or $9 + 10 - 2 = 17$.

6. *Critical region:* $|t| \geq 2.110$.

Since $N_1 \neq N_2$ and the population variances are assumed to be equal, we shall use Eq. (15.1) to estimate the standard error of the difference between means. The sum of squares for group 1 is

$$SS_1 = 996 - \frac{(88)^2}{9} = 135.56.$$

Similarly, the sum of squares for group 2 is

$$SS_2 = 2445 - \frac{(151)^2}{10} = 164.90.$$

The value of t in the present problem is

$$t = \frac{(\overline{X}_1 - \overline{X}_2) - (\mu_1 - \mu_2)}{\sqrt{\left(\frac{SS_1 + SS_2}{N_1 + N_2 - 2}\right)\left(\frac{1}{N_1} + \frac{1}{N_2}\right)}}$$

$$= \frac{(9.778 - 15.100) - 0}{\sqrt{\left(\frac{135.56 + 164.90}{17}\right)\left(\frac{1}{9} + \frac{1}{10}\right)}} = \frac{-5.322}{1.93} = -2.758.$$

Decision   Since the obtained t falls within the critical region (that is, $|-2.758| > 2.110$) we reject H_0. The negative t-ratio means that the mean for group 2 is greater than the mean for group 1. When referring to Table C, we ignore the sign of the obtained t-ratio. From these findings we would conclude that the traditional lecture method was more effective, since the mean of the group instructed by videotape (the experimental group) was significantly less than that of the group instructed by the traditional lecture method (the control group).

15.4 The t-Ratio and Homogeneity of Variance

The assumptions underlying the use of the t-distributions are as follows:

1. The sampling distribution of the difference between means is normally distributed.
2. Estimated $\sigma_{\overline{X}_1 - \overline{X}_2}$ (that is, $s_{\overline{X}_1 - \overline{X}_2}$) is based on the unbiased estimate of the population variance.
3. Both samples are drawn from populations whose variances are equal. This assumption is referred to as **homogeneity of variance.**

Occasionally, for reasons which may not be very clear, the scores of one group may be far more widely distributed than the scores of another group. This may indicate that we are sampling two different distributions, but the critical question becomes: Two different distributions of what? . . . means or variances?

To determine whether or not two variances differ significantly from one another, we must make reference to yet another distribution: the **F-distribution.** Named after R. A. Fisher, the statistician who first described it, the F-distribution is unlike any other we have encountered in the text; it is *tridimensional* in nature. To use the F Table (Table D_1, in Appendix C), we must begin at the entry stating the number of degrees of freedom of the group with the smaller variance, and move down the column until we find the entry for the number of degrees of freedom of the group with the larger variance. At that point, we will find the critical value of F required for rejecting the null hypothesis of no difference in variances. The F, itself, is defined as follows:

$$F = \frac{s^2 \text{ (larger variance)}}{s^2 \text{ (smaller variance)}} . \qquad (15.4)$$

The test is two-tailed, as we are interested in determining if either variance differs significantly from the other. It would be one-tailed *only* if we were interested in determining whether a specific variance is significantly greater than another. This is an unlikely comparison. Since Table D_1 is one-tailed, the use of the 0.025 critical value yields the 0.05 two-tailed significance level. In the preceding sample problem, the variance for group 1 is 135.56/8 or 16.94; and for group 2, 164.9/9 or 18.32. The F-ratio becomes

$$F = \frac{18.32}{16.94} = 1.08, \qquad df = 9/8.$$

Referring to Table D_1 under 9 and 8 df (df for groups 2 and 1, respectively), we find that an $F \geq 4.36$ is significant at the 0.05 level. We may therefore conclude that it is reasonable to assume that both samples were drawn from a population with the same variances.

What if we found a significant difference in variances? Would it have increased our likelihood of rejecting the null hypothesis of no difference between means? Very little, in all probability: as we have already pointed out, the *t*-ratio is a robust test. Our conclusions are not likely to be altered by any but extremely large departures from such assumptions as homogeneity of variances and normality.

Why, then, do we concern ourselves with an analysis of the variances? Frequently a significant difference in variances (particularly when the variance of the experimental group is significantly greater than that of the control group) is indicative of a *dual* effect of the experimental conditions. A larger variance indicates more extreme scores *at both ends* of a distribution. The alert researcher will seize upon these facts as a basis for probing into the possibility of dual effects.

15.5 Statistical Inference with Correlated Samples

When an experiment is conducted, data comparing two or more groups are obtained, a difference in some measure of central tendency is found, and then we raise the question: Is the difference of such magnitude that it is unlikely to be due to chance factors? As we have seen, a visual inspection of the data is not usually sufficient to answer this question. In many instances the experimental subjects themselves have widely varying aptitudes or proficiencies relative to the dependent or criterion measure. For example, college seniors in general are more proficient at test taking than are college freshmen due to differences in experience.

In an experiment, the score of any subject on the dependent variable may be thought to reflect at least three factors: (1) the subject's ability at the dependent task, (2) the effects of the independent or experimental variable, and (3) random error due to a wide variety of different causes; for example, minor variations from time to time in experimental procedures or conditions, momentary fluctuations in such things as attention span, motivation of the subjects, etc. There is little we can do about *random error* except to maintain as close control over experimental conditions as possible. The effects of the independent *variable* are, of course, what we are interested in assessing. In most studies the *individual differences among subjects* is, by and large, the most significant factor contributing to the scores and the variability of scores on the dependent variable. Anything we can do to take individual differences into account or statistically remove their effects will improve our ability to estimate the effects of the independent or experimental variable on the dependent variable scores. In essence, we want to isolate the effects of the independent variable by eliminating all other possible factors that might influence the score on the dependent variable. Up to this point in the chapter we have dealt only with independent samples; however, the use of non-independent or correlated samples is frequently necessary when conducting small-group research.

Two research situations are of particular interest:

Before-after design   A measurement of the *same* subjects is taken both before and after the introduction of the independent variable. It is presumed that each individual will remain relatively consistent. Thus there will be a correlation between the *before* sample and the *after* sample. Note that each subject may have been *selected* at random from a subject pool but, since he or she participates in both conditions, we cannot say that the subjects have been *assigned* randomly to experimental conditions.

A variation on the before-after design frequently used in survey research is the **panel design.** Respondents are interviewed at time 1 on a series of topics and the same respondents reinterviewed at a later time (time 2). In many instances the intial response to specific issues at time 1 will be very similar (and hence correlated) with the response at time 2.

Matched-group design The individuals available for participation in the study are matched on some variable known to be correlated with the dependent variable. The result is a set of paired individuals in which each member of a given pair obtains approximately the same score on the matching variable. Then one member of each pair is randomly assigned to the experimental condition and the other is assigned to the control group. Thus, if we wanted to determine the effect of a counseling program for drug addicts, we might match the participants on length of addiction, type of addiction, job skills, etc. Such a design has two advantages:

1. It ensures that the experimental groups are equivalent in initial ability.
2. It permits us to take advantage of the correlation based on initial ability and allows us in effect to remove one source of error from our measurements.

15.5.1 The Direct Difference Method: Student's *t*-Ratio

The direct difference method is a statistical technique used with correlated samples that allows for the direct calculation of the **standard error of the mean difference** ($s_{\overline{D}}$) which should not be confused with the standard error of the difference between means, which was discussed in Section 15.1.

In brief, the direct difference method consists of finding the differences between the dependent variable scores obtained by each pair of matched individuals and treating these differences as if they were raw scores. In effect, the direct–difference method transforms a two-sample into a one-sample case. We find one sample mean and a standard error based on one standard deviation. The null hypothesis is that the obtained mean of the difference scores ($\Sigma D/N$, symbolized as $\overline{D}$) comes from a population in which the mean difference ($\mu_{\overline{D}}$) is some specified value. The *t*-ratio employed to test H_0: $\mu_{\overline{D}} = 0$ is

$$t = \frac{\overline{D} - \mu_{\overline{D}}}{s_{\overline{D}}} = \frac{\overline{D}}{s_{\overline{D}}} \qquad (15.5)$$

where

$$s_{\overline{D}} = \sqrt{\frac{\Sigma d^2}{N(N - 1)}}. \qquad (15.6)$$

The raw score formula for calculating the sum of squares of the difference score is

$$\sum d^2 = \sum D^2 - \frac{(\Sigma D)^2}{N}, \qquad (15.7)$$

where D is the difference between paired scores, and d is the deviation of a difference score (D) from $\overline{D}$.

15.5.2 Sample Problem

Male police officers have traditionally worked with male partners; however, as more women become officers, the likelihood of having a partner of the opposite sex is increased. Traditionally, male officers have resisted the hiring of female officers and have also resisted working with a female partner. The police chief of a medium-sized metropolitan area was well aware of the biases on the part of male officers and assessed ten randomly selected male officers' attitudes toward having a female partner. These officers then worked with a female partner for a six-month period, at the end of which their attitudes toward female partners were reassessed. Their attitudes before and after having worked with a female partner are summarized in Table 15.2. It is anticipated that more favorable attitudes (that is, higher scores) will result from working with the female officers.

Let us set up this problem in formal statistical terms.

1. *Null hypothesis (H₀):* The difference in the attitudes of the male police officers before and after having worked with a female partner is equal to or greater than zero, that is $\mu_{\overline{D}} \geq 0$.

2. *Alternative hypothesis (H₁):* The attitudes of the male police officers will be more favorable after having worked with a female partner, that is, $\mu_{\overline{D}} < 0$. Note that our alternative hypothesis is directional; consequently, a one-tailed test of significance will be used.

3. *Statistical test:* Since we are using a panel or before-after design, the Student's *t*-ratio for correlated samples is appropriate.
4. *Significance level:* $\alpha = 0.01$.
5. *Sampling distribution:* The sampling distribution is the Student's *t*-distribution with df $= N - 1$, or $10 - 1 = 9$.
6. *Critical region:* $t_{0.01} \leq -2.821$. Since H_1 predicts that the scores in the *after* condition will be higher than those in the *before* condition, we expect the difference scores to be *negative*. Therefore, the critical region consists of all values of $t \leq -2.821$.

Table 15.2 presents the results of this study.

Table 15.2 Scores of ten police officers in a study using a panel design (hypothetical data)

Police Officer	BEFORE X_1	AFTER X_2	DIFFERENCE D	D^2
1	25	28	-3	9
2	23	19	4	16
3	30	34	-4	16
4	7	10	-3	9
5	3	6	-3	9
6	22	26	-4	16
7	12	13	-1	1
8	30	47	-17	289
9	5	16	-11	121
10	14	9	5	25
Σ	171	208	-37	511

The following steps are necessary in the direct difference method.

Step 1 The sum of squares of the difference scores applying Eq. (15.7) is

$$\sum d^2 = 511 - \frac{(-37)^2}{10} = 374.10.$$

Note: The after score is subtracted from the before score.

Step 2 The standard error of the mean difference applying Eq. (15.6) is

$$s_{\bar{D}} = \sqrt{374.10/10(9)} = 2.04.$$

Step 3 The value of $\bar{D} = \Sigma D/N = -37/10 = -3.70$. (To check the accuracy of ΣD we subtract ΣX_2 from ΣX_1, that is, $\Sigma X_1 - \Sigma X_2 = \Sigma D$, $171 - 208 = -37$.)

Step 4 The value of t in the present problem is

$$t = \frac{\overline{D}}{s_{\overline{D}}} = -\frac{3.70}{2.04} = -1.81.$$

Decision   Since the obtained t does not fall within the critical region (that is, $-1.81 > t_{0.01}$), we fail to reject H_0. Thus we must conclude that the male police officers did not significantly change their attitudes toward female partners.* Note, however, that eight of the ten officers became more favorable (see before and after columns of Table 15.2) after their experience with a female partner and that a larger sample size, given the same pattern of attitude change, would have resulted in rejection of the null hypothesis.

Summary

We have seen that if we take a number of pairs of samples either from the same population or from two different populations, then:

1. The distribution of differences between pairs of sample means tends to be normal.

2. The mean of these differences between means ($\mu_{\overline{X}_1 - \overline{X}_2}$) is equal to the difference between the population means, that is, $\mu_1 - \mu_2$.

3. The standard error of the difference between means ($\sigma_{\overline{X}_1 - \overline{X}_2}$) is equal to

$$\sqrt{\sigma_{\overline{X}_1}^2 + \sigma_{\overline{X}_2}^2}.$$

We presented equations for estimating $\sigma_{\overline{X}_1 - \overline{X}_2}$ from sample data. Using estimated $\sigma_{\overline{X}_1 - \overline{X}_2}$ (that is, $s_{\overline{X}_1 - \overline{X}_2}$), we demonstrated the use of Student's t to test hypotheses in the two-sample case.

An important assumption underlying the use of the t-distribution is that both samples are drawn from populations with equal variances. Although failure to find homogeneity of variance will probably not seriously affect our interpretations, the fact of heterogeneity of variance may have important theoretical implications.

We demonstrated the use of the direct difference method for determining the significance of the difference between the means of correlated samples.

---

* Statistical significance is discussed in Section 13.7.

Terms to Remember

Before-after design   A correlated-samples design in which each individual is measured on the dependent variable both before and after the introduction of the experimental conditions.

F-ratio   A ratio between sample variances.

Homogeneity of variance   The condition that exists when two or more sample variances have been drawn from populations with equal variances.

Matched-group design   A correlated-samples design in which pairs of individuals are matched on a variable correlated with the criterion measure. Each member of a pair receives different experimental conditions.

Panel design   A research design in which a sample is drawn and the same respondents are interviewed and then reinterviewed at a later time.

Standard error of the difference between means   Standard deviation of the sampling distribution of the difference between means

$$\sqrt{\sigma_{\overline{X}_1}^2 + \sigma_{\overline{X}_2}^2}.$$

Standard error of the mean difference   Standard error based upon a correlated-samples design. Used in the Student's t-ratio for correlated samples $(s_{\overline{D}})$.

Exercises

1. Two statistics classes of 25 students each obtained the following results on the final examination: $\overline{X}_1 = 82$, $SS_1 = 384.16$; $\overline{X}_2 = 77$, $SS_2 = 1536.64$. Test the hypothesis that the two classes are equal in ability, using $\alpha = 0.01$.

2. In a study of the effect of including a dollar with mailed questionnaires on the response rate, the following results were obtained:

GROUP RECEIVING DOLLAR	GROUP RECEIVING NO MONEY
$\Sigma X_1 = 324$	$\Sigma X_2 = 256$
$\Sigma X_1^2 = 6516$	$\Sigma X_2^2 = 4352$
$N_1 = 18$	$N_2 = 16$

Set up this study in formal terms, using $\alpha = 0.05$, and draw the appropriate conclusions concerning the effect of a dollar on the response rate.

3. A comparison of the number of visits to a dentist over the past five years by a group of upper-class children and a group of middle-class children is presented as follows.

UPPER CLASS CHILDREN	MIDDLE CLASS CHILDREN
9	6
6	7
8	7
8	9
9	8

Set up this comparison in formal statistical terms, using $\alpha = 0.05$, and draw the appropriate conclusions.

4. A study was undertaken to determine whether or not the acquisition of a response is influenced by a drug. The dependent variable was the number of trials required to master the task (X_1 is the experimental group and X_2 is the control group).

X_1	6	8	14	9	10	4	7	
X_2	4	5	3	7	4	2	1	3

a) Set up this study in formal statistical terms, and state the appropriate conclusions, using $\alpha = 0.01$.

b) Is there evidence of heterogeneity of variance?

5. Given two normal populations,

$$\mu_1 = 80, \quad \sigma_1 = 6; \quad \mu_2 = 77, \quad \sigma_2 = 6.$$

If a sample of 36 cases is drawn from population 1 and a sample of 36 cases from population 2, what is the probability that

a) $\overline{X}_1 - \overline{X}_2 \geq 5$?   b) $\overline{X}_1 - \overline{X}_2 \geq 0$?

c) $\overline{X}_1 - \overline{X}_2 \leq 0$?   d) $\overline{X}_1 - \overline{X}_2 \leq -5$?

6. Assuming the same two populations as in Exercise 5, calculate the probability that $\overline{X}_1 - \overline{X}_2 \geq 0$, when:

a) $N_1 = N_2 = 4$,   b) $N_1 = N_2 = 9$,

c) $N_1 = N_2 = 16$,   d) $N_1 = N_2 = 25$.

7. Graph these probabilities as a function of N. Can you formulate any generalization about the probability of finding a difference in the correct direction between sample means (that is, $\overline{X}_1 - \overline{X}_2 \geq 0$, when $\mu_1 > \mu_2$) as a function of N?

8. A college maintains that it has made vast strides in raising the standards of admission for its entering freshmen. It cites the fact that the mean high school grade average of 80 entering freshmen was 82.53 for last year with a standard deviation of 2.53. For the present year these statistics are 83.04 and 2.58, respectively, for 84 entering freshmen. Set up and test the appropriate null hypothesis, using $\alpha = 0.01$.

9. Research was conducted on the pupil responses of heterosexual and homosexual males when viewing pictures of men and women. The following table shows the change in pupil size of five heterosexual and five homosexual males when viewing pictures of a male.

SUBJECT	HETEROSEXUALS	SUBJECT	HOMOSEXUALS
1	−00.4	6	+18.8
2	−54.5	7	−04.6
3	+12.5	8	+18.9
4	+06.3	9	+18.2
5	−01.5	10	+15.8

Source: Hess, E. H., A. L. Seltzer, and J. M. Shlien, "Pupil response of hetero- and homosexual males to pictures of men and women: A pilot study," *J. Abn. Psych.* **70**, 1965, 165–168. Copyright 1965 by The American Psychological Association. Reprinted with permission.

Formulate H_0 and H_1, two-tailed test. Using the Student's t-ratio for independent samples, determine whether H_0 may be rejected at $\alpha = 0.05$. (*Hint:* To facilitate calculations when negative numbers are involved, algebraically add 55 to each score. This procedure eliminates all the negative values and makes use of the generalization shown in Section 2.3.)

To calculate the mean for each group:

$$\frac{\sum_{i=1}^{5} X_i}{N} = \frac{\sum_{i=1}^{5} X_i - 5(55)}{N}.$$

Note: Adding 55 to all scores will not change the difference between means and the standard error of the difference, since the relative differences among scores are maintained. However, the mean for each group will increase by 55.

10. The following table shows the change in pupil size of five heterosexual and five homosexual males when viewing pictures of a female.

SUBJECT	HETEROSEXUALS	SUBJECT	HOMOSEXUALS
1	+05.9	6	+11.2
2	−22.4	7	−38.0
3	+19.2	8	+18.1
4	+39.0	9	−05.6
5	+23.1	10	+21.5

Source: Hess, E. H., A. L. Seltzer, and J. M. Shlien, "Pupil response of hetero- and homosexual males to pictures of men and women: A pilot study," *J. Abn. Psych.* **70**, 1965, 165–168. Copyright 1965 by The American Psychological Association. Reprinted with permission.

Formulate H_0 and H_1, two-tailed test. Using the Student's t-ratio for independent samples, determine whether H_0 may be rejected at $\alpha = 0.05$. (*Hint:* To facilitate calculations when negative numbers are involved, algebraically add 39 to each score.)

11. You have been asked by the State Crime Commission to compare the robbery rates per 1000 residents in a group of eastern and a group of midwestern cities. Your boss believes that a person is more likely to be robbed in an eastern city than in a midwestern city. Determine the validity of this statement from the following robbery figures. Use Student's t-ratio for correlated samples.

POPULATION RANK IN RESPECTIVE GROUP	EASTERN CITIES	ROBBERY RATE	MIDWESTERN CITIES	ROBBERY RATE
1	New York	24	Chicago	29
2	Philadelphia	21	Detroit	37
3	Baltimore	35	Cleveland	27
4	Newark	23	St. Louis	19

Source: U.S. Law Enforcement Assistance Administration, *Criminal Victimization Surveys in Eight American Cities: 1971–72 vs. 1974–75;* and *Criminal Victimization Surveys in Chicago, Detroit, Los Angeles, New York, Philadelphia: A Comparison of 1972 and 1974 Findings.*

12. It has often been stated that women have a higher life expectancy than men. Using the data in the accompanying table and using $\alpha = 0.05$,

determine the validity of this statement for

    a) white Americans,   **b)** nonwhite Americans,

    c) white males compared to nonwhite females.

Expectation of life in the United States

Age	WHITE Male	WHITE Female	NONWHITE (CHIEFLY BLACK) Male	NONWHITE (CHIEFLY BLACK) Female	Age	WHITE Male	WHITE Female	NONWHITE (CHIEFLY BLACK) Male	NONWHITE (CHIEFLY BLACK) Female
0	67.5	74.4	60.9	66.5	11	58.6	65.2	53.5	58.7
1	68.2	74.8	62.9	68.0	12	57.6	64.2	52.6	57.7
2	67.3	73.9	62.1	67.2	13	56.7	63.2	51.6	56.7
3	66.4	73.0	61.2	66.3	14	55.7	62.2	50.7	55.7
4	65.4	72.0	60.3	65.4	15	54.7	61.3	49.7	54.8
5	64.4	71.0	59.3	64.5	16	53.8	60.3	48.8	53.8
6	63.5	70.1	58.4	63.5	17	52.8	59.3	47.8	52.8
7	62.5	69.1	57.4	62.5	18	51.9	58.3	46.9	51.9
8	61.6	68.1	56.5	61.6	19	51.0	57.4	46.0	50.9
9	60.6	67.1	55.5	60.6	20	50.1	56.4	45.1	50.0
10	59.6	66.2	54.5	59.6					

13. Imagine the following sampling experiment. You have two populations of means consisting of the following values:

Population 1: 3,4,5,5,6,6,6,7,7,8,9 *Population 2:* 0,1,2,2,3,3,3,4,4,5,6

You place slips of paper with these means into two separate hats—one for each population. You select a mean from population 1, replace it in the hat, and then select a mean from population 2 and replace it in the hat. You subtract mean 2 from mean 1 to obtain a difference between means. The cell entries in the following table show all possible differences between means that you could obtain.

		SELECTION FROM MEANS IN POPULATION 1 3	4	5	5	6	6	6	7	7	8	9
Selection from means in population 2	0	3	4	5	5	6	6	6	7	7	8	9
	1	2	3	4	4	5	5	5	6	6	7	8
	2	1	2	3	3	4	4	4	5	5	6	7
	2	1	2	3	3	4	4	4	5	5	6	7
	3	0	1	2	2	3	3	3	4	4	5	6
	3	0	1	2	2	3	3	3	4	4	5	6
	3	0	1	2	2	3	3	3	4	4	5	6
	4	−1	0	1	1	2	2	2	3	3	4	5
	4	−1	0	1	1	2	2	2	3	3	4	5
	5	−2	−1	0	0	1	1	1	2	2	3	4
	6	−3	−2	−1	−1	0	0	0	1	1	2	3

Note that the mean of the means of population 1 is 6; the corresponding mean of population 2 is 3.

a) Construct a frequency and probability distribution of differences between means $(\overline{X}_1 - \overline{X}_2)$.

b) Based on intuition, estimate what the mean of these differences should be. Now calculate the mean of the differences. How accurate was your estimate? Formulate a generalization about the mean of the sampling distribution of differences between means.

14. Let us vary the preceding sampling experiment slightly. Imagine we know that hat 1 contains the population of means that we have identified as population 1. However, the population of means in the second hat is a complete mystery. We select a sample mean from hat 1 and a sample mean from hat 2. By subtracting the second mean from the first mean, we obtain a difference between sample means.

a) What statistical hypothesis are we most likely to test?

b) What is the sampling distribution against which we should test H_0: $\mu_1 = \mu_2$?

c) Construct a frequency and probability distribution of differences between means under the hypothesis H_0: $\mu_1 = \mu_2$. (*Hint:* Assume that population 2 is identical to population 1, and find all possible differences between means.) This probability distribution is known as the sampling distribution of differences between means, in which the null hypothesis is that we are sampling from identical populations.

15. Using the sampling distribution under H_0: $\mu_1 = \mu_2$ (Exercise 14 above), find the probability of obtaining a difference in sample means:

a) $5 \le \overline{X}_1 - \overline{X}_2 \le -5$ **b)** $\overline{X}_1 - \overline{X}_2 \ge 6$

c) $\overline{X}_1 - \overline{X}_2 \le -3$ **d)** $6 \le \overline{X}_1 - \overline{X}_2 \le -6$

e) $\overline{X}_1 - \overline{X}_2 = -6$

16. Imagine that the *true* situation is the sampling distribution of differences between means found in response to Exercise 13. Find the probability of obtaining a difference in sample means:

a) $5 \le \overline{X}_1 - \overline{X}_2 \le -5$ **b)** $\overline{X}_1 - \overline{X}_2 \ge 6$

c) $\overline{X}_1 - \overline{X}_2 \le -3$ **d)** $6 \le \overline{X}_1 - \overline{X}_2 \le -6$

e) $\overline{X}_1 - \overline{X}_2 = -6$

17. Compare the probability values obtained in response to Exercises 15 and 16. See if you can formulate any broad observations about the different answers you obtained.

An Introduction to the Analysis of Variance

16

16.1 Multigroup comparisons

16.2 The Concept of Sums of Squares

16.3 Obtaining Variance Estimates

16.4 Fundamental Concepts of Analysis of Variance

16.5 Assumptions Underlying Analysis of Variance

16.6 An Example Involving Three Groups

16.7 The Interpretation of F

16.1 Multigroup Comparisons

In Chapters 14 and 15 we were concerned with comparisons between only two groups. Yet the complexity of the phenomena that the social scientist investigates does not allow us to restrict our observations to two groups. Events in nature rarely order themselves conveniently into two groups. More commonly the questions we pose are: Which type of juvenile detention facility is most successful in terms of the recidivism rate? Do the median incomes of the residents in several neighborhoods differ significantly? Are Southern Baptists significantly more orthodox than Lutherans and Methodists? Are persons over 60 more socially isolated than the middle-aged and the young? Which of three teaching techniques is most effective?

The research design necessary to provide answers to these questions would require comparison of more than two groups. You may wonder: But why should multigroup comparisons provide any obstacles? Can we not simply compare the mean of each group with the mean of every group and obtain a Student's *t*-ratio for each comparison? For example, if we had four groups, *A, B, C, D,* could we not calculate Student's *t*-ratios comparing *A* with *B, C,* and *D; B* with *C* and *D;* and *C* with *D?*

If you will think for a moment of the errors in inference, which we have so frequently discussed, you will recall that our greatest concern has been to avoid type I errors. When we establish the region of rejection at the 0.05 level, we are, in effect, acknowledging our willingness to take the risk of being wrong as often as 5% of the time in our rejection of the null hypothesis. Now, what happens when we have numerous comparisons to make? For an extreme example, let us imagine that we have conducted a study involving the calculation of 1000 separate Student's t-ratios. Would we be terribly impressed if, say, 50 of the t's proved to be significant at the 0.05 level? Of course not. Indeed, we would probably murmur something to the effect that, "With 1000 comparisons, we would be surprised if we don't obtain approximately 50 comparisons that are significant *by chance* (that is, due to predictable sampling error)."

The **analysis of variance** (sometimes abbreviated ANOVA) is a technique of statistical analysis that permits us to overcome the ambiguity involved in assessing significant differences when more than two group means are compared, by allowing for a single decision at a specified level of significance. It allows us to answer the question: Is there an overall indication that the independent variable is producing differences among the means of the various groups? Although the analysis of variance may be used in the two-sample case (in which event it yields precisely the same probability values as the Student's t-ratio), it is most commonly used when three or more categories or groups of the independent variable are involved. The focus of this chapter is **one-way analysis of variance,** which derives its name from the fact that various groups represent different categories or levels of a *single* independent variable (sometimes referred to as an *experimental stimulus,* or *treatment variable*). The procedures discussed may be generalized to any number of independent variables. We will now consider the two-group situation before examining the more complex three-group situation.

16.2 The Concept of Sums of Squares

Imagine that we have conducted a study (see Table 16.1) with two groups of students concerning their attitude toward the legalization of marijuana. None of the students in group 1 have ever used marijuana, and all of the students in group 2 use it at least once a month. A high score indicates the student strongly favors legalization. We will test the hypothesis that the two groups of students have been drawn from the same population (or that there is no difference in their attitude toward legalization of marijuana).

Table 16.1 Nonuser's and user's attitude toward the legalization of marijuana.

(NONUSERS) Group 1		(USERS) Group 2	
X_1	X_1^2	X_2	X_2^2
1	1	6	36
2	4	7	49
5	25	9	81
8	64	10	100
Σ 16	94	32	266

$N_1 = 4$, $\quad \overline{X}_1 = 4$ $\quad N_2 = 4$, $\quad \overline{X}_2 = 8$

$\Sigma X_{tot} = 48$, $\quad N = 8$, $\quad \overline{X}_{tot} = 6$.

The mean for group 1 is 4; the mean for group 2 is 8. The overall mean, $\overline{X}_{tot}$, is 48/8 or 6. Now, if we were to subtract the overall mean from each score and square, we would obtain the total sum of squares (SS_{tot})

$$SS_{tot} = \sum (X - \overline{X}_{tot})^2. \qquad (16.1)$$

For the data in Table 16.1, the total sum of squares using Eq. (16.1) is

$$SS_{tot} = (1 - 6)^2 + (2 - 6)^2 + (5 - 6)^2 + (8 - 6)^2 + (6 - 6)^2$$
$$+ (7 - 6)^2 + (9 - 6)^2 + (10 - 6)^2 = 25 + 16 + 1 + 4$$
$$+ 0 + 1 + 9 + 16 = 72.$$

Thus we see that the **total sum of squares** is the sum of the squared deviations from the overall mean and hence is a measure of the total variation in the data. We have already encountered the sum of squares in calculating the standard deviation, the variance, the standard error of the difference between means, and the total variation of a dependent variable when studying regression analysis. The alternative raw score equation for the total sum of squares is

$$SS_{tot} = \sum X_{tot}^2 - \frac{(\Sigma X_{tot})^2}{N}. \qquad (16.2)$$

For the data in Table 16.1, the total sum of squares is

$$SS_{tot} = 94 + 266 - \frac{(48)^2}{8}$$
$$= 360 - 288 = 72.$$

The **within-group sum of squares** (SS_w) is the sum of the sum of squares obtained within each group and is a measure of the variability within the groups. Figure 16.1 shows, for example, that the scores of the students in group 1 (nonusers) and group 2 (users) vary widely about each of their respective group means of 4 and 8. The source of the within-group variation includes all the variables influencing the dependent variable (attitude toward the legalization of marijuana) other than the independent variable. Could the variability we observed in Fig. 16.1 in the attitude toward marijuana legalization scores within group 1 or within group 2 have stemmed from the independent variable (nonuse versus use)? No, since everyone in group 1 was a nonuser and everyone in group 2 a user. Thus all within-group variability must stem from variables other than the specified independent variable.

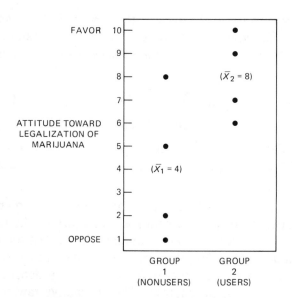

16.1 Scattergram of attitude toward legalization of marijuana scores for Nonusers and Users.

The within-group sum of squares, in this instance, measures the sum of the squared deviations of the scores in group 1 around $\overline{X}_1$ *plus* the sum of the squared deviations of the scores in group 2 around $\overline{X}_2$.

$$SS_w = SS_1 + SS_2$$

Where

$$SS_1 = \Sigma(X - \overline{X}_1)^2 \text{ and}$$
$$SS_2 = \Sigma(X - \overline{X}_2)^2.$$

For the data in Table 16.1, the within-group sum of squares is

$$\Sigma(X - \overline{X}_1)^2 + \Sigma(X - \overline{X}_2)^2 =$$

$(1 - 4)^2 + (2 - 4)^2 + (5 - 4)^2 + (8 - 4)^2 + (6 - 8)^2 +$
$(7 - 8)^2 + (9 - 8)^2 + (10 - 8)^2 = 9 + 4 + 1 + 16 + 4 +$
$1 + 1 + 4 = 40.$

Using the raw score equation (16.3) for the within-group sum of squares we also find that $SS_w = 40$.

$$SS_w = SS_1 + SS_2 \qquad (16.3)$$

$$SS_1 = \sum X_i^2 - \frac{(\Sigma X_1)^2}{N_1}$$

$$= 94 - \frac{(16)^2}{4} = 94 - 64 = 30,$$

$$SS_2 = \sum X_i^2 - \frac{(\Sigma X_2)^2}{N_2}$$

$$= 266 - \frac{(32)^2}{4} = 266 - 256 = 10,$$

$$SS_w = 30 + 10 = 40.$$

The **between-group sum of squares** (SS_{bet}) reflects the influence of the independent or treatment variable. It may be obtained by subtracting the overall mean from each group mean, squaring the result, multiplying by N in each group, and summing across all the groups. Thus

$$SS_{bet} = \sum N_i(\overline{X}_i - \overline{X}_{tot})^2, \qquad (16.4)$$

where N_i is the number in the ith group, and $\overline{X}_i$ is the mean of the ith group.

$$SS_{bet} = 4(4 - 6)^2 + 4(8 - 6)^2 = 32.$$

The raw score equation for calculating the between-group sum of squares is

$$SS_{bet} = \sum \frac{(\Sigma X_i)^2}{N_i} - \frac{(\Sigma X_{tot})^2}{N}, \qquad (16.5)$$

and

$$SS_{bet} = \frac{(16)^2}{4} + \frac{(32)^2}{4} - \frac{(48)^2}{8}$$
$$= 64 + 256 - 288$$
$$= 320 - 288 = 32.$$

It will be noted that the total sum of squares is equal to the sum of the between-group sum of squares and the within-group sum of squares. In other words,

$$SS_{tot} = SS_w + SS_{bet}. \qquad (16.6)$$

In this example, $SS_{tot} = 72$, $SS_w = 40$, and $SS_{bet} = 32$. Thus $72 = 40 + 32$. Therefore, we have partitioned the *total sum of squares* which is a measure of total variation in the dependent variable into two components: the within and between sum of squares. The *within sum of squares component* is a measure of the variability of the dependent variable scores *within* each of the categories or groups of the independent variable and hence is the variation in the dependent variable that cannot be attributed to the independent variable. The *between sum of squares component* is a measure of the variation *between* the categories or groups of the independent variable and hence is the variation in the dependent variable that is attributable to the independent or treatment variable.

16.3 Obtaining Variance Estimates

The analysis of variance consists of obtaining two independent **estimates of variance,** one based upon variability between groups (between-group variance) and the other based upon the variability within groups (within-group variance). The significance of the difference between these two variance estimates is provided by Fisher's F-distributions. (We are already familiar with F-distributions from our prior discussion of homogeneity of variance, Section 15.4). If the between-group variance estimate is large (that is, the difference between means is large) relative to the within-group variance estimate, the F-ratio is large. Conversely, if the between-group variance estimate is small relative to the within-group variance estimate, the F-ratio will be small.

A test of our null hypothesis that the two samples were drawn from the same population (H_0: $\mu_1 = \mu_2$) requires that we create a ratio of the between- and within-group sum of squares. You may have observed, however, that both values will increase as the sample size increases. Therefore as a "corrective" factor we control for sample size by dividing both values by their respective degrees of freedom.* This procedure allows us to arrive at variance estimates for the between- and within-group sum of squares. The *degrees of freedom of the between-group sum of squares* is the number of categories or groups (k) of the independent variable minus 1.

$$\text{df}_{\text{bet}} = k - 1. \qquad (16.7)$$

With two groups, $k = 2$. Therefore df $= 2 - 1 = 1$. Thus our between-group variance estimate for the problem at hand is

$$\hat{s}^2_{\text{bet}} = \frac{\text{SS}_{\text{bet}}}{\text{df}_{\text{bet}}} = \frac{32}{1} = 32, \qquad \text{df} = 1. \qquad (16.8)$$

The number of *degrees of freedom of the within-groups sum of squares* is the total N minus the number of categories within the independent variable. Thus

$$\text{df}_{\text{w}} = N - k. \qquad (16.9)$$

In the present problem, $\text{df}_{\text{w}} = 8 - 2 = 6$ and our within-group variance estimate becomes

$$\hat{s}^2_{\text{w}} = \frac{\text{SS}_{\text{w}}}{\text{df}_{\text{w}}} = \frac{40}{6} = 6.67, \qquad \text{df} = 6. \qquad (16.10)$$

Now, all we have left to do is to calculate the F-ratio and determine whether or not our two variance estimates could have reasonably been drawn from the same population. If not, we will conclude that the significantly larger between-group variance is due to the operation of the group differences. In other words, we shall conclude that the groups' differences in marijuana usage produced a significant difference in means. The **F-ratio**, in analysis of variance, is the between-

---

* The rationale for the degrees of freedom concept was presented in Section 14.5. The same rationale applies here. Degrees of freedom represent the number of values that are free to vary once we have placed certain restrictions on our data.

group variance estimate divided by the within-group variance estimate. Symbolically,

$$F = \frac{\hat{s}^2_{bet}}{\hat{s}^2_w}.$$

(16.11)

For this problem our F-ratio is

$$F = \frac{32}{6.67} = 4.80, \qquad df = 1/6.$$

Looking up the F-ratio under 1 and 6 degrees of freedom in Table D, Appendix C, we find that an F-ratio of 5.99 or larger is required for significance at the 0.05 level. For the present problem, then, we cannot reject the null hypothesis. You should note that Table D provides two-tailed values. The analysis of variance test for significance is automatically two-tailed since *any* difference among the sample means will enlarge the entire value of F, not just the difference in which the researcher is interested.

In sum, we must conclude that the nonusers and users we have studied do not differ significantly in terms of their attitudes toward the legalization of marijuana and that they were drawn from the same population.

16.4 Fundamental Concepts of Analysis of Variance

In these few pages, we have examined all the basic concepts necessary to understand simple analysis of variance. Before proceeding with an example involving three groups, let us briefly review these fundamental concepts.

1. We have seen that in a study involving two or more groups it is possible to identify two different bases for estimating the population variance: the between-group and the within-group.

    a) The between-group variance estimate reflects the magnitude of the difference between and or among the group means. The larger the difference between means, the larger the between-group variance.

    b) The within-group variance estimate reflects the dispersion of scores within each treatment group. The within-group variance is analogous to $s_{\overline{X}_1 - \overline{X}_2}$ in the Student's t-ratio. It is often referred to as the *error term*.

2. The null hypothesis is that the samples were drawn from the same population, or that $\mu_1 = \mu_2 = \cdots = \mu_k$.

3. The alternative hypothesis is that the samples were not drawn from the same population, that is, $\mu_1 \neq \mu_2 \ldots \neq \mu_k$.

4. The F-ratio consists of the between-group variance estimate divided by the within-group variance estimate. By consulting Table D of the distribution of F we can determine whether or not the null hypothesis of equal population means can reasonably be entertained. In the event of a significant F-ratio, we may conclude that the group means are not all estimates of a common population mean.

5. In the two-sample case, the F-ratio yields probability values identical to those of the Student's t-ratio. Indeed, in the one-degree-of-freedom situation (that is, $k = 2$), $t = \sqrt{F}$ or $t^2 = F$. You may check this statement by calculating the Student's t-ratio for the sample problem we have just completed.

16.5 Assumptions Underlying Analysis of Variance

When discussing the Student's t-ratio in Chapter 15, we noted that a fundamental assumption underlying the use of the Student's t-ratio is that the within-group population variance for all groups must be homogeneous, that is, drawn from the same population of variances. The same assumption holds true for the analysis of variance. In other words, a basic assumption underlying the analysis of variance is that the within-group variances (which, when summed together, make up $\hat{s}_w^2$) are homogeneous and is referred to as **homogeneity of variance**. As with the Student's t-ratio, there is a test for determining whether or not the hypothesis of identical variances is tenable. However, it is beyond the scope of this introductory text to delve into Bartlett's test of homogeneity of variances. Application of this test is described in Kirk (1968).

In addition, the analysis of variance requires the assumptions of normality within groups, random and independent sampling, interval scaling for the dependent variable, and nominal scaling for the independent variable(s).

16.6 An Example Involving Three Groups

Imagine we have just completed a study to assess attitudes toward welfare payments, and we have three samples of nine respondents each: group 1 is a middle-class sample, group 2 is a sample of persons presently on welfare, and group 3 is a lower-middle-class sample. A high score indicates strong support of welfare payments. Our null hypothesis is that these samples were drawn from the same population or that the groups do not differ in their attitudes toward welfare. If we reject the null hypothesis we would conclude that one or more of the group means is significantly different from one or both of the other group means. The results of this study are presented in Table 16.2.

Table 16.2   Attitude toward welfare scores of three groups of respondents in a hypothetical study.

(MIDDLE-CLASS SAMPLE)		(WELFARE SAMPLE)		(LOWER-MIDDLE-CLASS SAMPLE)	
Group 1		Group 2		Group 3	
X_1	X_1^2	X_2	X_2^2	X_3	X_3^2
4	16	12	144	1	1
5	25	8	64	3	9
4	16	10	100	4	16
3	9	5	25	6	36
6	36	7	49	8	64
10	100	9	81	5	25
1	1	14	196	3	9
8	64	9	81	2	4
5	25	4	16	2	4
Σ 46	292	78	756	34	168

$N_1 = 9,$ $\overline{X}_1 = 5.11$ $N_2 = 9,$ $\overline{X}_2 = 8.67$ $N_3 = 9,$ $\overline{X}_3 = 3.78$

$$\Sigma X_{tot} = 46 + 78 + 34 = 158,$$
$$\Sigma X_{tot}^2 = 292 + 756 + 168 = 1216,$$
$$N = 27.$$

The following steps are followed in a three-group analysis of variance:

Step 1. Using Eq. (16.2), the total sum of squares is

$$SS_{tot} = 1216 - \frac{(158)^2}{27} = 291.41.$$

Step 2. Using Eq. (16.5) for three groups, the between-group sum of squares is

$$SS_{bet} = \frac{(46)^2}{9} + \frac{(78)^2}{9} + \frac{(34)^2}{9} - \frac{(158)^2}{27} = 114.96.$$

Step 3. The within-group sum of squares may be obtained by employing Eq. (16.3) for three groups or by subtracting SS_{bet} from SS_{tot}:

$$SS_w = \left(292 - \frac{(46)^2}{9}\right) + \left(756 - \frac{(78)^2}{9}\right) + \left(168 - \frac{(34)^2}{9}\right) = 176.45.$$

$$SS_w = SS_{tot} - SS_{bet} = 291.41 - 114.96 = 176.45.$$

Step 4. The between-group variance estimate is

$$df_{bet} = k - 1 = 2, \quad \hat{s}^2_{bet} = \frac{114.96}{2} = 57.48.$$

Step 5. The within-group variance estimate is

$$df_w = N - k = 24, \quad \hat{s}^2_w = \frac{176.45}{24} = 7.35.$$

Step 6. Using Eq. (16.11) we find that the value of *F* is

$$F = \frac{57.48}{7.35} = 7.82, \quad df = 2/24.$$

To summarize these steps, we use the format shown in Table 16.3.

Table 16.3 Summary table for representing the relevant statistics in analysis of variance problems

SOURCE OF VARIATION	SUM OF SQUARES	DEGREES OF FREEDOM	VARIANCE ESTIMATE*	F
Between groups	114.96	2	57.48	7.82
Within groups	176.45	24	7.35	
Total	291.41	26		

*In many texts, the term *mean square* appears in this column. However, we prefer the term *variance estimate* since this term accurately describes the nature of the entries in the column.

By following the format recommended in Table 16.3, you have a final check upon your calculation of sum of squares and your assignment of degrees of freedom. Thus $SS_{bet} + SS_w$ must equal SS_{tot}. The degrees of freedom of the total are found by

$$df_{tot} = N - 1. \qquad (16.12)$$

In the present example, the number of degrees of freedom for the total is

$$df_{tot} = 27 - 1 = 26.$$

16.7 The Interpretation of *F*

When we look up the *F* required for significance with 2 and 24 degrees of freedom, we find that an *F* of 3.40 or larger is significant at the 0.05 level. Since our *F* of 7.82 exceeds this value, we may conclude that the three-group means are not all estimates of a common population mean or that the groups' attitudes toward welfare differ. Now do we stop at this point? After all, are we not interested in determining whether or not one of the three samples is more prowelfare than the other two? The answer to the first question is negative, and the answer to the second is affirmative.

The truth of the matter is that our finding an overall significant *F*-ratio now permits us to investigate the following specific hypotheses which will allow us to determine exactly which means are significantly different.

$$H_0: \mu_1 = \mu_2$$
$$H_0: \mu_1 = \mu_3$$
$$H_0: \mu_2 = \mu_3$$

In the absence of a significant *F*-ratio, any significant differences between specific comparisons would have to be regarded as representing a chance difference.

Over the past several years statisticians have developed a large number of tests that permit the researcher to investigate specific hypotheses concerning population parameters. Two broad classes of such tests exist:

1. *A priori* **or planned comparisons:** When comparisons are planned in advance of the investigation, an *a priori* test is appropriate. For *a priori* tests, it is not necessary that the overall *F*-ratio be significant.

2. *A posteriori* **comparisons:** When the comparisons are not planned in advance, an *a posteriori* test is appropriate.

In the present example we shall illustrate the use of an *a posteriori* test for making pairwise comparisons among means. Tukey (1953) has developed such a test, which he named the HSD (honestly significant difference) test. To use this test, the overall *F*-ratio must be significant. A difference between two means is significant at a given α-level if it equals or exceeds HSD, which is:

$$\text{HSD} = q_\alpha \sqrt{\frac{\hat{s}_w^2}{N}}, \qquad (16.13)$$

where

$\hat{s}_w^2$ = the within-group variance estimate,
N = number of subjects in each group,
q_α = tabled value for a given α-level found in Table K for df_w and k (number of means).

16.7.1 A Worked Example

Let us use the data from Section 16.5 to illustrate the application of the HSD test. We shall assume $\alpha = 0.05$ for testing the significance of the difference between each pair of means.

Step 1. Prepare a matrix showing the mean of each group and the differences between pairs of means. This is shown in Table 16.4.

Table 16.4   Differences among means

	$\overline{X}_1$	$\overline{X}_2$	$\overline{X}_3$
	$\overline{X}_1 = 5.11$	$\overline{X}_2 = 8.67$	$\overline{X}_3 = 3.78$
$\overline{X}_1 = 5.11$	. . .	3.56	1.33
$\overline{X}_2 = 8.67$	. . .	. . .	4.89
$\overline{X}_3 = 3.78$	. . .	. . .	. . .

Step 2. We must now determine the value of q. Given that the within sum of squares degrees of freedom equals 24 ($N - k = 27 - 3 = 24$), $k = 3$, and $\alpha = 0.05$, we find by referring to Table K that $q_{0.05} = 3.53$.

Step 3. Find HSD by multiplying $q_{0.05}$ by $\sqrt{\hat{s}_w^2/N}$. The quantity $\hat{s}_w^2$ is found in Table 16.3 under within-group variance estimate. The N per group is 9. Thus

$$\text{HSD} = 3.53 \sqrt{\frac{7.35}{9}} = 3.53(0.90) = 3.18.$$

Step 4. Referring to Table 16.4, we find that the differences between $\overline{X}_1$ versus $\overline{X}_2$ and $\overline{X}_2$ versus $\overline{X}_3$ both exceed HSD = 3.18. We may therefore conclude that these differences are statistically significant at $\alpha = 0.05$. Since the group 2 mean (welfare sample) is significantly higher than the mean of the other two samples, we may conclude that the respondents' experience with welfare has influenced their attitude toward welfare in such a manner as to result in a high score relative to the other two samples.

Summary

We began this chapter with the observation that the researcher is frequently interested in conducting studies that are more extensive than the classical two-group design. However, when more than two groups are involved in a study, we increase the risk of making a type I error if we accept as significant any comparison that falls within the rejection region. In multigroup studies it is desirable to know whether or not there is an indication of an overall effect of the independent variable before we investigate specific hypotheses. The analysis of variance technique provides such a test.

In this chapter, we presented an introduction to the complexities of analysis of variance. We showed that total sums of squares can be partitioned into two component sums of squares: the within-group and the between-group. These two component sums of squares provide us in turn with independent estimates of the population variance. A between-group variance estimate that is large, relative to the within-group variance, suggests that the independent variable or experimental treatments are responsible for the large differences among the group means. The significance of the difference in variance estimates is obtained by reference to the F-table (Table D, Appendix C).

When the overall F-ratio is found to be statistically significant, we are free to investigate specific hypotheses, employing a multiple-comparison test.

Terms to Remember

Analysis of variance   A method, described initially by R. A. Fisher, for partitioning the sum of squares for experimental or survey data into known components of variation.

A posteriori **comparisons**   Comparisons not planned in advance to investigate specific hypotheses concerning population parameters.

A priori **or planned comparisons**   Comparisons planned in advance to investigate specific hypotheses concerning population parameters.

Between-group variance   Estimate of variance based upon variability between groups.

F-**ratio**   The between-group variance estimate divided by the within-group variance estimate.

Homogeneity of variance   The condition that exists when two or more sample variances have been drawn from populations with equal variances.

One-way analysis of variance   Statistical analysis of various categories, groups, or levels of one independent (experimental) variable.

Sum of squares   Deviations from the mean squared and summed.

Variance estimate   Sum of the squared deviations from the mean divided by degrees of freedom.

Within-group variance   Estimate of variance based upon the variability within groups.

Exercises

1. Using the following data derived from the ten-year period 1955–1964, determine whether there is a significant difference at the 0.01 level in death rate among the various seasons. (*Note:* Assume death rates for any given year to be independent.)

YEAR	WINTER	SPRING	SUMMER	FALL
1955	9.8	9.0	8.8	9.4
1956	9.9	9.3	8.7	9.4
1957	9.8	9.3	8.8	10.3
1958	10.6	9.2	8.6	9.8
1959	9.9	9.4	8.7	9.4
1960	10.7	9.1	8.3	9.6
1961	9.7	9.2	8.8	9.5
1962	10.2	8.9	8.8	9.6
1963	10.9	9.3	8.7	9.5
1964	10.0	9.3	8.9	9.4

2. Conduct an HSD test, comparing the death rates of each season with every other season. Use the 0.01 level, two-tailed test for each comparison.

3. Professor Stevens negotiates contracts with ten different independent research organizations to compare the effectiveness of his marital satisfaction assessment instrument with that designed by another sociologist ten years earlier. A significant difference (0.05 level) in favor of Professor Stevens's instrument is found in one of ten studies. He subsequently claims that independent research has demonstrated the superiority of his instrument over the instrument developed earlier. Criticize this conclusion.

4. A random sample of four interviewers associated with a large public opinion firm have completed four interviews each, and the time in minutes necessary to complete the interviews is indicated in the accompanying table. Test the null hypothesis that all the interviewers are drawn from a common population.

INTERVIEWER A	INTERVIEWER B	INTERVIEWER C	INTERVIEWER D
28	34	29	22
19	23	24	31
30	20	33	18
25	16	21	24

5. A labor organization randomly selected three medium-sized businesses for a study of worker grievances. The number of official grievances during the past five calendar years was tabulated. Set up and test an appropriate null hypothesis which will assess the extent to which a particular business has significantly more grievances than the others. Con-

duct an HSD test comparing each business with every other business, using $\alpha = 0.05$.

BUSINESS A	BUSINESS B	BUSINESS C
42	52	38
36	48	44
47	43	33
43	49	35
38	51	32

6. If the F-ratio is less than 1.00, what do we conclude?

7. Determine whether there is a significant difference in the number of major work stoppages among the three geographical areas given (hypothetical data).

NEW ENGLAND		MID-ATLANTIC		FAR WEST	
Maine	8	New York	7	Washington	9
New Hampshire	7	New Jersey	7	Oregon	7
Vermont	8	Pennsylvania	8	Nevada	6
Massachusetts	6.5	Delaware	7	California	7
Rhode Island	8	Maryland	7	Alaska	8
Connecticut	8	District of Columbia	7	Hawaii	5

8. Suppose a family stability measure were administered in four public high schools to a sample of 24 students who attend the schools. The following scores were obtained:

SCHOOL A	SCHOOL B	SCHOOL C	SCHOOL D
50	55	50	70
50	60	65	80
55	65	75	65
60	55	55	70
45	70	60	75
55	65	65	60

Note: (Low score = family instability)

Is there a significant difference in the family stability scores among the four groups of students?

9. A nutritional expert divides a sample of bicyclists into three groups. Group B is given a vitamin supplement and group C is given a diet of health foods. Group A is instructed to eat as they normally do. The expert subsequently records the number of minutes it takes each person to ride six miles:

A	B	C
15	14	13
16	13	12
14	15	11
17	16	14
15	14	11

Set up the appropriate hypothesis and conduct an analysis of variance.

10. For Exercise 9, determine which diet or diets are superior.

11. Banks A and B use two different forms for recording checks written. The police found that the following number of checks had bounced for 15 customers during the last 10 years:

BANK A	BANK B
4	2
8	0
3	1
0	2
3	1
5	1
3	3
4	3
0	4
5	3
2	1
4	4
6	0
2	5
0	0

Determine whether there is a significant difference between the number of checks bounced at each bank. Use $\alpha = 0.05$.

12. In 1971 three cities had the following numbers of speeches made by their top administrators:

CHICAGO	HOUSTON	CINCINNATI
2	9	13
16	2	39
28	1	25
2	10	9
21	13	3
2	1	0
2	12	13
19	0	27
0	1	5
4	2	1
8	7	2
6	7	

Test the hypothesis that all three cities were drawn from the same population with respect to speeches by top administrators.

Inferential Statistics: Nonparametric Tests of Significance

17 Power and Efficiency of a Statistical Test

18 Statistical Inference with Categorical Variables:
        Chi Square and Related Measures

Power and Efficiency of a Statistical Test

17

17.1 The Concept of Power

17.2 Calculation of Power: One-Sample Case

17.3 The Effect of Sample Size on Power

17.4 The Effect of α-Level on Power

17.5 The Effect of the Nature of H_1 on Power

17.6 Parametric versus Nonparametric Tests: Power

17.7 Calculation of Power: Two-Sample Case

17.8 The Effect of Correlated Measures on Power

17.9 Power, Type I, and Type II Errors

17.10 Power Efficiency of a Statistical Test

17.1 The Concept of Power

Throughout the first two parts of this book, only fleeting references were made to the power and the power efficiency of a statistical test (although they were not identified as such). Before proceeding into nonparametric tests of significance, it is desirable to examine these concepts in more detail.

While discussing type I and type II errors in Section 13.6, we pointed out that the basic conservatism of scientists causes them to set up a rejection level sufficiently low to make type I errors less frequently than type II errors. In other words, scientists have preferred to make the mistake of failing to reject a false null hypothesis than the mistake of rejecting a true one. However this conservatism should not be construed to mean that the scientist is happy about the prospect of making type II errors. To the contrary, it is quite likely that many promising research projects have been abandoned because of the failure of the experimenter to reject the null hypothesis when it was actually false. There have also been many studies that wasted time and resources by using a sample that was much too large. Both of these mistakes (failing to reject the null hypothesis when it should have been rejected and using too large a sample) could be avoided by examining power prior to conducting the study.

Now up to this point in the book, our concern has been to establish a level of significance that will reduce the likelihood of falsely rejecting the null hypothesis. In other words, we have been primarily concerned with avoiding type I rather than type II errors. However, it should be recognized that the ideal statistical test is one that effects some sort of *balance* between these two types of error. Ideally, we should specify in advance of our study the probability of making both a type I and a type II error. In practice, however, most researchers content themselves with stating only the *p*-value they will use to reject the null hypothesis. As we have seen, this *p*-value represents the probability of a type I error (that is, α).

When we begin to concern ourselves with effecting a balance between type I and type II errors, we are dealing with the concept of the **power of a test.** The power of a test is defined as the probability of rejecting the null hypothesis when it is in fact false. Symbolically, power is defined as follows:

Power = 1 − probability of a type II error.

If we let β represent the probability of a type II error, the definition of power becomes:

$$\text{Power} = 1 - \beta. \qquad (17.1)$$

We can calculate the power of a test only when H_0 is false and we are given the true value of the population mean under H_1. However, it must be recognized that the necessary information to calculate the power of a test is *almost never* available to us. Yet theoretically we know the effects of various conditions on the power of a test. These conditions include the sample size, the α-level, the nature of the alternative hypothesis (H_1), the nature of the statistical test itself, and correlated measures and are discussed along with three hypothetical examples to alert you to the factors that researchers must consider when designing a research project. In the vast majority of instances, researchers can only assume or estimate the parameters necessary to compute the power of a test.

But the importance of considering power cannot be underestimated. A doctoral student designed a research project to test the sociolinguistic impact of two teaching strategies on school children (H_0: Outcome of strategy I = Outcome of strategy II). In the course of the study, there was a high attrition rate among the teachers (and hence, their students) due to illness, pregnancy, and lack of administrative cooperation. As a result, the sample size was reduced and the study threatened.

Because it was no longer likely that the tests of significance were powerful enough to detect real differences in such a small sample, a failure to reject the null hypothesis could not have been interpreted. It might mean that the two experimental treatments (the teaching strategies) resulted in the same outcome (which is a valid interpretation if we have no reason to question the power of the statistical test) or it might mean that there could be real differences between the strategies but that the test was not powerful enough to detect the difference.

In this particular case, the statistical analysis did show a significant difference in student gains on the sociolinguistic measure and H_0 was rejected. Hence, it was assumed that the sample size was adequate (albeit small) and that the test was powerful enough. But it was also assumed that the difference between the outcomes of the two strategies was large; if the difference had been smaller, H_0 could not have been rejected even though it was false because the tests used in the analysis were not powerful enough to detect small differences given the small sample size.

Compare this to inspecting what appears, with unaided sight, to be two identical small insects (H_0: $Insect_1$ = $Insect_2$). If, in fact, the insects are different, that is, H_0 is false, but the unaided eye cannot detect the differences, then we have to assume that either they are the same type of insect, or that our sight is inadequate to detect the difference. In short, we don't know very much about the insects in question. On the other hand, if a school child's magnifying glass reveals that one has small spots on its back and the other does not, a relatively obvious difference, even a weak instrument is sufficient to detect the difference. However, if the difference is discernible only under a powerful laboratory microscope, the weaker instrument could not lead us to reject a false null hypothesis.

Before proceeding with a formal discussion of power, let us review a series of problems that appeared in the end-of-chapter exercises for Chapter 15 (Exercises 13 through 17). You will recall that we imagined drawing all possible pairs of samples from two populations of means.

Population 1 means: 3, 4, 5, 5, 6, 6, 6, 7, 7, 8, 9: μ_1 = 6:
Population 2 means: 0, 1, 2, 2, 3, 3, 3, 4, 4, 5, 6; μ_2 = 3.

We obtained the sampling distribution of differences between means shown in Table 17.1.

Then we pretended not to know the characteristics of the second population. We formulated the null hypothesis that the first and second populations are the same (μ_1 = μ_2). This sampling distribution of differences between means is shown in Table 17.2.

Table 17.1   Frequency and sampling distributions of differences between means when all possible sample means are selected, with replacement, from two populations of means in which $\mu_1 = 6$ and $\mu_2 = 3$. This represents the true sampling distribution of differences between means for the indicated populations of means.

$\overline{X}_1 - \overline{X}_2$	f	$p(\overline{X}_1 - \overline{X}_2)$
9	1	0.0083
8	2	0.0165
7	5	0.0413
6	10	0.0826
5	14	0.1157
4	18	0.1488
3	21	0.1736
2	18	0.1488
1	14	0.1157
0	10	0.0826
−1	5	0.0413
−2	2	0.0165
−3	1	0.0083
$N_{\overline{X}_1 - \overline{X}_2} = 121$		$\Sigma p(\overline{X}_1 - \overline{X}_2) = 1.000$

Table 17.2   Sampling distribution of the differences between means under H_0: $\mu_1 = \mu_2$. (*Note:* We have assumed that population 2 is identical to population 1, that is, $\mu_1 = \mu_2$, and found all possible differences between means.)

$\overline{X}_1 - \overline{X}_2$	f	$p(\overline{X}_1 - \overline{X}_2)$	
6	1	0.0083	⎫ 0.0248
5	2	0.0165	⎭
4	5	0.0413	
3	10	0.0826	
2	14	0.1157	
1	18	0.1488	
0	21	0.1736	
−1	18	0.1488	
−2	14	0.1157	
−3	10	0.0826	
−4	5	0.0413	
−5	2	0.0165	⎫ 0.0248
−6	1	0.0083	⎭
$N_{\overline{X}_1 - \overline{X}_2} = 121$		$\Sigma p(\overline{X}_1 - \overline{X}_2) = 1.0000$	

Note that under the null hypotheses of no difference between population means, a difference equal to or greater than 5 or equal to or less than -5 would lead to the rejection of H_0 at $\alpha = 0.05$ (0.0248 + 0.0248 = 0.0496), two-tailed test.

Now we return to the *true* sampling distribution of differences between sample means (Table 17.1) and ask, "Under $H_0 : \mu_1 = \mu_2$, how often would we have rejected this false null hypothesis when $\alpha = 0.05$, two-tailed test? How often would we have *failed* to reject this false H_0?" Stated another way, how often would we have made a type II error?

Let us look at each of these questions in turn. Under $H_0 : \mu_1 = \mu_2$, we would reject the null hypothesis at $\alpha = 0.05$, two-tailed test, whenever we obtained a difference in sample means equal to or greater than 5 or equal to or less than -5. Looking at Table 17.1, we see that, under the *true* sampling distribution of differences, we would obtain a difference in means equal to or greater than 5 about 26% of the time (0.1157 + 0.0826 + 0.0413 + 0.0165 + 0.0083 = 0.2644). Under the *true* distribution, we would never obtain a difference in means equal to or less than -5. Thus the probability of correctly rejecting H_0 is 0.2644.

How often would we fail to reject H_0? Any difference in means less than 5 would *not* cause us to reject the null hypothesis. Since we know H_0 to be false, this entire region would represent a type II error. The probability of a type II error is therefore $1.0000 - 0.2644 = 0.7356$. In other words, we would *fail* to reject a false null hypothesis almost 74% of the time. The power of the test is 0.2644, which means that we would correctly reject a false null hypothesis only about 26% of the time.

Figure 17.1 graphically presents probability histograms for both the null and its *true* distributions of $\overline{X}_1 - \overline{X}_2$ for the sampling problem we have been discussing.

We might note that failure to reject H_0 when it is false is not only a statistical and research problem; it is also an economic one. Research costs both time and money. One must question the advisability of undertaking a project when the probability of making a type II error is high. In this chapter, we shall discuss several strategies for reducing the risk of a type II error.

The following section illustrates the calculation of power. While the following research situation is hypothetical, it is presented along with the example in Section 17.7 to emphasize the importance of several conditions (sample size, etc.) on the power of a test.

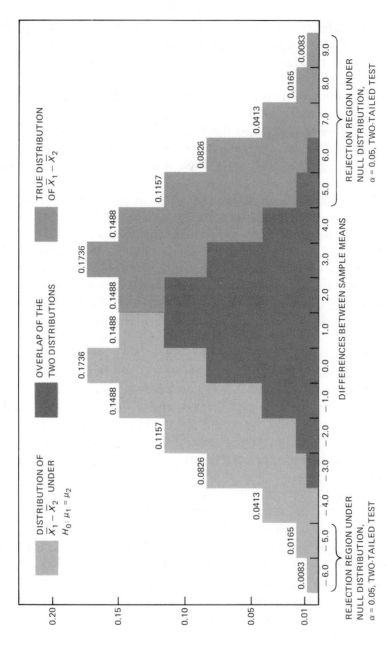

17.1 Probability histograms for true distributions of $\overline{X}_1 - \overline{X}_2$ for the sampling example given in the text.

17.2 Calculation of Power: One-Sample Case

A sociologist working for a state welfare office has constructed two aptitude scales which he administers interchangeably to incoming groups of social work trainees. He estimates from considerable experience that the mean performance on scale A is 70, and on scale B is 72. Both scales have a standard deviation of 5. He is embarrassed to discover that his assistant failed to record which scale was administered to a group of 16 trainees. Scanning the data and noticing a number of low scores, he believes that this sample came from a population in which $\mu = 70$ (that is, scale A was administered to this group), or $H_0: \mu = \mu_0 = 70$.

As a matter of fact, however, scale B was administered. Thus since we know that H_0 is false and we know the true value of μ under H_1 ($\mu = \mu_1 = 72$), we may calculate the power of the test (i.e., the probability that he will correctly reject the false null hypothesis).

Let us set up this problem in formal statistical terms.

1. *Null hypothesis (H_0):* The mean of the population from which this sample was drawn equals 70, that is $\mu = \mu_0 = 70$.

2. *Alternative hypothesis (H_1):* The mean of the population from which this sample was drawn equals 72, that is $\mu = \mu_1 = 72$.

3. *Statistical test:* Since σ is known, $z = (\overline{X} - \mu_0)/\sigma_{\overline{x}}$ is the appropriate test statistic.

4. *Significance level:* $\alpha = 0.01$ (one-tailed test).

5. *Sampling distribution:* The sampling distribution of the mean is known to be a normal distribution.

6. *Critical region:* $z_{0.01} \geq +2.33$. Since we are using a one-tailed test, the critical region consists of all values of $z = (\overline{X} - \mu_0)/\sigma_{\overline{x}} \geq 2.33$.

Therefore the critical value of the sample statistic (the minimum value of $\overline{X}$ leading to rejection of H_0) is

$$\overline{X} = (2.33)\sigma_{\overline{x}} + \mu_0.$$

Thus the power equals the probability of obtaining this critical value in the distribution under H_1. The following steps are necessary.

Step 1. Calculate the value of $\sigma_{\overline{x}}$:

$$\sigma_{\overline{x}} = \frac{\sigma}{\sqrt{N}} = \frac{5}{\sqrt{16}} = 1.25.$$

Step 2. Determine the critical value of $\overline{X}$ ($\alpha = 0.01$, one-tailed test):

$$\overline{X} = (2.33)(1.25) + 70 = 72.91.$$

Step 3. Determine the probability of obtaining this critical value in the *true* sampling distribution under H_1. The critical value of $\overline{X}$ has a z score, in the distribution under H_1, of

$$z = \frac{72.91 - \mu_1}{\sigma_{\overline{X}}} = \frac{72.91 - 72.00}{1.25} = 0.73.$$

Referring to column C, Table A, Appendix C, we see that the probability of correctly rejecting H_0 is 0.2327. This probability is $1 - \beta$, or the power of the test. Incidentally, the probability of making a type II error (β) is 0.7673.

Figure 17.2 clarifies these relationships by indicating the region of rejection for H_0 in the *true* distribution under H_1, $1 - \beta$ or power. The shaded area indicates β, which is the probability of falsely accepting H_0.

17.2 Region of rejection for H_0 in the distribution under H_1.

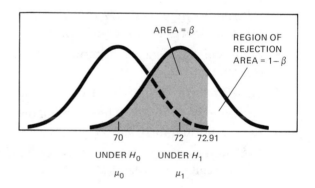

AREA = β

REGION OF REJECTION
AREA = $1 - \beta$

70   72   72.91

UNDER H_0 UNDER H_1

μ_0 μ_1

17.3 The Effect of Sample Size on Power

Power varies as a function of several different factors which were noted earlier in the chapter. Let us first examine the effect of varying the size of the sample on the power of the test. For example, let us assume $N = 25$ in the problem described in the previous section and see what effect this has on the power of the test.

Using the same procedures as just described, we shall test the hypothesis:

$$H_0 : \mu = \mu_0 = 70,$$

given that the true hypothesis is

$$H_1 : \mu = \mu_1 = 72.$$

Since, with $\alpha = 0.01$ (one-tailed test), the critical region consists of all values of $z = (\overline{X} - \mu_0)/\sigma_{\overline{X}} \geq 2.33$, the critical value of $\overline{X} = (2.33)\sigma_{\overline{X}} + \mu_0$. Thus

Step 1. The value of $\sigma_{\overline{X}}$ is

$$\sigma_{\overline{X}} = \frac{\sigma}{\sqrt{N}} = \frac{5}{\sqrt{25}} = 1.00.$$

Step 2. The critical value of $\overline{X}$ is

$$\overline{X} = (2.33)(1.00) + 70 = 72.33.$$

Step 3. The critical value of $\overline{X}$ has a z score in the distribution under H_1 of

$$z = \frac{72.33 - 72.00}{1.00} = 0.33.$$

Referring to column C, Table A, we see that the power of the test is 0.3707.

We have seen in our illustrative problem that when $N = 16$, then the power is 0.2327. When we increased our N to 25, power increased to 0.3707. Had we determined power for $N = 100$ for this example, we would find that the power $= 0.9525$. Thus we may conclude that the power of a test is a function of N; however, the rate at which the power of a test increases with an increasing sample size is not always uniform.

17.4 The Effect of α-Level on Power

In our previous discussion on type I and type II errors (Section 13.5), we indicated that the lower we set α, the lesser the likelihood of a type I error and the greater the likelihood of a type II error. Since β is the probability that a type II error will occur, and power $= 1 - \beta$, the higher the α-level chosen, the greater the power of the test. One can readily demonstrate this relationship between α and power by substituting a different α-level in the preceding problems and observing the change in power.

For example, using $\alpha = 0.05$ (one-tailed test) with $N = 16$, we find that the critical region consists of all values of

$$z = \frac{\overline{X} - \mu_0}{\sigma_{\overline{X}}} \geq 1.65.*$$

Therefore the critical value of $\overline{X} = (1.65)\sigma_{\overline{X}} + \mu_0 = 72.06$. Thus power $= 0.4801$ when $\alpha = 0.05$, as compared to 0.2327 when $\alpha = 0.01$ in the example from Section 17.2.

17.5 The Effect of the Nature of H_1 on Power

The power of a test is also a function of the nature of the alternative hypothesis. In the event that H_0 is actually false, the directional or one-tailed H_1 is more powerful than the two-tailed test so long as the parameter is in the predicted direction.

Inspection of Table 17.3 reveals that the higher the α-level, the lower the absolute value of z required to reject H_0. We have already seen that power increases with increasing α. It follows that power increases as the critical value of z decreases. Table 17.3 shows that for any given α-level, the critical value of z is lower for a one-tailed test than for a two-tailed test. Therefore an obtained z which is not significant for a two-tailed test may be significant for a one-tailed test. Thus the one-

Table 17.3   Critical values of z required to reject H_0 at various α-levels as a function of the nature of H_1

	NATURE OF H_1	
Directional (one-tailed test)		Nondirectional (two-tailed test)
$\alpha = 0.005$	$z = 2.58$	$z = \pm 2.81$
$\alpha = 0.01$	$z = 2.33$	$z = \pm 2.58$
$\alpha = 0.025$	$z = 1.96$	$z = \pm 2.24$
$\alpha = 0.05$	$z = 1.65$	$z = \pm 1.96$

* When $\alpha = 0.05$ (one-tailed test), the critical value of z is exactly halfway between 1.64 and 1.65. We shall use $z = 1.65$ as the critical value so that $p < 0.05$ rather than $z = 1.64$, which results in $p > 0.05$.

tailed test is more powerful than its two-tailed alternative, unless the parameter happens to lie in a direction opposite to the one predicted. In this case, the one-tailed test will be less powerful.

17.6 Parametric versus Nonparametric Tests: Power

Another factor determining the power of a statistical test is the nature of the test itself. We can state as a general rule that for any given N, the parametric tests are more powerful than their nonparametric counterparts. It is primarily for this reason that we have deferred the discussion of statistical power until the present section of the text. For any given N, the parametric tests of significance (those assuming normally distributed populations with the same variance) entail less risk of a type II error. They are more likely to reject H_0 when H_0 is false. Thus given the choice between a nonparametric and a parametric test of significance, the parametric test should be used so long as its underlying assumptions are fulfilled. However, as we shall see in the following chapter, there are numerous situations in which the very nature of our data excludes the possibility of a parametric test of significance. We shall therefore be forced to use less powerful nonparametric tests.*

Why do nonparametric tests have less power? Briefly stated, the answer is that parametric statistical tests (as opposed to nonparametric tests) make maximum use of all the information that is available in the data when the populations are normally distributed.

Let us look at a simple illustration. Imagine that we have obtained the following scores in the course of conducting a study: 50, 34, 21, 12, 10. Now, if we were to convert these scores into ranks, we would obtain 1, 2, 3, 4, 5. Note that all the information concerning the *magnitudes* of the scores is lost when we convert to ranks. The difference between the scores of 50 and 34 becomes "equivalent" when expressed as ranks to the difference between, say, 12 and 10. This greater sensitivity of the parametric tests to the magnitudes of scores makes them a more accurate basis for arriving at probability values when the basic measurement assumptions are met.

---

* It must be reiterated that the parametric tests are more powerful only when the assumptions underlying their use are valid. When the assumptions are not met, a nonparametric treatment may be as powerful as the parametric.

17.7 Calculation of Power: Two-Sample Case

So far, we have examined the effect of various factors on power, using the one-sample case. All the conclusions drawn apply equally to the two-sample case.

At this point, we present a sample problem in which we calculate the power of a test for the two-sample case. Assume that we wish to determine if the residents' mean ages in two nursing homes for the elderly differ significantly. The two nursing home populations have the following age parameters:

$$\mu_1 = 80, \qquad \mu_2 = 75,$$
$$\sigma_1 = 6, \qquad \sigma_2 = 6.$$

If we draw a sample of nine cases from each of the two populations ($N_1 = 9$, $N_2 = 9$), we may test for the significance of the difference between the two sample means obtained. First, let us set up this problem in formal statistical terms.

1. *Null hypothesis (H_0):* The two samples were drawn from nursing homes (populations) with equal means, that is $\mu_1 = \mu_2$.
2. *Alternative hypothesis (H_1)* The two samples were drawn from populations with different means, that is, $\mu_1 \neq \mu_2$.
3. *Statistical test:* Since we are comparing two sample means drawn from normally distributed populations with known variances, z is the appropriate test statistic.
4. *Significance level:* $\alpha = 0.01$.
5. *Sampling distribution:* The sampling distribution of the statistic ($\overline{X}_1 - \overline{X}_2$) is known to be a normal distribution (see Section 15.1).
6. *Critical region:* $|z_{0.01}| \geq 2.58$. Since H_1 is nondirectional, the critical region consists of all values of $z \geq 2.58$ and $z \leq -2.58$.

In other words, when

$$|z| = \left| \frac{(\overline{X}_1 - \overline{X}_2) - (\mu_1 - \mu_2)}{\sigma_{\overline{X}_1 - \overline{X}_2}} \right| \geq 2.58.$$

we will reject H_0. Since H_0 means that $\mu_1 - \mu_2 = 0$, the lower critical value of ($\overline{X}_1 - \overline{X}_2$) = $(-2.58)\sigma_{\overline{X}_1 - \overline{X}_2}$, and the upper critical value of ($\overline{X}_1 - \overline{X}_2$) = $(+2.58)\sigma_{\overline{X}_1 - \overline{X}_2}$.

Now, since we know H_0 to be false (that is, $\mu_1 - \mu_2 \neq 0$), the power of the test is equal to the probability of obtaining these critical values. Any obtained sample difference which is less than these critical values will lead to a type II error (that is, acceptance of a false H_0).

We use the following steps to calculate power:

Step 1. Calculate the value of $\sigma_{\bar{x}_1 - \bar{x}_2}$:

$$\sigma_{\bar{x}_1} = \frac{\sigma_1}{\sqrt{N_1}} = 2, \qquad \sigma_{\bar{x}_2} = \frac{\sigma_2}{\sqrt{N_2}} = 2.$$

$$\sigma_{\bar{x}_1 - \bar{x}_2} = \sqrt{\sigma_{\bar{x}_1}^2 + \sigma_{\bar{x}_2}^2} = 2.828.$$

Step 2. Determine the critical values of $(\bar{X}_1 - \bar{X}_2)$, $\alpha = 0.01$, two-tailed test. The lower critical value is $(\bar{X}_1 - \bar{X}_2) = (-2.58)(2.828) = -7.296$. The upper critical value is $(\bar{X}_1 - \bar{X}_2) = 7.296$.

Step 3. Determine the probability of obtaining these critical values in the *true* sampling distribution under H_1. The upper critical value of $(\bar{X}_1 - \bar{X}_2)$ has a z score, in the distribution under H_1, of

$$z = \frac{7.296 - \mu_{\bar{x}_1 - \bar{x}_2}}{\sigma_{\bar{x}_1 - \bar{x}_2}} = \frac{7.296 - 5.0}{2.828} = 0.81.$$

Referring to column C, Table A, we see that the area beyond a z of 0.81 is 0.2090. The lower critical value of $(\bar{X}_1 - \bar{X}_2)$ has a z score of

$$z = \frac{-7.296 - 5.0}{2.828} = -4.35.$$

A z of -4.35 is so large that only a negligible proportion of area falls beyond it (< 0.00003). Thus the power $= 0.2090$.

17.8   The Effect of Correlated Measures on Power

In Section 17.2, we indicated that when people have been successfully matched on a variable correlated with the dependent variable, a statistical test that takes this correlation into account provides a more powerful test than one that does not. This may be readily demonstrated.

Using the data in the preceding problem, let us assume that the nine people drawn from population 1 are matched on a related variable with the nine people drawn from population 2 and that the correlation between these two variables is 0.80.

Since, with $\alpha = 0.01$ (two-tailed test), the critical region consists of all values of

$$|z| = \left| \frac{(\overline{X}_1 - \overline{X}_2) - (\mu_1 - \mu_2)}{\sigma_{\overline{x}_1 - \overline{x}_2}} \right| \geq 2.58,$$

the critical values of $(\overline{X}_1 - \overline{X}_2) = (\pm 2.58)\sigma_{\overline{x}_1 - \overline{x}_2}$.

Step 1. The value of $\sigma_{\overline{x}_1 - \overline{x}_2}$ is

$$\sigma_{\overline{x}_1 - \overline{x}_2} = \sqrt{\sigma_{\overline{x}_1}^2 + \sigma_{\overline{x}_2}^2 - 2r\sigma_{\overline{x}_1}\sigma_{\overline{x}_2}} = 1.26.$$

Step 2. The critical values of $(\overline{X}_1 - \overline{X}_2)$ are

$$(\overline{X}_1 - \overline{X}_2) = (\pm 2.58)(1.26) = \pm 3.25.$$

Step 3. The lower critical value of $(\overline{X}_1 - \overline{X}_2)$ has a z score, in the distribution under H_1, of

$$z = \frac{(-3.25 - 5.00)}{1.26} = -6.55.$$

Referring to Table A, we find that the area beyond a z of -6.55 is negligible. Therefore power will be determined according to the upper critical value.

The upper critical value of $(\overline{X}_1 - \overline{X}_2)$ has a z of

$$z = \frac{3.25 - 5.00}{1.26} = -1.39.$$

To find the probability of obtaining $(\overline{X}_1 - \overline{X}_2) \geq 3.25$, we refer to Table A to find the area *above* a z of -1.39. Thus power $= 1.0000 - 0.0823 = 0.9177$. Since power $= 1 - \beta$, the probability of a type II error (β) is 0.0823.

17.9 Power, Type I, and Type II Errors

Let us take a moment to tie together some of our observations about power and type I and type II errors. To begin, we must emphasize the fact that there are only two possibilities with respect to the null hypothesis, that is, either it is true (for example, $\mu_1 = \mu_2$) or it is not true (for example, $\mu_1 \neq \mu_2$). These are two mutually exclusive situations. Now, since a type I error is defined as the probability of rejecting H_0 when it is true, two points should be noted.

1. If H_0 is *false*, the probability of a type I error is zero.
2. It is only when we *reject H_0* that any possibility exists for a type I error. Such an error will be made only when H_0 is true, in which case the probability of a type I error is α.

   Further, since a type II error is defined as the probability of accepting H_0 when it is false, we arrive at the following conclusions.

1. If H_0 is *true*, the probability of a type II error is zero.
2. It is only when we *accept H_0* that any possibility exists for a type II error. Such an error will be made only when H_0 is false, in which case the probability of a type II error is β. Therefore, the concept of power, which is defined in terms of a type II error $(1 - \beta)$, applies only when H_0 is not true.

   Table 17.4 summarizes the probabilities associated with acceptance or rejection of H_0 depending on the true state of affairs. (Recall that this table was presented in Section 13.6.2.)

Table 17.4   The type of error made as a function of the true status of H_0 and the statistical decision we have made. To illustrate, if H_0 is true (column 1) and we have rejected H_0 (row 2), we have made a type I error. If H_0 is false (column 2) and we have rejected H_0, we have made a correct decision.

		TRUE STATUS OF H_0	
		H_0 *True*	H_0 *False*
Decision	Accept H_0	Correct	Type II error
		$1 - \alpha$	β
	Reject H_0	Type I error	Correct
		α	$1 - \beta$

17.10 Power Efficiency of a Statistical Test

In Section 17.6, we pointed out that when the underlying assumptions can be considered valid, parametric tests are more powerful than nonparametric tests for any given N. However, it is also true that when nonparametric tests are to be utilized, we can make any specific nonparametric test as powerful as a parametric test by using a larger sample size. Thus test A may be more powerful than test B when the N's are equal, but B may be as powerful as A when an N of, say, 40 is used compared to an N of 30 with test A.

The concept of **power efficiency** is concerned with the increase in sample size required to make one test as powerful as a competing test. Let us assume that test A is the most powerful for the type of data that we are analyzing. Let us assume that test B is equal in power to test A when their N's are 40 and 30 respectively. We shall let N_b represent the N required to make it as powerful as test A when N_a is used. The power efficiency of test B may now be stated:

$$\text{Power efficiency of test } B = 100\frac{N_a}{N_b}\%. \qquad (17.2)$$

Thus in the preceding example, the power efficiency of test B relative to test A is 100(30/40) or 75%. Therefore, given that all the assumptions for employing test A are met, we shall have to use four cases of test B for every three cases of test A to achieve equal power. If the assumptions underlying test A are not met, the concept of power efficiency has no meaning, since test A should not be used.

Summary

In this chapter, we discussed two important concepts: power and power efficiency. Power is defined as the probability of rejecting H_0 when it is actually false, that is, *power* $= 1 - \beta$. We demonstrated the calculation of power for the one-sample and the two-sample cases when H_0 is known to be false and the true value of the parameter under H_1 is known. The calculation of power requires that we compute: the standard error of the sampling distribution under both H_0 and H_1, the critical value of the sample statistic [$\overline{X}$ in the one-sample case: ($\overline{X}_1 - \overline{X}_2$) in the two-sample case], the probability of obtaining this critical value in the sampling distribution under H_1. This probability is the power of the test. We showed that power varies as a function of: sample size, α-level, the nature of H_1, the nature of the statistical test, and the use of correlated measures.

Power efficiency is concerned with the increase in sample size of a given test necessary to make it as powerful as another test with a smaller N. Symbolically, the power efficiency of test B relative to test A may be represented as

$$\text{Power efficiency of test } B \ = \ (100)\frac{N_a}{N_b}\%.$$

Terms to Remember

Power efficiency   The increase in sample size required to make one test as powerful as a competing test.

Power of a test   The probability of rejecting the null hypothesis when it is, in fact, false.

Exercises

1. Test A has a power efficiency of 80% relative to test B. If in test B we included a total of 24 cases, what is the N required to achieve equal power with test A?
2. Given the sample problem in Section 17.3, calculate power when $N = 100$.
3. Using the sample problem in Section 17.7, demonstrate the effect of the nature of H_1 on power by calculating the power when $\alpha = 0.01$, *one-tailed* test, that is, $H_1 : \mu_1 > \mu_2$.
4. Given $\alpha = 0.01$, two-tailed test, with the following two normal populations:

$$\mu_1 = 100, \quad \mu_2 = 90,$$
$$\sigma_1 = 10, \quad \sigma_2 = 10.$$

a) If a sample of 25 cases is drawn from each population, find

1) the probability of a type I error,
2) the probability of a type II error,
3) the power of the test.

b) If two samples of 25 cases each are drawn from population 1, find

1) the probability of a type I error,
2) the probability of a type II error,
3) the power of the test.

5. In Chapter 15, Exercise 5, you were given two normal populations:

$$\mu_1 = 80, \sigma_1 = 6 \text{ and } \mu_2 = 77, \sigma_2 = 6,$$

and asked to calculate the probability of the following four examples. Assume that a sample of 36 cases is drawn from population 1 and a sample of 36 cases from population 2. Calculate the power for each of the following four examples, using $\alpha = 0.01$, one-tailed test. Which of the factors influencing power do these examples illustrate?

a) $\overline{X}_1 - \overline{X}_2 \geq 5$?
b) $\overline{X}_1 - \overline{X}_2 \geq 0$?
c) $\overline{X}_1 - \overline{X}_2 \leq 0$?
d) $\overline{X}_1 - \overline{X}_2 \leq -5$?

6. Explain why power is an economic as well as a research and statistical problem?

7. Why doesn't the concept of power apply whenever the null hypothesis is true?

8. Explain why a type I error cannot be made whenever the null hypothesis is false.

9. Assume that test A is the most powerful test for the data we are analyzing. What is the power efficiency of test B when $N_a = 20$ and the following values of N_b are required to make test B equally powerful?

a) 25   b) 30   c) 35   d) 40   e) 50

10. Summarize the various factors affecting the power of a test. Enumerate the various ways that power may be increased.

11. Discuss the relationship between Type II errors and the power of a test.

12. Explain how the power of a test can be utilized for statistical purposes.

Statistical Inference with Categorical Variables: Chi Square and Related Measures

18

18.1 Introduction

18.2 The χ^2 One-Variable Case

18.3 The χ^2 Test of the Independence of Categorical Variables

18.4 Limitations in the Use of χ^2

18.5 Nominal Measures of Association Based on χ^2

18.1 Introduction

In recent years there has been a broadening in both scope and penetration of research in the social sciences. Much provocative and stimulating research has been initiated in such diverse areas as socialization, welfare rights, public opinion polling, group processes, and social stratification. New variables have been added to the arsenal of the researcher, many of which do not lend themselves to traditional parametric statistical treatment, either because of the scales of measurement used or because of flagrant violations of the assumptions of these parametric tests.* For these reasons, many new statistical techniques have been developed. In this chapter, we present a few techniques from among over 40 nonparametric procedures, none of which require that we make any assumptions about the shape of the population distribution.

Parametric techniques are usually preferable because of their greater sensitivity. This generalization is not true, however, when the underlying assumptions are seriously violated. Indeed, under certain circumstances (e.g., badly skewed distributions, particularly with small

---

* Parametric statistics requires that the population from which the data have been drawn be normally distributed.

sample sizes) a nonparametric test may well be as powerful as its parametric counterpart.* Consequently the researcher is frequently faced with the difficult choice of a statistical test appropriate for his or her data.

18.2 The χ^2 One-Variable Case

Let us suppose you are a demographer conducting research on the number of children preferred by American couples. Research five years ago on the national level indicated that 16% of the couples polled preferred no children, 65% preferred one or two children, and 19% preferred three or four children (hypothetical data). Realizing that attitudes may have changed over the past five years, you conduct a national study of 600 couples selected by an accepted sampling technique. Their preferred number of children is indicated in Table 18.1.

Table 18.1 Contingency table showing the preferred number of children (observed frequencies).

	OBSERVED FREQUENCIES			
	Preferred number of children			Marginal total
	0	*1 or 2*	*3 or 4*	
Number of couples preferring	120	400	80	600

This is the type of problem for which the χ^2† **one-variable test** is ideally suited. In single-variable applications, the χ^2 test has been described as a "goodness-of-fit" technique: It permits us to determine whether or not a significant difference exists between the *observed* number of cases falling into each category and the *expected* number of cases, based on the null hypothesis. In other words, it permits us to

---

* Numerous investigators have demonstrated the robustness of the t and F tests; i.e., even substantial departures from the assumptions underlying parametric tests do not seriously affect the validity of statistical inferences. For articles dealing with this topic, see Haber, Runyon, and Badia, *Readings in Statistics.* Reading, Mass.: Addison-Wesley Publishing Co., Inc., 1970.

† The symbol χ^2 will be used to denote the test of significance as well as the quantity obtained from applying the test to observed frequencies, whereas the word *chi square* will refer to the theoretical chi-square distribution.

answer the question, "How well does our observed distribution fit the theoretical distribution?"

What we require, then, is a null hypothesis that allows us to specify the expected frequencies in each category and, subsequently, a test of this null hypothesis. The null hypothesis may be tested by

$$\chi^2 = \sum_{i=1}^{k} \frac{(f_o - f_e)^2}{f_e} , \qquad (18.1)$$

where

f_o = the observed number in a given category,

f_e = the expected number in that category, and

$\sum_{i=1}^{k}$ = directs us to sum this ratio over all k categories.

In this example, the null hypothesis could be that the preference for the number of children has not changed over the past five years. Thus based on the study five years ago, we expect that 16% or 96 of the present sample of 600 to prefer no children, 65% or 390 persons to prefer one or two children, and 19% or 114 persons to prefer three or four children, as is shown in the expected frequency table (Table 18.2):

Table 18.2   Contingency table showing the preferred number of children (expected frequency).

	EXPECTED FREQUENCIES			
	Preferred number of children			Marginal total
	0	1 or 2	3 or 4	
Number of couples preferring	96	390	114	600

As is readily apparent from Eq. (18.1), if there is close agreement between the observed frequencies and expected frequencies, the resulting χ^2 will be small, leading to a failure to reject the null hypothesis. In fact, if the observed and expected values for all cells in the table are equal, χ^2 will reach its lower limit of 0.* As the discrepancy

---

* The upper limit of χ^2 is $N(k - 1)$ where N is the total number of cases and k the number of rows or columns in the table, whichever is smaller. For example, in a 3 × 3 table with 100 cases, the maximum value of χ^2 would be 200 because $100(3 - 1) = 200$.

$(f_o - f_e)$ increases, the value of χ^2 increases. The larger the χ^2, the more likely we are to reject the null hypothesis.

Substituting the observed and expected values into Eq. (18.1), we find χ^2 to be 16.40 as shown in Table 18.3.

Table 18.3 Computation of chi square

f_o	f_e	$f_o - f_e$	$(f_o - f_e)^2$	$(f_o - f_e)^2/f_e$
120	96	24	576	6.00
400	390	10	100	0.26
80	114	-34	1156	10.14
600	600			$\chi^2 = 16.40$

In studying the Student's t-ratio (Section 14.5), we saw that the sampling distributions of t varied as a function of degrees of freedom. The same is true for χ^2. However, assignments of degrees of freedom with the Student's t-ratio are based on N, whereas for χ^2 the degrees of freedom are a function of the number of cells (k). In the one-variable case, df $= k - 1$ since the marginal total is fixed, only $k - 1$ cells are free to vary. Given that the total is 600, as soon as two cells or categories are filled, the third is completely determined. Thus there are only two degrees of freedom: once we know the first and second cells to be 135 and 400, the third *must* be 65.*

In the one-variable case, df $= k - 1$. Table B, Appendix C, lists the critical values of χ^2 for various α-levels. If the obtained χ^2 value *exceeds* the critical value at a given probability level, the null hypothesis may be rejected at that level of significance. Hence, we are only concerned with the right-hand tail of the distribution as all χ^2 tests are one-tailed.

In the preceding example, $k = 3$; therefore df $= 2$. Assuming $\alpha = 0.01$, we find that Table B indicates that an χ^2 value of 9.21 or greater is required for significance. Since our calculated value of 16.40 is greater than 9.21, we may reject the null hypothesis (see Fig. 18.1) and assert that the preferred number of children has changed over the past five years.

---

* This logic applies only to 1 × N and 2 × 2 tables.

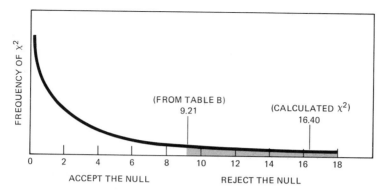

18.1 Sampling distribution of X^2 for two degrees of freedom.

Before proceeding, it should be noted that the shape of the theoretical χ^2 distribution is determined by the degrees of freedom. Figure 18.2 shows the distribution for one, two, and ten degrees of freedom.

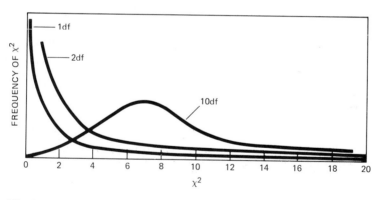

18.2 X^2 distribution for 1, 2 and 10 degrees of freedom.

As is apparent in Fig. 18.2, for all degrees of freedom, the χ^2 distribution is skewed to the right, with those curves associated with large degrees of freedom reaching far to the right before approaching the horizontal axis, which designates the value of χ^2. For example, from

Table B we see that the critical region for a test, given $\alpha = 0.01$ and degrees of freedom equal to 1, begins at 6.635. By following the 0.01 column down you will see that as we deal with greater degrees of freedom (that is, more cells), the corresponding value of χ^2 from Table B that must be exceeded to reject the null hypothesis increases until it reaches 50.892 for 30 degrees of freedom.

18.3 The χ^2 Test of the Independence of Variables

So far in this chapter, we have been concerned with the one-variable case. In practice, we do not encounter the one-variable case very frequently when using categorical variables. More often we ask questions concerning the interrelationships between and among variables. For example, we may ask: "Is there a difference in the crime rate of children coming from different socioeconomic backgrounds? (In other words, is crime rate independent of socioeconomic background, or does it depend in part on their backgrounds?)" Or, "If we are conducting an opinion poll, can we determine whether there is a difference between the opinions of males and females about a given issue? (Stated another way, does the opinion on a given issue depend to any extent on the gender of the respondent?)"

These are but two examples of problems for which the χ^2 technique can be used. You could undoubtedly extend this list to include many campus activities, such as attitudes of fraternity and sorority members versus nonfraternity and nonsorority students toward certain basic issues (for example, cheating on exams), or differences in grading practices among professors in various academic departments. All of these problems have certain characteristics in common: (1) They deal with two or more variables in which (2) the data consist of a frequency count that is tabulated and placed in the appropriate cells. These examples also share an additional and more important characteristic: (3) There is no immediately obvious way to assign expected frequency values to each category. However, as we shall point out shortly, we must consequently base our expected frequencies on the observed frequencies themselves.

Let us take a look at another example. Political scientists, political sociologists, and politicians have long been interested in the relationship between religious affiliation and political party affiliation. Traditionally, Catholics have aligned themselves with the Democratic party, whereas Protestants were more likely to support Republican candidates. The following study of 350 registered voters in California will allow us to test the null hypothesis that Catholics are no more likely to affiliate with either party than are Protestants (see Table 18.4).

Table 18.4   2 × 2 contingency table showing the relationship between religious and political affiliation

POLITICAL AFFILIATION	RELIGIOUS AFFILIATION		Total
	c_1 Catholic	c_2 Protestant	
Democrat r_1	a 125	b 225	350
Republican r_2	c 75	d 125	200
Total	200	350	550

We will now apply a test of significance. In formal statistical terms,

1. *Null hypothesis (H_0):* There is no difference in political party affiliation among Catholics and Protestants.
2. *Alternative hypothesis (H_1):* There is no difference in political party affiliation among Catholics and Protestants; consequently, a one-tailed test of significance will be used.
3. *Statistical test:* Since we are assuming that religious and political affiliation are independent and the data are in terms of frequencies in discrete categories, the χ^2 test of independence is the appropriate statistical test.
4. *Significance level:* $\alpha = 0.05$.
5. *Sampling distribution:* The sampling distribution is the chi-square distribution with df $= (r - 1)(c - 1)$. Since marginal totals are fixed, the frequency of only one cell is free to vary; that is, once one cell value is known in a 2 × 2 table, the other three cell values are determined. Therefore, we have a one-degree-of-freedom situation. The general rule for finding df in the two-variable case is $(r - 1)(c - 1)$, in which $r =$ number of rows and $c =$ number of columns. Thus in the present example, df $= (2 - 1)(2 - 1) = 1$.
6. *Critical region:* Table B in Appendix C shows that for df $= 1$, $\alpha = 0.05$, the critical region consists of all values of $\chi^2 \geq 3.841$. Thus χ^2 is

calculated from Eq. (18.1).

$$\chi^2 = \sum_{i=1}^{k} \frac{(f_o - f_e)^2}{f_e}.$$ *(18.1)*

The main problem now is to decide on a basis for determining the expected cell frequencies. Recall from Chapter 12, Section 12.4.4 that "the probability of the simultaneous or successive occurrence of two events is the product of the separate probabilities of each event." In Table 18.4, the probability of a respondent in the total sample being Catholic is 200/550 or 0.364 and the probability of a respondent being a Democrat is 350/550 or 0.636; therefore, the joint probability of being a Catholic and a Democrat (see cell *a*) is (0.364)(0.636) or 0.2315. Multiplying 0.2315 by our total sample size of 550 results in an expected frequency in cell *a* of 127.3. You could proceed further in this manner and calculate the expected frequency for the remaining cells. However, there is a simple rule that may be followed in determining expected frequency of a given cell: you multiply the two marginal frequencies common to that cell and divide by *N*.

Let us again concentrate on cell *a*. The two marginal totals common to cell *a* are row 1 marginal and column 1 marginal. If the null hypothesis is correct and political affiliation is independent of religious affiliation, we would expect the same proportion of respondents who are Catholics to affiliate with the Democratic party as would those who are Protestants. Since 200 of the total sample of 550 are Catholic, we would expect that (200 × 350)/550 would be found in cell *a*. This figure comes to 127.3, a value identical to the expected frequency we calculated utilizing the more cumbersome *joint-probability method.*

Table 18.5   Relationship between religious and political affiliation (*expected* frequencies within parenthesis)

POLITICAL AFFILIATION	RELIGIOUS AFFILIATION		Total
	Catholic	*Protestant*	
Democrat	*a* 125 (127.3)	*b* 225 (222.7)	350
Republican	*c* 75 (72.7)	*d* 125 (127.3)	200
Total	200	350	550

Table 18.5 presents the observed data, with the expected cell frequencies in the lower right-hand corner of each cell. Incidentally, you may have noted that the four expected frequencies total 550 and that the expected frequencies for a given column or row sum to the respective observed column or row marginal.

It can be seen in Table 18.5 that 125 of the 200 Catholics claimed an affiliation with the Democratic party and that 225 of the 350 Protestants were also affiliated with the Democrats. Do these data indicate that political affiliation is independent of religious affiliation? Table 18.6 summarizes the necessary calculations.

Table 18.6 Computation of chi square

CELL	f_0	f_e	$f_0 - f_e$	$(f_0 - f_e)^2$	$(f_0 - f_e)^2/f_e$
a	125	127.3	-2.3	5.29	0.04
b	225	222.7	2.3	5.29	0.02
c	75	72.7	2.3	5.29	0.07
d	125	127.3	-2.3	5.29	0.04
	550	550.0			$\chi^2 = 0.17$

Since the χ^2 value of 0.17 is less than the value of 3.84 (1df, $\alpha = 0.05$) from Table B required for significance at the 0.05 level, we may not reject H_0. In other words, we must conclude that political party affiliation and religious affiliation are independent; that is, there is no difference in political party affiliation between Catholics and Protestants.

In research we often find that we have more than two subgroups within a nominal class. For example, we might have three categories in one scale and four in another, resulting in a 3 × 4 contingency table. The procedure for obtaining the expected frequencies is the same as the one for the 2 × 2 contingency table. Of course, the degrees of freedom will be greater than 1 (for example, 3 × 4 contingency table, df = 6).

18.4 Limitations in the Use of χ^2

A fundamental assumption in the use of χ^2 is that each observation or frequency is independent of all other observations. Consequently one may not make several observations on the same individual and treat each as though it were independent of all the other observations. Such an error produces what is referred to as an **inflated** N since you are counting or including some people more than once; that is, you are treating the data as though you had a greater number of independent observations than you actually have. This error is extremely serious and may easily lead to the rejection of the null hypothesis when it is in fact true.

Consider the following hypothetical example. Imagine that you are a student in a sociology course and as a class project you decide to poll the student body to determine whether male and female students differ in their opinions on some issue of contemporary significance. Each of 15 members of the class is asked to obtain replies from 10 respondents, 5 male and 5 female. The results are listed in Table 18.7 in which the expected cell frequencies are included in parentheses.

Table 18.7   Response to question

RESPONSE TO QUESTION	SEX		
	Male	Female	Total
Approve	30 (40)	50 (40)	80
Disapprove	45 (35)	25 (35)	70
Total	75	75	150
	$\chi^2 = 10.71$		

Assuming $\alpha = 0.05$, we find that the critical region consists of all the values of $\chi^2 \geq 3.84$. Since the obtained χ^2 of 10.71 > 3.84, you reject the null hypothesis of no difference in the opinions of male and female students on the issue in question. You conclude instead that approval of the issue is dependent on the sex of the respondent.

After completing the study, you discover that a number of students were inadvertently polled as many as two or three times by different members of the class. Consequently the frequencies within the cells are not independent since some individuals had contributed as many as two or three responses. In a reanalysis of the data, in which only one frequency per respondent was permitted, we obtained the results shown in Table 18.8.

Note that now the obtained χ^2 of 2.72 < 3.84; thus you must accept H_0. The failure to achieve independence of responses resulted in a serious error in the original conclusion. Incidentally, you should note

Table 18.8   Response to question

		SEX	
RESPONSE TO QUESTION	Male	Female	Total
Approve	28 (32.5)	32 (27.5)	60
Disapprove	37 (32.5)	23 (27.5)	60
Total	65	55	120

$$\chi^2 = 2.72$$

that the requirement of independence within a cell or condition is basic to *all* statistical tests. We have mentioned this specifically in connection with the χ^2 test because violations may be very subtle and not easily recognized.

An equally important limitation of χ^2 stems from the fact that the value of χ^2 is proportional to the sample size. For example, if the sample in Table 18.8 were doubled to 240 and the observed frequencies were increased proportionately, χ^2 would equal 5.44 rather than 2.72. Consequently, in spite of the fact that the relationship has *not* changed, we would reject rather than accept the H_0. For this reason, many social scientists prefer to avoid χ^2 when dealing with large samples because the results can be very misleading. Indeed, it is difficult not to reject the H_0 when the sample size is over 1000 cases!

With small N's or when the expected proportion in any cell is small, the approximation of the sample statistics to the chi-square distribution may not be very close. A rule which has been generally adopted, in the one-degree-of-freedom situation, is that the *expected frequency* in all cells should be equal to or greater than 5. When df > 1, the expected frequency should be equal to or greater than 5 in at least 80% of the cells. When these requirements are not met, other statistical tests are available. (See Siegel, 1956.)

When a contingency table is larger than 2 × 2 it is possible to collapse rows or columns in such a manner as to increase the frequencies within the resulting cells. Such a practice is used widely in the social sciences with the understanding that the categories that are combined logically fit together and that the procedure is not intended to increase or decrease the value of χ^2.

18.5 Nominal Measures of Association Based on χ^2

We learned in Section 18.4 that the value of χ^2 is directly proportional to the sample size. This characteristic of χ^2 plus the observation that $\chi^2 = 0$ when two variables are statistically independent led to the development of the **phi coefficient**, which is used primarily with 2 × 2 tables.

$$\phi = \sqrt{\frac{\chi^2}{N}} \qquad\qquad (18.2)$$

or

$$\phi^2 = \frac{\chi^2}{N}. \qquad\qquad (18.3)$$

For any table in which either the number of rows or columns equals two, phi ranges from 0 when the variables are independent to 1 when the variables are perfectly related.* Phi is sometimes referred to as a measure of the degree of diagonal concentration; hence, when two diagonally opposite cells are both empty as in the example below, phi equals 1, as is demonstrated in Tables 18.9 and 18.10.

Table 18.9

	SEX		
VOTE	*Male*	*Female*	*Total*
Yes	50	0	50
No	0	50	50
Total	50	50	N = 100

$$\chi^2 = \frac{(50 - 25)^2}{25} + \frac{(0 - 25)^2}{25} + \frac{(0 - 25)^2}{25} + \frac{(50 - 25)^2}{25}$$

$$= 25 + 25 + 25 + 25$$

$$= 100$$

$$\phi = \sqrt{\frac{100}{100}} = 1 \quad \text{and} \quad \phi^2 = \frac{100}{100} = 1.$$

---

* The maximum value of phi exceeds 1 when a table contains more than two rows and two columns.

Table 18.10

		SEX	
VOTE	*Male*	*Female*	*Total*
Yes	0	50	50
No	50	0	50
Total	50	50	$N = 100$

$$\chi^2 = \frac{(0 - 25)^2}{25} + \frac{(50 - 25)^2}{25} + \frac{(50 - 25)^2}{25} + \frac{(0 - 25)^2}{25}$$

$$= 25 + 25 + 25 + 25$$

$$= 100$$

$$\phi = \sqrt{\frac{100}{100}} = 1 \quad \text{and} \quad \phi^2 = \frac{100}{100} = 1.$$

Some social scientists prefer ϕ^2 for a 2×2 table as its value equals Goodman's and Kruskal's tau (see Chapter 8) in this particular instance. Furthermore, as is the case with the other two measures of association to be discussed, phi only requires nominal-level data and provides us with a measure of the strength, but not the direction, of association since the sign of phi will always be positive. By examining the percentage distribution of cases in the cells, it is possible, of course, to determine the pattern of the relationship.

The **contingency coefficient** was developed by Karl Pearson primarily for use with square tables having more than two rows and columns, for example, 3×3 or 4×4.

$$C = \sqrt{\frac{\chi^2}{\chi^2 + N}}. \tag{18.4}$$

The contingency coefficient equals 0 when the variables are independent; however, its maximum value is always less than 1 and is determined by the number of rows and columns in the table.* C has been calculated for the data in Table 18.11.

In this instance $C = 0.55$, and we can conclude that there is a moderately strong relationship between educational attainment and political activism since the maximum value for C with a 3×3 table is

---

*   The maximum value for a square table (for example, 2×2, 3×3, etc.) is calculated using the equation $\sqrt{(k - 1)/k}$ when k equals the number of rows or columns, whichever is less. For a 2×2 table, the maximum value would be $\sqrt{(2 - 1)/2} = \sqrt{1/2} = 0.707$, and for a 3×3 table, the maximum value would be $\sqrt{(3 - 1)/3} = 0.816$.

Table 18.11 Educational attainment by political activism

| POLITICAL ACTIVISM | EDUCATIONAL ATTAINMENT | | | |
	Low	Moderate	High	Total
High	5	10	35	50
Moderate	15	30	5	50
Low	30	10	10	50
Total	50	50	50	N = 150

$$\chi^2 = \frac{(5 - 16.7)^2}{16.7} + \frac{(10 - 16.7)^2}{16.7} + \frac{(35 - 16.7)^2}{16.7} + \frac{(15 - 16.7)^2}{16.7} +$$

$$\frac{(30 - 16.7)^2}{16.7} + \frac{(5 - 16.7)^2}{16.7} + \frac{(30 - 16.7)^2}{16.7} + \frac{(10 - 16.7)^2}{16.7} +$$

$$\frac{(10 - 16.7)^2}{16.7}$$

$$= 8.2 + 2.7 + 20.1 + 0.2 + 10.6 + 8.2 + 10.6 + 2.7 + 2.7$$

$$= 66.0$$

$$C = \sqrt{\frac{\chi^2}{\chi^2 + N}} = \sqrt{\frac{66.0}{66.0 + 150}} = 0.55$$

0.816. By examining Table 18.11 we can also see that the respondents who have completed higher levels of education tend to be more politically active than those respondents who have had less education.

A disadvantage of the contingency coefficient, in addition to its interpretability when the variables are neither independent nor perfectly related, is the difficulty of comparing contingency coefficients for tables of unequal size (differing numbers of rows and columns). The fluctuation of its maximum value has led to the development of **Cramer's** V, a third measure of association for nominal level data based on χ^2.

Cramer's V can be used with square and nonsquare tables of any size, and it ranges from 0 when the variables are independent to 1 when they are perfectly related.

$$V = \sqrt{\frac{\chi^2}{N \cdot \text{Minimum } (r - 1 \text{ or } c - 1)}}. \qquad (18.5)$$

The denominator of Cramer's V is calculated by multiplying N times the smaller of the two quantities, $r - 1$ or $c - 1$. Calculating V for Table 18.11, we find it to equal 0.47 versus 0.55 for C. This example should help clarify the problems of interpreting ϕ, C, and V. They are not equivalent measures with one limited exception.*

$$V = \sqrt{\frac{66.0}{150(3 - 1)}} = \sqrt{0.220} = 0.47.$$

Cramer's V is the most versatile of the three measures of association since its range is always 0 to 1 and it can be used with a table of any dimension. It does share limitations with the phi and contingency coefficients. Cramer's V does not allow for a PRE interpretation, nor does it provide an indication of the direction of relationship.†

Summary

In this chapter we have discussed two tests of significance used with categorical variables, that is, the χ^2 one-variable test and the χ^2 two-variable test.

The χ^2 one-variable test has been described as a "goodness-of-fit" technique, permitting us to determine whether or not a significant difference exists between the *observed* number of cases appearing in each category and the expected number of cases specified under the null hypothesis.

The χ^2 test of independence of variables may be used to determine whether two variables are related or independent. If the χ^2 value is significant, we may conclude that the variables are interdependent or related.

We discussed three limitations on the use of the χ^2 test. In the one-degree-of-freedom situation, the expected frequency should equal or exceed 5 to permit the use of the χ^2 test. When df > 1, the expected frequency in 80% of the cells should equal or exceed 5. A second, and most important, restriction is that the frequency counts must be independent of one another. Failure to meet this requirement results in an error known as the inflated N and may well lead to the rejection of the null hypothesis when it is true (type I error). A third limitation is that χ^2 is directly proportional to N and hence can be misleading.

---

* $V^2 = \phi^2$ in a 2 $\times$ k table.

†   Phi square does allow for a PRE interpretation with a 2 $\times$ 2 table as its value is equivalent to Goodman's and Kruskal's tau, which was discussed in Chapter 8.

Three nominal measures of association were discussed, including the phi coefficient, the contingency coefficient, and Cramer's V.

Terms to Remember

Contingency coefficient   Nominal measure of association based on χ^2 and used primarily for square tables.

Cramer's V   Nominal measure of association based on χ^2 suitable for tables of any dimension.

Inflated N   An error produced whenever several observations are made on the same individual and treated as though they were independent observations.

One-variable test ("goodness-of-fit" technique)   Test of whether or not a significant difference exists between the *observed* number of cases falling into each category and the *expected* number of cases, based on the null hypothesis.

Phi coefficient   Nominal measure of association based on χ^2 used with 2×2 tables.

Exercises

1. Marital status and depression has long been a topic of interest for social scientists. Given the following data, calculate χ^2, an appropriate measure of association and interpret the results if $\alpha = 0.05$.

	MARITAL STATUS		
DEPRESSION	*Married*	*Never Married*	*Formerly Married*
High	20	15	15
Moderate	26	12	10
Low	18	9	8

2. A researcher has reported that a very strong, positive relationship exists between religious affiliation and race. Critique. (*Hint*: Consider the level of measurement.)

3. Professor Stevens has the following data that do not meet the assumptions of χ^2 and must be altered. Combine categories in a defensible manner and calculate χ^2.

EDUCATIONAL ATTAINMENT	COUNTRY OF BIRTH United States	Germany	England	Vietnam	Thailand	Korea
Graduate school	6	4	3	3	1	1
College	8	4	3	2	2	2
High school	9	2	5	4	3	0
Grade school	4	3	2	1	3	4

4. A study was conducted to determine if there is a relationship between socioeconomic status and attitudes toward a new urban renewal program. The results are listed in the accompanying table.

ATTITUDE	SOCIOECONOMIC STATUS Lower	Middle
Approve	90	60
Disapprove	200	100

Set up this study in formal statistical terms and draw the appropriate conclusion.

5. Why does $\chi^2 = 0$ when two variables are independent?

6. Discuss the ways in which ϕ, C, and V are unique with respect to each other and similar to each other.

7. In a study concerned with preferences of schools for their children, 100 people in a high-income group and 200 people in a lower-income group were interviewed. The results of their choices follow:

PREFERENCE STATED	UPPER-INCOME GROUP	LOWER-INCOME GROUP
Private, nonreligious	36	84
Private, religious	39	51
Public	16	44
Have no preference	9	21

What conclusions would you draw from these data?

8. Professor Norman Yetman of the University of Kansas has concerned himself with possible economic discrimination against black athletes by the media. (*Source*: Gary Lehman, Associated Press Writer, August 22, 1975.) He has analyzed their opportunities to appear on commercials, make guest appearances, and obtain off-season jobs. The following table summarizes the results.

APPEARANCE IN COMMERCIALS	RACE OF ATHLETES	
	White	*Nonwhite*
Opportunity	8	2
No opportunity	3	11

Test the null hypothesis that opportunity to appear in commercials is independent of racial background (that is, whites and nonwhites have an equal opportunity to appear).

9. Recent evidence suggests that the ownership of a dog may be therapeutic. In a study of status with respect to ownership of a pet dog and survival record after suffering coronary heart disease, the following data were reported. Test the null hypothesis that ownership of a dog is unrelated to survival. (*Source*: Cited in a syndicated column by Dr. Neil Solomon in *The Los Angeles Times,* Jan. 25, 1979.)

		STATUS OF DOG OWNERSHIP		
		Owners	*Not owners*	*Total*
SURVIVAL STATUS	*Survived one year*	50	28	78
	Did not survive	3	11	14
	Total	53	39	92

10. In the preceding exercise, we found a significant relationship between dog ownership and survival following coronary heart disease. Critique the conclusion, "Dog ownership leads to better survival rates of coronary victims," in the light of what we know about correlation and causation.

Appendixes

A   Review of Basic Mathematics

B   Glossary of Symbols

C   Tables

D   Glossary of Terms

E   References

Review of Basic
Mathematics

Arithmetic Operations

You already know that addition is indicated by the sign +, subtraction by the sign −, multiplication in one of three ways, 2×4, $2(4)$, or $2 \cdot 4$, and division by a slash, /, an overbar, —, or the symbol ÷. However, it is not unusual to forget the rules concerning addition, subtraction, multiplication, and division, particularly when these operations occur in a single problem.

Addition and Subtraction

When numbers are added together, the order of adding the numbers has no influence on the sum. Thus we may add $2 + 5 + 3$ in any of the following ways:

$$2 + 5 + 3, \quad 5 + 2 + 3, \quad 2 + 3 + 5,$$
$$5 + 3 + 2, \quad 3 + 2 + 5, \quad 3 + 5 + 2.$$

When a series of numbers containing both positive and negative signs are added, the order of adding the numbers has no influence on the sum. It is often desirable, however, to group together the numbers preceded by positive signs, group together the numbers preceded by negative signs, add each group separately, and subtract the latter sum from the former. Thus

$$-2 + 3 + 5 - 4 + 2 + 1 - 8$$

may best be added by grouping in the following ways:

$$
\begin{array}{cc}
+3 & \\
+5 & -2 \\
+2 & -4 \\
\underline{+1} & \underline{-8} \\
+11 & -14 = -3.
\end{array}
$$

Incidentally, to subtract a larger numerical value from a smaller numerical value, as in the above example (11 − 14), we ignore the signs, subtract the smaller number from the larger, and affix the sign of the larger to the sum. Thus $-14 + 11 = -3$.

Multiplication

The order in which numbers are multiplied has no effect on the product. In other words,

$$2 \times 3 \times 4 = 2 \times 4 \times 3 = 3 \times 2 \times 4$$
$$= 3 \times 4 \times 2 = 4 \times 2 \times 3 = 4 \times 3 \times 2 = 24.$$

When addition, subtraction, and multiplication occur in the same expression, we must develop certain procedures governing *which* operations are to be performed first.

In the expression

$$2 \times 4 + 7 \times 3 - 5,$$

multiplication is performed first. Thus the above expression is equal to

$$2 \times 4 = 8,$$
$$7 \times 3 = 21,$$
$$-5 = -5$$

and

$$8 + 21 - 5 = 24.$$

We may *not* add first and then multiply. Thus $2 \times 4 + 7$ is *not* equal to $2(4 + 7)$ or 22.

If a problem involves finding the product of one term multiplied by a second expression that includes two or more terms either added or subtracted, we may multiply first and then add, or add first and then multiply. Thus the solution to the following problem becomes

$$8(6 - 4) = 8 \times 6 - 8 \times 4$$
$$= 48 - 32$$
$$= 16$$

or

$$8(6 - 4) = 8(2)$$
$$= 16.$$

In most cases, however, it is more convenient to reduce the expression within the parentheses first. Thus, generally speaking, the second solution appearing above will be more frequently used.

Finally, if numbers having like signs are multiplied, the product is always positive; for example, $(+2) \times (+4) = +8$ and $(-2) \times (-4) = +8$. If numbers bearing unlike signs are multiplied, the product is always negative; for example, $(+2) \times (-4) = -8$ and $(-2) \times (+4) = -8$. The same rule applies also to division. When we obtain the quotient of two numbers of like signs, it is always positive; when the numbers differ in sign, the quotient is always negative.

Multiplication as successive addition   Many students tend to forget that multiplication is a special form of successive addition. Thus

$$15 + 15 + 15 + 15 + 15 = 5(15)$$

and

$$(15 + 15 + 15 + 15 + 15) + (16 + 16 + 16 + 16) = 5(15) + 4(16).$$

This formulation is useful in understanding the advantages of "grouping" scores into what is called a frequency distribution. In obtaining the sum of an array of scores, some of which occur a number of times, it is desirable to multiply each score by the frequency with which it occurs and then add the products. Thus, if we were to obtain the following distribution of scores,

12, 13, 13, 13, 14, 14, 14, 14, 15, 15, 15, 15,
15, 15, 15, 16, 16, 16, 17, 17, 17, 17, 18,

and wanted the sum of these scores, it would be advantageous to form the following frequency distribution:

X	f	fX
12	1	12
13	3	39
14	4	56
15	7	105
16	3	48
17	4	68
18	1	18
$N = 23$		$\sum fX = 346$

Algebraic Operations

Transposing

To transpose a term from one side of an equation to another, you merely have to *change the sign* of the transposed term. All the following are equivalent statements:

$$a + b = c,$$
$$a = c - b,$$
$$b = c - a,$$
$$0 = c - a - b,$$
$$0 = c - (a + b).$$

Solving Equations Involving Fractions

Much of the difficulty encountered in solving equations that involve fractions can be avoided by remembering one important mathematical principle:

Equals multiplied by equals are equal.

Let us look at a few sample problems.

1. Solve the following equation for x:

$$b = \frac{a}{x}.$$

In solving for x, we want to express the value of x in terms of a and b. In other words, we want our final equation to read, $x = \underline{\hspace{1cm}}$. Note that we may multiply both sides of the equation by x/b and obtain the following:

$$\cancel{b} \cdot \frac{x}{\cancel{b}} = \frac{a}{\cancel{x}} \cdot \frac{\cancel{x}}{b}.$$

This reduces to

$$x = \frac{a}{b}.$$

2. Solve the same equation for a. Similarly, if we wanted to solve the equation in terms of a, we could multiply both sides of the equation by x. Thus

$$b \cdot x = \frac{a}{\cancel{x}} \cdot \cancel{x}$$

becomes $bx = a$, or $a = bx$.

In each of the preceding solutions, you will note that the net effect of multiplying by a constant has been to rearrange the terms in the numerator and the denominator of the equations. In fact, we may state two general rules that will permit us to solve the above problems without having to use multiplication by equals (although multiplication by equals is implicit in the arithmetic operations):

a) A term that is in the denominator on one side of the equation may be moved to the other side of the equation by multiplying it by the numerator on that side. Thus

$$\frac{x}{a} = b$$

becomes

$$x = ab.$$

b) A term in the numerator on one side of an equation may be moved to the other side of the equation by dividing the numerator on that side by it. Thus

$$ab = x$$

may become

$$a = \frac{x}{b} \quad \text{or} \quad b = \frac{x}{a}.$$

Thus we have seen that all of the following are equivalent statements:

$$b = \frac{a}{x}, \qquad a = bx, \qquad x = \frac{a}{b}.$$

Similarly,

$$\frac{\Sigma X}{N} = \overline{X}, \qquad \Sigma X = N\overline{X}, \qquad \frac{\Sigma X}{\overline{X}} = N.$$

Dividing by a sum or a difference   It is true that

$$\frac{x + y}{z} = \frac{x}{z} + \frac{y}{z} \quad \text{and} \quad \frac{x - y}{z} = \frac{x}{z} - \frac{y}{z}.$$

We cannot, however, simplify the following expressions as easily:

$$\frac{x}{y + z} \quad \text{or} \quad \frac{x}{y - z}.$$

Thus

$$\frac{x}{y + z} \neq \frac{x}{y} + \frac{x}{z},$$

in which $\neq$ means "not equal to."

Reducing Fractions to Simplest Expressions

A corollary to the rule stating that equals multiplied by equals are equal is

Unequals multiplied by equals remain proportional.

Thus, if we were to multiply 1/4 by 8/8, the product, 8/32, is in the same proportion as 1/4. This corollary is useful in reducing the complex fractions to their simplest expression. Let us look at an example.

Example

Reduce

$$\frac{a/b}{c/d} \quad \text{or} \quad \frac{a}{b} \div \frac{c}{d}$$

to its simplest expression.

Note that, if we multiply both the numerator and the denominator by

$$\frac{bd/1}{bd/1},$$

we obtain

$$\frac{(a/b) \cdot (bd/1)}{(c/d) \cdot (bd/1)},$$

which becomes ad/bc.

However, we could obtain the same result if we were to *invert the divisor* and multiply. Thus

$$\frac{a/b}{c/d} = \frac{a}{b} \cdot \frac{d}{c} = \frac{ad}{bc}.$$

We may now formulate a general rule for dividing one fraction into another fraction. In dividing fractions, we *invert the divisor and multiply*. Thus

$$\frac{x/y}{a^2/b} \quad \text{becomes} \quad \frac{x}{y} \cdot \frac{b}{a^2},$$

which equals

$$\frac{bx}{a^2 y}.$$

To illustrate: if $a = 5$, $b = 2$, $x = 3$, and $y = 4$, the preceding expressions become

$$\frac{3/4}{5^2/2} = \frac{3}{4} \cdot \frac{2}{5^2} = \frac{2 \cdot 3}{4 \times 5^2} = \frac{6}{100}.$$

A general practice you should follow when substituting numerical values into fractional expressions is to reduce the expression to its simplest form *prior* to substitution.

Multiplication and Division of Terms Having Exponents

An exponent indicates how many times a number is to be multiplied by itself. For example, X^5 means that X is to be multiplied by itself 5 times, or

$$X^5 = X \cdot X \cdot X \cdot X \cdot X.$$

If $X = 3$,

$$X^5 = 3 \cdot 3 \cdot 3 \cdot 3 \cdot 3 = 243$$

and

$$\left(\frac{1}{X}\right)^5 = \frac{1^5}{X^5} = \frac{1 \cdot 1 \cdot 1 \cdot 1 \cdot 1}{3 \cdot 3 \cdot 3 \cdot 3 \cdot 3} = \frac{1}{243}.$$

To multiply X to the ath power (X^a) times X raised to the bth power, you simply *add the exponents,* thus raising X to the $(a + b)$th power. The reason for the addition of exponents may be seen from the following illustration.

If $a = 3$ and $b = 5$, then

$$X^a \cdot X^b = X^3 X^5 = (X \cdot X \cdot X)(X \cdot X \cdot X \cdot X \cdot X),$$

which equals X^8.

Now, if $X = 5$, $a = 3$, and $b = 5$, then

$$X^a \cdot X^b = X^{a+b} = X^{3+5} = X^8 = 5^8 = 390{,}625.$$

If $X = 1/6$, $a = 2$, and $b = 3$,

$$X^a \cdot X^b = X^{a+b} = \left(\frac{1}{6}\right)^{2+3} = \left(\frac{1}{6}\right)^5 = \frac{1^5}{6^5} = \frac{1}{7776}.$$

To divide X raised to the ath power by X raised to the bth power, you simply *subtract* the exponent in the denominator from the exponent in the numerator.* The reason for the subtraction is made clear in the

---

* This leads to an interesting exception to the rule that an exponent indicates the number of times a number is multiplied by itself; that is,

$$\frac{X^N}{X^N} = X^{N-N} = X^0;$$

however,

$$\frac{X^N}{X^N} = 1; \qquad \text{therefore} \qquad X^0 = 1.$$

Any number raised to the zero power is equal to 1.

following illustration:

$$\frac{X^a}{X^b} = \frac{X^5}{X^2} = \frac{X \cdot X \cdot X \cdot X \cdot X}{X \cdot X} = X^3 = 3^3 = 27.$$

If $X = 5/6$, $a = 4$, $b = 2$, then

$$\frac{X^a}{X^b} = X^{a-b} = X^{4-2} = X^2.$$

Substituting 5/6 for X we have

$$X^2 = \left(\frac{5}{6}\right)^2 = \frac{5^2}{6^2} = \frac{25}{36} \; .$$

Extracting Square Roots

The square root of a number is the value that, when multiplied by itself, equals that number. Appendix C contains a table of square roots.

The usual difficulty encountered in calculating square roots is the decision as to how many digits precede the decimal, for example $\sqrt{25,000,000} = 5000$, not 500 or 50,000; that is, there are four digits before the decimal. In order to calculate the number of digits preceding the decimal, simply count the number of *pairs* to the left of the decimal: number of pairs = number of digits.

However, if there is an odd number of digits, then the number of digits preceding the decimal equals the number of pairs + 1. The following examples illustrate this point:

$$\frac{50.0}{\sqrt{2500.00}}, \qquad \frac{5.0}{\sqrt{25.00}}$$

and

$$\frac{15.8}{\sqrt{250.00}}, \qquad \frac{1.58}{\sqrt{2.5000}}.$$

Glossary of Symbols

B

Following are definitions of the symbols that appear in the text, followed by the page number showing the first reference to the symbol.

English letters and Greek letters are listed separately in their approximate alphabetical order. Mathematical operators are also listed separately.

Symbol	Definition	Page

Mathematical operators

Symbol	Definition	Page
$\neq$	Not equal to	122
$a < b$	a is less than b	28
$a > b$	a is greater than b	28
$\leq$	Less than or equal to	275
$\geq$	Greater than or equal to	290
$\sqrt{\ }$	Square root	20
X^a	X raised to the ath power	21
$\lvert X \rvert$	Absolute value of X	117
Σ	Sum all quantities or scores that follow	20
$\displaystyle\sum_{i=1}^{N} X_i$	Sum all quantities X_1 through X_N: $X_1 + X_2 + \cdots + X_N$	21

Greek letters

α Probability of a type I error, probability of rejecting H_0 when it is true 318

β
1. Probability of a type II error, probability of accepting H_0 when it is false 318
2. Standardized multiple regression coefficient 259

χ^2 Chi square 435

ϵ Epsilon: percentage difference in a contingency table 159

λ Lambda: an asymmetric or symmetric measure of association for nominal-level contingency table data 161

μ Population mean 132

μ_0 Value of the population mean under H_0 338

μ_1 Value of the population mean under H_1 374

$\mu_{\bar{X}}$ Mean of the distribution of sample means 334

$\mu_{\bar{X}_1 - \bar{X}_2}$ Mean of the distribution of the difference between pairs of sample means 372

$\mu_{\bar{D}}$ Mean of the difference between paired scores 379

ϕ Phi coefficient 444

σ^2 Population variance 118

σ_P Standard error of the proportion 357

σ Population standard deviation 132

$\sigma_{\bar{X}}^2 = \dfrac{\sigma^2}{N}$ Variance of the sampling distribution of the mean 336

$\sigma_{\bar{X}} = \dfrac{\sigma}{\sqrt{N}}$ True standard error of the mean given random samples of a fixed N 335

$\sigma_{\bar{X}_1 - \bar{X}_2}$ True standard error of the difference between means 373

τ Tau: Goodman and Kruskal's asymmetric measure of association for nominal-level contingency table data   166

English letters

a Constant term in a regression equation (Y-intercept)   254

b_y Unstandardized slope of a line relating values of Y to values of X 218

c Number of columns in a contingency table   439

C
1. Centile: Equivalent to a percentile and is a percentage rank that divides a distribution into equal parts   87
2. Concordant pairs in a contingency table   174
3. Contingency coefficient   445

cum f Cumulative frequency   56

cum f_{ll} Cumulative frequency at the lower real limit of the interval containing X 85

cum % Cumulative percent   56

D
1. Rank on X-variable — rank on Y-variable (r_s equation)   203
2. Score on X-variable — score on Y-variable ($X - Y$)   203
3. Decile: a percentage rank that divides a distribution into 10 equal parts   87
4. Discordant pairs in a contingency table   174

d_{yx} Somer's d for ordinal-level data that incorporates tied pairs   179

$\overline{D}$ Mean of the differences between the paired scores   379

df Degrees of freedom: number of values free to vary after certain restrictions have been placed on the data   346

F	A ratio of the between-group variance estimate to the within-group variance estimate	169
f	Frequency	56
f_i	Number of cases within the interval containing X	85
f_e	Expected number in a given category	435
f_o	Observed number in a given category	435
fX	A score multiplied by its corresponding frequency	121
Gamma	Goodman and Kruskal's symmetric measure of association for ordinal-level contingency table data, sometimes written γ.	174
H_0	The null hypothesis; hypothesis actually tested	315
H_1	The alternative hypothesis; hypothesis entertained if H_0 is rejected	315
i	Width of the class interval	21
k	Number of groups or categories or cells	435
Md_n	Median	100
M_o	Mode	103
MD	Mean deviation	117
N	1. Number of pairs	20
	2. Number in either sample	20
	3. Total number of scores or quantities	20
$N_{\bar{x}}$	Total number of means obtained in a sampling experiment	277
PRE	Proportional reduction in error	174
p	Probability	274
$p(A)$	Probability of event A	274
$p(B\mid A)$	Probability of B given that A has occurred	283
P	Probability of the occurrence of an event	280

Q $\begin{cases} \textbf{1.} \text{ Probability of the nonoccurrence} \\ \quad \text{of an event} \hfill 280 \\ \textbf{2.} \text{ Yule's special case of gamma for} \\ \quad 2 \times 2 \text{ tables} \hfill 174 \end{cases}$

Q_1 First quartile, 25th percentile   115

Q_3 Third quartile, 75th percentile   115

r $\begin{cases} \textbf{1.} \text{ Pearson product-moment correla-} \\ \quad \text{tion coefficient (Pearson's } r) \hfill 190 \\ \textbf{2.} \text{ Number of rows in a contingency} \\ \quad \text{table} \hfill 439 \end{cases}$

r^2 Coefficient of determination   231

r_s Spearman rank-order correlation co-efficient (Spearman's rho)   203

$r_{XY \cdot Z}$ Partial correlation coefficient   250

R Multiple correlation coefficient   259

R^2 Coefficient of multiple determination   260

$1 - R^2$ Coefficient of non-determination   260

$s^2 = \dfrac{\sum (X - \overline{X})^2}{N}$ Variance of a sample   119

$s = \sqrt{\dfrac{\sum (X - \overline{X})^2}{N}}$ Standard deviation of a sample   119

$\hat{s}^2 = \dfrac{\sum (X - \overline{X})^2}{N - 1}$ Unbiased estimate of the population variance   343

$s_{\overline{X}}^2$ Estimated variance of the sampling distribution of the mean   343

$s_{\overline{X}} = \dfrac{\hat{s}}{\sqrt{N}} = \dfrac{s}{\sqrt{N-1}}$ Estimated standard error of the mean   344

$s_{\overline{X}_1 - \overline{X}_2}$ Estimated standard error of the difference between means   373

$s_{\overline{D}}$ Estimated standard error of the difference between means, direct-difference method   380

$\hat{s}_{\text{bet}}^2$ Between-group variance estimate   397

$\hat{s}_w^2$ Within-group variance estimate   397

$s_{\text{est } y}$ Standard error of estimate when predictions are made from X to Y 227

$\displaystyle\sum (X - \overline{X})^2$	Sum of squares, sum of the squared deviations from the mean	99
SS_{tot}	Total sum of squares, sum of the squared deviations of each score (X) from the overall mean ($\overline{X}_{tot}$)	393
SS_W	Within-group sum of squares, sum of the squared deviations of each score (X) from the mean of its own group ($\overline{X}_i$)	393
SS_{bet}	Between-group sum of squares, sum of the squared deviations of each group mean ($\overline{X}_i$) from the overall mean ($\overline{X}_{tot}$), multiplied by the N in each group	395
t	Statistic employed to test hypotheses when σ is unknown	345
T_y	Tied pairs with different X values but which share the same Y value. Used with Somer's d.	179
X, Y	Variables; quantities or scores of variables	20
X_i, Y_i	Specific quantities indicated by the subscript i	21
$\overline{X}, \overline{Y}$	Arithmetic means of a sample	95
$\overline{X}_i$	Mean of the ith group	395
$\overline{X}_{tot} = \dfrac{\Sigma X_{tot}}{N}$	Overall mean	393
$(X - \overline{X})$	Deviation of a score from its mean	98
ΣX^2	Sum of the squares of the raw scores	122
$(\Sigma X)^2$	Sum of the raw scores, the quantity squared	122
X_{ll}	Score at lower real limit of interval containing X	85
X_m	Midpoint of an interval	97
Y'	Scores predicted by regression equations (also Y)	217
Y_T	Interval around the regression line within which the true value occurs	228

V Cramer's V 446

z $\begin{cases} \textbf{1.} \text{ Deviation of a specific score from} \\ \quad \text{the mean, expressed in standard} \\ \quad \text{deviation units} \\ \textbf{2.} \text{ Statistic employed to test hypothe-} \\ \quad \text{ses when } \sigma \text{ is known} \end{cases}$

     132

     136

$z_{0.01} = \pm 2.58$ Critical value of z, minimum z required to reject H_0 at the 0.01 level of significance, two-tailed test   339

$z_{0.05} = \pm 1.96$ Minimum value of z required to reject H_0 at the 0.05 level of significance, two-tailed test   339

z_y Y' expressed in terms of a z-score   220

Z Control variable in a multivariate context   249

Tables

C

A   Proportions of area under the normal curve

B   Table of χ^2

C   Critical values of t

D₁   Critical values of F that cut off the upper and lower 2.5% of the F-distribution

E   Transformation of r to z_r

F   Critical values of r_s

G   Functions of r

H   Binominal coefficients

I   Critical values of x or $N - x$ (whichever is larger) at 0.05 and 0.01 levels when $p = Q = 1/2$

J   Critical values of x for various stages of P and Q when $N \leqslant 49$

K   Percentage points of the Studentized range

L   Squares, square roots, and reciprocals of numbers from 1 to 1000

M   Random digits

The use of Table A

The use of Table A requires that the raw score be transformed into a z-score and that the variable be normally distributed.

The values in Table A represent the proportion of area in the standard normal curve, which has a mean of 0, a standard deviation of 1.00, and a total area also equal to 1.00.

Since the normal curve is symmetrical, it is sufficient to indicate only the areas corresponding to positive z-values. Negative z-values will have precisely the same proportions of area as their positive counterparts.

Column B represents the proportion of area between the mean and a given z.

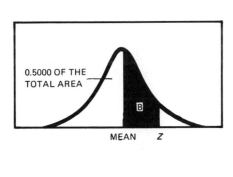

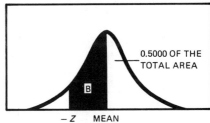

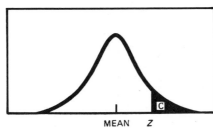

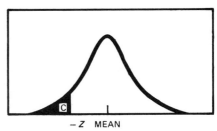

Table A   Proportions of area under the normal curve

(A) z	(B) AREA BETWEEN MEAN AND z	(C) AREA BEYOND z	(A) z	(B) AREA BETWEEN MEAN AND z	(C) AREA BEYOND z	(A) z	(B) AREA BETWEEN MEAN AND z	(C) AREA BEYOND z
0.00	0.0000	0.5000	0.55	0.2088	0.2912	1.10	0.3643	0.1357
0.01	0.0040	0.4960	0.56	0.2123	0.2877	1.11	0.3665	0.1335
0.02	0.0080	0.4920	0.57	0.2157	0.2843	1.12	0.3686	0.1314
0.03	0.0120	0.4880	0.58	0.2190	0.2810	1.13	0.3708	0.1292
0.04	0.0160	0.4840	0.59	0.2224	0.2776	1.14	0.3729	0.1271
0.05	0.0199	0.4801	0.60	0.2257	0.2743	1.15	0.3749	0.1251
0.06	0.0239	0.4761	0.61	0.2291	0.2709	1.16	0.3770	0.1230
0.07	0.0279	0.4721	0.62	0.2324	0.2676	1.17	0.3790	0.1210
0.08	0.0319	0.4681	0.63	0.2357	0.2643	1.18	0.3810	0.1190
0.09	0.0359	0.4641	0.64	0.2389	0.2611	1.19	0.3830	0.1170
0.10	0.0398	0.4602	0.65	0.2422	0.2578	1.20	0.3849	0.1151
0.11	0.0438	0.4562	0.66	0.2454	0.2546	1.21	0.3869	0.1131
0.12	0.0478	0.4522	0.67	0.2486	0.2514	1.22	0.3888	0.1112
0.13	0.0517	0.4483	0.68	0.2517	0.2483	1.23	0.3907	0.1093
0.14	0.0557	0.4443	0.69	0.2549	0.2451	1.24	0.3925	0.1075
0.15	0.0596	0.4404	0.70	0.2580	0.2420	1.25	0.3944	0.1056
0.16	0.0636	0.4364	0.71	0.2611	0.2389	1.26	0.3962	0.1038
0.17	0.0675	0.4325	0.72	0.2642	0.2358	1.27	0.3980	0.1020
0.18	0.0714	0.4286	0.73	0.2673	0.2327	1.28	0.3997	0.1003
0.19	0.0753	0.4247	0.74	0.2704	0.2296	1.29	0.4015	0.0985
0.20	0.0793	0.4207	0.75	0.2734	0.2266	1.30	0.4032	0.0968
0.21	0.0832	0.4168	0.76	0.2764	0.2236	1.31	0.4049	0.0951
0.22	0.0871	0.4129	0.77	0.2794	0.2206	1.32	0.4066	0.0934
0.23	0.0910	0.4090	0.78	0.2823	0.2177	1.33	0.4082	0.0918
0.24	0.0948	0.4052	0.79	0.2852	0.2148	1.34	0.4099	0.0901
0.25	0.0987	0.4013	0.80	0.2881	0.2119	1.35	0.4115	0.0885
0.26	0.1026	0.3974	0.81	0.2910	0.2090	1.36	0.4131	0.0869
0.27	0.1064	0.3936	0.82	0.2939	0.2061	1.37	0.4147	0.0853
0.28	0.1103	0.3897	0.83	0.2967	0.2033	1.38	0.4162	0.0838
0.29	0.1141	0.3859	0.84	0.2995	0.2005	1.39	0.4177	0.0823
0.30	0.1179	0.3821	0.85	0.3023	0.1977	1.40	0.4192	0.0808
0.31	0.1217	0.3783	0.86	0.3051	0.1949	1.41	0.4207	0.0793
0.32	0.1255	0.3745	0.87	0.3078	0.1922	1.42	0.4222	0.0778
0.33	0.1293	0.3707	0.88	0.3106	0.1894	1.43	0.4236	0.0764
0.34	0.1331	0.3669	0.89	0.3133	0.1867	1.44	0.4251	0.0749
0.35	0.1368	0.3632	0.90	0.3159	0.1841	1.45	0.4265	0.0735
0.36	0.1406	0.3594	0.91	0.3186	0.1814	1.46	0.4279	0.0721
0.37	0.1443	0.3557	0.92	0.3212	0.1788	1.47	0.4292	0.0708
0.38	0.1480	0.3520	0.93	0.3238	0.1762	1.48	0.4306	0.0694
0.39	0.1517	0.3483	0.94	0.3264	0.1736	1.49	0.4319	0.0681
0.40	0.1554	0.3446	0.95	0.3289	0.1711	1.50	0.4332	0.0668
0.41	0.1591	0.3409	0.96	0.3315	0.1685	1.51	0.4345	0.0655
0.42	0.1628	0.3372	0.97	0.3340	0.1660	1.52	0.4357	0.0643
0.43	0.1664	0.3336	0.98	0.3365	0.1635	1.53	0.4370	0.0630
0.44	0.1700	0.3300	0.99	0.3389	0.1611	1.54	0.4382	0.0618
0.45	0.1736	0.3264	1.00	0.3413	0.1587	1.55	0.4394	0.0606
0.46	0.1772	0.3228	1.01	0.3438	0.1562	1.56	0.4406	0.0594
0.47	0.1808	0.3192	1.02	0.3461	0.1539	1.57	0.4418	0.0582
0.48	0.1844	0.3156	1.03	0.3485	0.1515	1.58	0.4429	0.0571
0.49	0.1879	0.3121	1.04	0.3508	0.1492	1.59	0.4441	0.0559
0.50	0.1915	0.3085	1.05	0.3531	0.1469	1.60	0.4452	0.0548
0.51	0.1950	0.3050	1.06	0.3554	0.1446	1.61	0.4463	0.0537
0.52	0.1985	0.3015	1.07	0.3577	0.1423	1.62	0.4474	0.0526
0.53	0.2019	0.2981	1.08	0.3599	0.1401	1.63	0.4484	0.0516
0.54	0.2054	0.2946	1.09	0.3621	0.1379	1.64	0.4495	0.0505

Table A   (continued)

(A) z	(B) AREA BETWEEN MEAN AND z	(C) AREA BEYOND z	(A) z	(B) AREA BETWEEN MEAN AND z	(C) AREA BEYOND z	(A) z	(B) AREA BETWEEN MEAN AND z	(C) AREA BEYOND z
1.65	0.4505	0.0495	2.22	0.4868	0.0132	2.79	0.4974	0.0026
1.66	0.4515	0.0485	2.23	0.4871	0.0129	2.80	0.4974	0.0026
1.67	0.4525	0.0475	2.24	0.4875	0.0125	2.81	0.4975	0.0025
1.68	0.4535	0.0465	2.25	0.4878	0.0122	2.82	0.4976	0.0024
1.69	0.4545	0.0455	2.26	0.4881	0.0119	2.83	0.4977	0.0023
1.70	0.4554	0.0446	2.27	0.4884	0.0116	2.84	0.4977	0.0023
1.71	0.4564	0.0436	2.28	0.4887	0.0113	2.85	0.4978	0.0022
1.72	0.4573	0.0427	2.29	0.4890	0.0110	2.86	0.4979	0.0021
1.73	0.4582	0.0418	2.30	0.4893	0.0107	2.87	0.4979	0.0021
1.74	0.4591	0.0409	2.31	0.4896	0.0104	2.88	0.4980	0.0020
1.75	0.4599	0.0401	2.32	0.4898	0.0102	2.89	0.4981	0.0019
1.76	0.4608	0.0392	2.33	0.4901	0.0099	2.90	0.4981	0.0019
1.77	0.4616	0.0384	2.34	0.4904	0.0096	2.91	0.4982	0.0018
1.78	0.4625	0.0375	2.35	0.4906	0.0094	2.92	0.4982	0.0018
1.79	0.4633	0.0367	2.36	0.4909	0.0091	2.93	0.4983	0.0017
1.80	0.4641	0.0359	2.37	0.4911	0.0089	2.94	0.4984	0.0016
1.81	0.4649	0.0351	2.38	0.4913	0.0087	2.95	0.4984	0.0016
1.82	0.4656	0.0344	2.39	0.4916	0.0084	2.96	0.4985	0.0015
1.83	0.4664	0.0336	2.40	0.4918	0.0082	2.97	0.4985	0.0015
1.84	0.4671	0.0329	2.41	0.4920	0.0080	2.98	0.4986	0.0014
1.85	0.4678	0.0322	2.42	0.4922	0.0078	2.99	0.4986	0.0014
1.86	0.4686	0.0314	2.43	0.4925	0.0075	3.00	0.4987	0.0013
1.87	0.4693	0.0307	2.44	0.4927	0.0073	3.01	0.4987	0.0013
1.88	0.4699	0.0301	2.45	0.4929	0.0071	3.02	0.4987	0.0013
1.89	0.4706	0.0294	2.46	0.4931	0.0069	3.03	0.4988	0.0012
1.90	0.4713	0.0287	2.47	0.4932	0.0068	3.04	0.4988	0.0012
1.91	0.4719	0.0281	2.48	0.4934	0.0066	3.05	0.4989	0.0011
1.92	0.4726	0.0274	2.49	0.4936	0.0064	3.06	0.4989	0.0011
1.93	0.4732	0.0268	2.50	0.4938	0.0062	3.07	0.4989	0.0011
1.94	0.4738	0.0262	2.51	0.4940	0.0060	3.08	0.4990	0.0010
1 95	0.4744	0.0256	2.52	0.4941	0.0059	3.09	0.4990	0.0010
1 96	0.4750	0.0250	2.53	0.4943	0.0057	3.10	0.4990	0.0010
1.97	0.4756	0.0244	2.54	0.4945	0.0055	3.11	0.4991	0.0009
1.98	0.4761	0.0239	2.55	0.4946	0.0054	3.12	0.4991	0.0009
1.99	0.4767	0.0233	2.56	0.4948	0.0052	3.13	0.4991	0.0009
2.00	0.4772	0.0228	2.57	0.4949	0.0051	3.14	0.4992	0.0008
2.01	0.4778	0.0222	2.58	0.4951	0.0049	3.15	0.4992	0.0008
2.02	0.4783	0.0217	2.59	0.4952	0.0048	3.16	0.4992	0.0008
2.03	0.4788	0.0212	2.60	0.4953	0.0047	3.17	0.4992	0.0008
2.04	0.4793	0.0207	2.61	0.4955	0.0045	3.18	0.4993	0.0007
2.05	0.4798	0.0202	2.62	0.4956	0.0044	3.19	0.4993	0.0007
2.06	0.4803	0.0197	2.63	0.4957	0.0043	3.20	0.4993	0.0007
2.07	0.4808	0.0192	2.64	0.4959	0.0041	3.21	0.4993	0.0007
2.08	0.4812	0.0188	2.65	0.4960	0.0040	3.22	0.4994	0.0006
2.09	0.4817	0.0183	2.66	0.4961	0.0039	3.23	0.4994	0.0006
2.10	0.4821	0.0179	2.67	0.4962	0.0038	3.24	0.4994	0.0006
2.11	0.4826	0.0174	2.68	0.4963	0.0037	3.25	0.4994	0.0006
2.12	0.4830	0.0170	2.69	0.4964	0.0036	3.30	0.4995	0.0005
2.13	0.4834	0.0166	2.70	0.4965	0.0035	3.35	0.4996	0.0004
2.14	0.4838	0.0162	2.71	0.4966	0.0034	3.40	0.4997	0.0003
2.15	0.4842	0.0158	2.72	0.4967	0.0033	3.45	0.4997	0.0003
2.16	0.4846	0.0154	2.73	0.4968	0.0032	3.50	0.4998	0.0002
2.17	0.4850	0.0150	2.74	0.4969	0.0031	3.60	0.4998	0.0002
2.18	0.4854	0.0146	2.75	0.4970	0.0030	3.70	0.4999	0.0001
2.19	0.4857	0.0143	2.76	0.4971	0.0029	3.80	0.4999	0.0001
2.20	0.4861	0.0139	2.77	0.4972	0.0028	3.90	0.49995	0.00005
2.21	0.4864	0.0136	2.78	0.4973	0.0027	4.00	0.49997	0.00003

Tabled values are two-tailed.

Table B   Table of χ^2

DEGREES OF FREEDOM df	0.10	0.05	0.02	0.01
1	2.706	3.841	5.412	6.635
2	4.605	5.991	7.824	9.210
3	6.251	7.815	9.837	11.341
4	7.779	9.488	11.668	13.277
5	9.236	11.070	13.388	15.086
6	10.645	12.592	15.033	16.812
7	12.017	14.067	16.622	18.475
8	13.362	15.507	18.168	20.090
9	14.684	16.919	19.679	21.666
10	15.987	18.307	21.161	23.209
11	17.275	19.675	22.618	24.725
12	18.549	21.026	24.054	26.217
13	19.812	22.362	25.472	27.688
14	21.064	23.685	26.873	29.141
15	22.307	24.996	28.259	30.578
16	23.542	26.296	29.633	32.000
17	24.769	27.587	30.995	33.409
18	25.989	28.869	32.346	34.805
19	27.204	30.144	33.687	36.191
20	28.412	31.410	35.020	37.566
21	29.615	32.671	36.343	38.932
22	30.813	33.924	37.659	40.289
23	32.007	35.172	38.968	41.638
24	33.196	36.415	40.270	42.980
25	34.382	37.652	41.566	44.314
26	35.563	38.885	42.856	45.642
27	36.741	40.113	44.140	46.963
28	37.916	41.337	45.419	48.278
29	39.087	42.557	46.693	49.588
30	40.256	43.773	47.962	50.892

For any given df, Table C shows the values of *t* corresponding to various levels of probability. Obtained *t* is significant at a given level if it is equal to or *greater than* the value shown in the table.

Table C Critical values of *t*

	LEVEL OF SIGNIFICANCE FOR ONE-TAILED TEST					
	0.10	0.05	0.025	0.01	0.005	0.0005
	LEVEL OF SIGNIFICANCE FOR TWO-TAILED TEST					
df	0.20	0.10	0.05	0.02	0.01	0.001
1	3.078	6.314	12.706	31.821	63.657	636.619
2	1.886	2.920	4.303	6.965	9.925	31.598
3	1.638	2.353	3.182	4.541	5.841	12.941
4	1.533	2.132	2.776	3.747	4.604	8.610
5	1.476	2.015	2.571	3.365	4.032	6.859
6	1.440	1.943	2.447	3.143	3.707	5.959
7	1.415	1.895	2.365	2.998	3.499	5.405
8	1.397	1.860	2.306	2.896	3.355	5.041
9	1.383	1.833	2.262	2.821	3.250	4.781
10	1.372	1.812	2.228	2.764	3.169	4.587
11	1.363	1.796	2.201	2.718	3.106	4.437
12	1.356	1.782	2.179	2.681	3.055	4.318
13	1.350	1.771	2.160	2.650	3.012	4.221
14	1.345	1.761	2.145	2.624	2.977	4.140
15	1.341	1.753	2.131	2.602	2.947	4.073
16	1.337	1.746	2.120	2.583	2.921	4.015
17	1.333	1.740	2.110	2.567	2.898	3.965
18	1.330	1.734	2.101	2.552	2.878	3.922
19	1.328	1.729	2.093	2.539	2.861	3.883
20	1.325	1.725	2.086	2.528	2.845	3.850
21	1.323	1.721	2.080	2.518	2.831	3.819
22	1.321	1.717	2.074	2.508	2.819	3.792
23	1.319	1.714	2.069	2.500	2.807	3.767
24	1.318	1.711	2.064	2.492	2.797	3.745
25	1.316	1.708	2.060	2.485	2.787	3.725
26	1.315	1.706	2.056	2.479	2.779	3.707
27	1.314	1.703	2.052	2.473	2.771	3.690
28	1.313	1.701	2.048	2.467	2.763	3.674
29	1.311	1.699	2.045	2.462	2.756	3.659
30	1.310	1.697	2.042	2.457	2.750	3.646
40	1.303	1.684	2.021	2.423	2.704	3.551
60	1.296	1.671	2.000	2.390	2.660	3.460
120	1.289	1.658	1.980	2.358	2.617	3.373
∞	1.282	1.645	1.960	2.326	2.576	3.291

Note: Table C is taken from Table III (page 46) of Fisher and Yates *Statistical Tables for Biological, Agricultural and Medical Research*, published by Longman Group Ltd., London (previously published by Oliver and Boyd, Edinburgh), and by permission of the authors and publishers.

The obtained F is significant at a given level if it is equal to or *greater than* the value shown in Table D. 0.05 (light row) and 0.01 (dark row) points for the distribution of F.

The values shown are the right tail of the distribution obtained by dividing the larger variance estimate by the smaller variance estimate. To find the complementary left or lower tail for a given df and α level, reverse the degrees of freedom and find the reciprocal of that value in the F-table. For example, the value cutting off the top 5% of the area for df 7 and 12 is 2.85. To find the cutoff point of the bottom 5% of the area, find the tabled value at the $\alpha = 0.05$ level for 12 and 7 df. This is found to be 3.57. The reciprocal is $1/3.57 = 0.28$. Thus 5% of the area falls *at or below an* $F = 0.28$.

Table D   Critical values of F

DEGREES OF FREEDOM FOR NUMERATOR

Each cell shows the 5% critical value (upper) and the 1% critical value (lower, bold).

df (denom.)	1	2	3	4	5	6	7	8	9	10	11	12	14	16	20	24	30	40	50	75	100	200	500	∞
1	161 / **4052**	200 / **4999**	216 / **5403**	225 / **5625**	230 / **5764**	234 / **5859**	237 / **5928**	239 / **5981**	241 / **6022**	242 / **6056**	243 / **6082**	244 / **6106**	245 / **6142**	246 / **6169**	248 / **6208**	249 / **6234**	250 / **6258**	251 / **6286**	252 / **6302**	253 / **6323**	253 / **6334**	254 / **6352**	254 / **6361**	254 / **6366**
2	18.51 / **98.49**	19.00 / **99.01**	19.16 / **99.17**	19.25 / **99.25**	19.30 / **99.30**	19.33 / **99.33**	19.36 / **99.34**	19.37 / **99.36**	19.38 / **99.38**	19.39 / **99.40**	19.40 / **99.41**	19.41 / **99.42**	19.42 / **99.43**	19.43 / **99.44**	19.44 / **99.45**	19.45 / **99.46**	19.46 / **99.47**	19.47 / **99.48**	19.47 / **99.48**	19.48 / **99.49**	19.49 / **99.49**	19.49 / **99.49**	19.50 / **99.50**	19.50 / **99.50**
3	10.13 / **34.12**	9.55 / **30.81**	9.28 / **29.46**	9.12 / **28.71**	9.01 / **28.24**	8.94 / **27.91**	8.88 / **27.67**	8.84 / **27.49**	8.81 / **27.34**	8.78 / **27.23**	8.76 / **27.13**	8.74 / **27.05**	8.71 / **26.92**	8.69 / **26.83**	8.66 / **26.69**	8.64 / **26.60**	8.62 / **26.50**	8.60 / **26.41**	8.58 / **26.30**	8.57 / **26.27**	8.56 / **26.23**	8.54 / **26.18**	8.54 / **26.14**	8.53 / **26.12**
4	7.71 / **21.20**	6.94 / **18.00**	6.59 / **16.69**	6.39 / **15.98**	6.26 / **15.52**	6.16 / **15.21**	6.09 / **14.98**	6.04 / **14.80**	6.00 / **14.66**	5.96 / **14.54**	5.93 / **14.45**	5.91 / **14.37**	5.87 / **14.24**	5.84 / **14.15**	5.80 / **14.02**	5.77 / **13.93**	5.74 / **13.83**	5.71 / **13.74**	5.70 / **13.69**	5.68 / **13.61**	5.66 / **13.57**	5.65 / **13.52**	5.64 / **13.48**	5.63 / **13.46**
5	6.61 / **16.26**	5.79 / **13.27**	5.41 / **12.06**	5.19 / **11.39**	5.05 / **10.97**	4.95 / **10.67**	4.88 / **10.45**	4.82 / **10.27**	4.78 / **10.15**	4.74 / **10.05**	4.70 / **9.96**	4.68 / **9.89**	4.64 / **9.77**	4.60 / **9.68**	4.56 / **9.55**	4.53 / **9.47**	4.50 / **9.38**	4.46 / **9.29**	4.44 / **9.24**	4.42 / **9.17**	4.40 / **9.13**	4.38 / **9.07**	4.37 / **9.04**	4.36 / **9.02**
6	5.99 / **13.74**	5.14 / **10.92**	4.76 / **9.78**	4.53 / **9.15**	4.39 / **8.75**	4.28 / **8.47**	4.21 / **8.26**	4.15 / **8.10**	4.10 / **7.98**	4.06 / **7.87**	4.03 / **7.79**	4.00 / **7.72**	3.96 / **7.60**	3.92 / **7.52**	3.87 / **7.39**	3.84 / **7.31**	3.81 / **7.23**	3.77 / **7.14**	3.75 / **7.09**	3.72 / **7.02**	3.71 / **6.99**	3.69 / **6.94**	3.68 / **6.90**	3.67 / **6.88**
7	5.59 / **12.25**	4.74 / **9.55**	4.35 / **8.45**	4.12 / **7.85**	3.97 / **7.46**	3.87 / **7.19**	3.79 / **7.00**	3.73 / **6.84**	3.68 / **6.71**	3.63 / **6.62**	3.60 / **6.54**	3.57 / **6.47**	3.52 / **6.35**	3.49 / **6.27**	3.44 / **6.15**	3.41 / **6.07**	3.38 / **5.98**	3.34 / **5.90**	3.32 / **5.85**	3.29 / **5.78**	3.28 / **5.75**	3.25 / **5.70**	3.24 / **5.67**	3.23 / **5.65**
8	5.32 / **11.26**	4.46 / **8.65**	4.07 / **7.59**	3.84 / **7.01**	3.69 / **6.63**	3.58 / **6.37**	3.50 / **6.19**	3.44 / **6.03**	3.39 / **5.91**	3.34 / **5.82**	3.31 / **5.74**	3.28 / **5.67**	3.23 / **5.56**	3.20 / **5.48**	3.15 / **5.36**	3.12 / **5.28**	3.08 / **5.20**	3.05 / **5.11**	3.03 / **5.06**	3.00 / **5.00**	2.98 / **4.96**	2.96 / **4.91**	2.94 / **4.88**	2.93 / **4.86**
9	5.12 / **10.56**	4.26 / **8.02**	3.86 / **6.99**	3.63 / **6.42**	3.48 / **6.06**	3.37 / **5.80**	3.29 / **5.62**	3.23 / **5.47**	3.18 / **5.35**	3.13 / **5.26**	3.10 / **5.18**	3.07 / **5.11**	3.02 / **5.00**	2.98 / **4.92**	2.93 / **4.80**	2.90 / **4.73**	2.86 / **4.64**	2.82 / **4.56**	2.80 / **4.51**	2.77 / **4.45**	2.76 / **4.41**	2.73 / **4.36**	2.72 / **4.33**	2.71 / **4.31**
10	4.96 / **10.04**	4.10 / **7.56**	3.71 / **6.55**	3.48 / **5.99**	3.33 / **5.64**	3.22 / **5.39**	3.14 / **5.21**	3.07 / **5.06**	3.02 / **4.95**	2.97 / **4.85**	2.94 / **4.78**	2.91 / **4.71**	2.86 / **4.60**	2.82 / **4.52**	2.77 / **4.41**	2.74 / **4.33**	2.70 / **4.25**	2.67 / **4.17**	2.64 / **4.12**	2.61 / **4.05**	2.59 / **4.01**	2.56 / **3.96**	2.55 / **3.93**	2.54 / **3.91**
11	4.84 / **9.65**	3.98 / **7.20**	3.59 / **6.22**	3.36 / **5.67**	3.20 / **5.32**	3.09 / **5.07**	3.01 / **4.88**	2.95 / **4.74**	2.90 / **4.63**	2.86 / **4.54**	2.82 / **4.46**	2.79 / **4.40**	2.74 / **4.29**	2.70 / **4.21**	2.65 / **4.10**	2.61 / **4.02**	2.57 / **3.94**	2.53 / **3.86**	2.50 / **3.80**	2.47 / **3.74**	2.45 / **3.70**	2.42 / **3.66**	2.41 / **3.62**	2.40 / **3.60**
12	4.75 / **9.33**	3.88 / **6.93**	3.49 / **5.95**	3.26 / **5.41**	3.11 / **5.06**	3.00 / **4.82**	2.92 / **4.65**	2.85 / **4.50**	2.80 / **4.39**	2.76 / **4.30**	2.72 / **4.22**	2.69 / **4.16**	2.64 / **4.05**	2.60 / **3.98**	2.54 / **3.86**	2.50 / **3.78**	2.46 / **3.70**	2.42 / **3.61**	2.40 / **3.56**	2.36 / **3.49**	2.35 / **3.46**	2.32 / **3.41**	2.31 / **3.38**	2.30 / **3.36**
13	4.67 / **9.07**	3.80 / **6.70**	3.41 / **5.74**	3.18 / **5.20**	3.02 / **4.86**	2.92 / **4.62**	2.84 / **4.44**	2.77 / **4.30**	2.72 / **4.19**	2.67 / **4.10**	2.63 / **4.02**	2.60 / **3.96**	2.55 / **3.85**	2.51 / **3.78**	2.46 / **3.67**	2.42 / **3.59**	2.38 / **3.51**	2.34 / **3.42**	2.32 / **3.37**	2.28 / **3.30**	2.26 / **3.27**	2.24 / **3.21**	2.22 / **3.18**	2.21 / **3.16**
14	4.60 / **8.86**	3.74 / **6.51**	3.34 / **5.56**	3.11 / **5.03**	2.96 / **4.69**	2.85 / **4.46**	2.77 / **4.28**	2.70 / **4.14**	2.65 / **4.03**	2.60 / **3.94**	2.56 / **3.86**	2.53 / **3.80**	2.48 / **3.70**	2.44 / **3.62**	2.39 / **3.51**	2.35 / **3.43**	2.31 / **3.34**	2.27 / **3.26**	2.24 / **3.21**	2.21 / **3.14**	2.19 / **3.11**	2.16 / **3.06**	2.14 / **3.02**	2.13 / **3.00**
15	4.54 / **8.68**	3.68 / **6.36**	3.29 / **5.42**	3.06 / **4.89**	2.90 / **4.56**	2.79 / **4.32**	2.70 / **4.14**	2.64 / **4.00**	2.59 / **3.89**	2.55 / **3.80**	2.51 / **3.73**	2.48 / **3.67**	2.43 / **3.56**	2.39 / **3.48**	2.33 / **3.36**	2.29 / **3.29**	2.25 / **3.20**	2.21 / **3.12**	2.18 / **3.07**	2.15 / **3.00**	2.12 / **2.97**	2.10 / **2.92**	2.08 / **2.89**	2.07 / **2.87**

DEGREES OF FREEDOM FOR DENOMINATOR

Table D (continued)

DEGREES OF FREEDOM FOR NUMERATOR

df (denom.)	1	2	3	4	5	6	7	8	9	10	11	12	14	16	20	24	30	40	50	75	100	200	500	∞
16	4.49 / 8.53	3.63 / 6.23	3.24 / 5.29	3.01 / 4.77	2.85 / 4.44	2.74 / 4.20	2.66 / 4.03	2.59 / 3.89	2.54 / 3.78	2.49 / 3.69	2.45 / 3.61	2.42 / 3.55	2.37 / 3.45	2.33 / 3.37	2.28 / 3.25	2.24 / 3.18	2.20 / 3.10	2.16 / 3.01	2.13 / 2.96	2.09 / 2.89	2.07 / 2.86	2.04 / 2.80	2.02 / 2.77	2.01 / 2.75
17	4.45 / 8.40	3.59 / 6.11	3.20 / 5.18	2.96 / 4.67	2.81 / 4.34	2.70 / 4.10	2.62 / 3.93	2.55 / 3.79	2.50 / 3.68	2.45 / 3.59	2.41 / 3.52	2.38 / 3.45	2.33 / 3.35	2.29 / 3.27	2.23 / 3.16	2.19 / 3.08	2.15 / 3.00	2.11 / 2.92	2.08 / 2.86	2.04 / 2.79	2.02 / 2.76	1.99 / 2.70	1.97 / 2.67	1.96 / 2.65
18	4.41 / 8.28	3.55 / 6.01	3.16 / 5.09	2.93 / 4.58	2.77 / 4.25	2.66 / 4.01	2.58 / 3.85	2.51 / 3.71	2.46 / 3.60	2.41 / 3.51	2.37 / 3.44	2.34 / 3.37	2.29 / 3.27	2.25 / 3.19	2.19 / 3.07	2.15 / 3.00	2.11 / 2.91	2.07 / 2.83	2.04 / 2.78	2.00 / 2.71	1.98 / 2.68	1.95 / 2.62	1.93 / 2.59	1.92 / 2.57
19	4.38 / 8.18	3.52 / 5.93	3.13 / 5.01	2.90 / 4.50	2.74 / 4.17	2.63 / 3.94	2.55 / 3.77	2.48 / 3.63	2.43 / 3.52	2.38 / 3.43	2.34 / 3.36	2.31 / 3.30	2.26 / 3.19	2.21 / 3.12	2.15 / 3.00	2.11 / 2.92	2.07 / 2.84	2.02 / 2.76	2.00 / 2.70	1.96 / 2.63	1.94 / 2.60	1.91 / 2.54	1.90 / 2.51	1.88 / 2.49
20	4.35 / 8.10	3.49 / 5.85	3.10 / 4.94	2.87 / 4.43	2.71 / 4.10	2.60 / 3.87	2.52 / 3.71	2.45 / 3.56	2.40 / 3.45	2.35 / 3.37	2.31 / 3.30	2.28 / 3.23	2.23 / 3.13	2.18 / 3.05	2.12 / 2.94	2.08 / 2.86	2.04 / 2.77	1.99 / 2.69	1.96 / 2.63	1.92 / 2.56	1.90 / 2.53	1.87 / 2.47	1.85 / 2.44	1.84 / 2.42
21	4.32 / 8.02	3.47 / 5.78	3.07 / 4.87	2.84 / 4.37	2.68 / 4.04	2.57 / 3.81	2.49 / 3.65	2.42 / 3.51	2.37 / 3.40	2.32 / 3.31	2.28 / 3.24	2.25 / 3.17	2.20 / 3.07	2.15 / 2.99	2.09 / 2.88	2.05 / 2.80	2.00 / 2.72	1.96 / 2.63	1.93 / 2.58	1.89 / 2.51	1.87 / 2.47	1.84 / 2.42	1.82 / 2.38	1.81 / 2.36
22	4.30 / 7.94	3.44 / 5.72	3.05 / 4.82	2.82 / 4.31	2.66 / 3.99	2.55 / 3.76	2.47 / 3.59	2.40 / 3.45	2.35 / 3.35	2.30 / 3.26	2.26 / 3.18	2.23 / 3.12	2.18 / 3.02	2.13 / 2.94	2.07 / 2.83	2.03 / 2.75	1.98 / 2.67	1.93 / 2.58	1.91 / 2.53	1.87 / 2.46	1.84 / 2.42	1.81 / 2.37	1.80 / 2.33	1.78 / 2.31
23	4.28 / 7.88	3.42 / 5.66	3.03 / 4.76	2.80 / 4.26	2.64 / 3.94	2.53 / 3.71	2.45 / 3.54	2.38 / 3.41	2.32 / 3.30	2.28 / 3.21	2.24 / 3.14	2.20 / 3.07	2.14 / 2.97	2.10 / 2.89	2.04 / 2.78	2.00 / 2.70	1.96 / 2.62	1.91 / 2.53	1.88 / 2.48	1.84 / 2.41	1.82 / 2.37	1.79 / 2.32	1.77 / 2.28	1.76 / 2.26
24	4.26 / 7.82	3.40 / 5.61	3.01 / 4.72	2.78 / 4.22	2.62 / 3.90	2.51 / 3.67	2.43 / 3.50	2.36 / 3.36	2.30 / 3.25	2.26 / 3.17	2.22 / 3.09	2.18 / 3.03	2.13 / 2.93	2.09 / 2.85	2.02 / 2.74	1.98 / 2.66	1.94 / 2.58	1.89 / 2.49	1.86 / 2.44	1.82 / 2.36	1.80 / 2.33	1.76 / 2.27	1.74 / 2.23	1.73 / 2.21
25	4.24 / 7.77	3.38 / 5.57	2.99 / 4.68	2.76 / 4.18	2.60 / 3.86	2.49 / 3.63	2.41 / 3.46	2.34 / 3.32	2.28 / 3.21	2.24 / 3.13	2.20 / 3.05	2.16 / 2.99	2.11 / 2.89	2.06 / 2.81	2.00 / 2.70	1.96 / 2.62	1.92 / 2.54	1.87 / 2.45	1.84 / 2.40	1.80 / 2.32	1.77 / 2.29	1.74 / 2.23	1.72 / 2.19	1.71 / 2.17
26	4.22 / 7.72	3.37 / 5.53	2.98 / 4.64	2.74 / 4.14	2.59 / 3.82	2.47 / 3.59	2.39 / 3.42	2.32 / 3.29	2.27 / 3.17	2.22 / 3.09	2.18 / 3.02	2.15 / 2.96	2.10 / 2.86	2.05 / 2.77	1.99 / 2.66	1.95 / 2.58	1.90 / 2.50	1.85 / 2.41	1.82 / 2.36	1.78 / 2.28	1.76 / 2.25	1.72 / 2.19	1.70 / 2.15	1.69 / 2.13
27	4.21 / 7.68	3.35 / 5.49	2.96 / 4.60	2.73 / 4.11	2.57 / 3.79	2.46 / 3.56	2.37 / 3.39	2.30 / 3.26	2.25 / 3.14	2.20 / 3.06	2.16 / 2.98	2.13 / 2.93	2.08 / 2.83	2.03 / 2.74	1.97 / 2.63	1.93 / 2.55	1.88 / 2.47	1.84 / 2.38	1.80 / 2.33	1.76 / 2.25	1.74 / 2.21	1.71 / 2.16	1.68 / 2.12	1.67 / 2.10
28	4.20 / 7.64	3.34 / 5.45	2.95 / 4.57	2.71 / 4.07	2.56 / 3.76	2.44 / 3.53	2.36 / 3.36	2.29 / 3.23	2.24 / 3.11	2.19 / 3.03	2.15 / 2.95	2.12 / 2.90	2.06 / 2.80	2.02 / 2.71	1.96 / 2.60	1.91 / 2.52	1.87 / 2.44	1.81 / 2.35	1.78 / 2.30	1.75 / 2.22	1.72 / 2.18	1.69 / 2.13	1.67 / 2.09	1.65 / 2.06
29	4.18 / 7.60	3.33 / 5.42	2.93 / 4.54	2.70 / 4.04	2.54 / 3.73	2.43 / 3.50	2.35 / 3.33	2.28 / 3.20	2.22 / 3.08	2.18 / 3.00	2.14 / 2.92	2.10 / 2.87	2.05 / 2.77	2.00 / 2.68	1.94 / 2.57	1.90 / 2.49	1.85 / 2.41	1.80 / 2.32	1.77 / 2.27	1.73 / 2.19	1.71 / 2.15	1.68 / 2.10	1.65 / 2.06	1.64 / 2.03
30	4.17 / 7.56	3.32 / 5.39	2.92 / 4.51	2.69 / 4.02	2.53 / 3.70	2.42 / 3.47	2.34 / 3.30	2.27 / 3.17	2.21 / 3.06	2.16 / 2.98	2.12 / 2.90	2.09 / 2.84	2.04 / 2.74	1.99 / 2.66	1.93 / 2.55	1.89 / 2.47	1.84 / 2.38	1.79 / 2.29	1.76 / 2.24	1.72 / 2.16	1.69 / 2.13	1.66 / 2.07	1.64 / 2.03	1.62 / 2.01

DEGREES OF FREEDOM FOR DENOMINATOR

Table D (*continued*)   0.05 (light row) and 0.01 (dark row) points for the distribution of F

DEGREES OF FREEDOM FOR NUMERATOR

df (denom.)	1	2	3	4	5	6	7	8	9	10	11	12	14	16	20	24	30	40	50	75	100	200	500	∞
32	4.15/7.50	3.30/5.34	2.90/4.46	2.67/3.97	2.51/3.66	2.40/3.42	2.32/3.25	2.25/3.12	2.19/3.01	2.14/2.94	2.10/2.86	2.07/2.80	2.02/2.70	1.97/2.62	1.91/2.51	1.86/2.42	1.82/2.34	1.76/2.25	1.74/2.20	1.69/2.12	1.67/2.08	1.64/2.02	1.61/1.98	1.59/1.96
34	4.13/7.44	3.28/5.29	2.88/4.42	2.65/3.93	2.49/3.61	2.38/3.38	2.30/3.21	2.23/3.08	2.17/2.97	2.12/2.89	2.08/2.82	2.05/2.76	2.00/2.66	1.95/2.58	1.89/2.47	1.84/2.38	1.80/2.30	1.74/2.21	1.71/2.15	1.67/2.08	1.64/2.04	1.61/1.98	1.59/1.94	1.57/1.91
36	4.11/7.39	3.26/5.25	2.86/4.38	2.63/3.89	2.48/3.58	2.36/3.35	2.28/3.18	2.21/3.04	2.15/2.94	2.10/2.86	2.06/2.78	2.03/2.72	1.98/2.62	1.93/2.54	1.87/2.43	1.82/2.35	1.78/2.26	1.72/2.17	1.69/2.12	1.65/2.04	1.62/2.00	1.59/1.94	1.56/1.90	1.55/1.87
38	4.10/7.35	3.25/5.21	2.85/4.34	2.62/3.86	2.46/3.54	2.35/3.32	2.26/3.15	2.19/3.02	2.14/2.91	2.09/2.82	2.05/2.75	2.02/2.69	1.96/2.59	1.92/2.51	1.85/2.40	1.80/2.32	1.76/2.22	1.71/2.14	1.67/2.08	1.63/2.00	1.60/1.97	1.57/1.90	1.54/1.86	1.53/1.84
40	4.08/7.31	3.23/5.18	2.84/4.31	2.61/3.83	2.45/3.51	2.34/3.29	2.25/3.12	2.18/2.99	2.12/2.88	2.07/2.80	2.04/2.73	2.00/2.66	1.95/2.56	1.90/2.49	1.84/2.37	1.79/2.29	1.74/2.20	1.69/2.11	1.66/2.05	1.61/1.97	1.59/1.94	1.55/1.88	1.53/1.84	1.51/1.81
42	4.07/7.27	3.22/5.15	2.83/4.29	2.59/3.80	2.44/3.49	2.32/3.26	2.24/3.10	2.17/2.96	2.11/2.86	2.06/2.77	2.02/2.70	1.99/2.64	1.94/2.54	1.89/2.46	1.82/2.35	1.78/2.26	1.73/2.17	1.68/2.08	1.64/2.02	1.60/1.94	1.57/1.91	1.54/1.85	1.51/1.80	1.49/1.78
44	4.06/7.24	3.21/5.12	2.82/4.26	2.58/3.78	2.43/3.46	2.31/3.24	2.23/3.07	2.16/2.94	2.10/2.84	2.05/2.75	2.01/2.68	1.98/2.62	1.92/2.52	1.88/2.44	1.81/2.32	1.76/2.24	1.72/2.15	1.66/2.06	1.63/2.00	1.58/1.92	1.56/1.88	1.52/1.82	1.50/1.78	1.48/1.75
46	4.05/7.21	3.20/5.10	2.81/4.24	2.57/3.76	2.42/3.44	2.30/3.22	2.22/3.05	2.14/2.92	2.09/2.82	2.04/2.73	2.00/2.66	1.97/2.60	1.91/2.50	1.87/2.42	1.80/2.30	1.75/2.22	1.71/2.13	1.65/2.04	1.62/1.98	1.57/1.90	1.54/1.86	1.51/1.80	1.48/1.76	1.46/1.72
48	4.04/7.19	3.19/5.08	2.80/4.22	2.56/3.74	2.41/3.42	2.30/3.20	2.21/3.04	2.14/2.90	2.08/2.80	2.03/2.71	1.99/2.64	1.96/2.58	1.90/2.48	1.86/2.40	1.79/2.28	1.74/2.20	1.70/2.11	1.64/2.02	1.61/1.96	1.56/1.88	1.53/1.84	1.50/1.78	1.47/1.73	1.45/1.70
50	4.03/7.17	3.18/5.06	2.79/4.20	2.56/3.72	2.40/3.41	2.29/3.18	2.20/3.02	2.13/2.88	2.07/2.78	2.02/2.70	1.98/2.62	1.95/2.56	1.90/2.46	1.85/2.39	1.78/2.26	1.74/2.18	1.69/2.10	1.63/2.00	1.60/1.94	1.55/1.86	1.52/1.82	1.48/1.76	1.46/1.71	1.44/1.68
55	4.02/7.12	3.17/5.01	2.78/4.16	2.54/3.68	2.38/3.37	2.27/3.15	2.18/2.98	2.11/2.85	2.05/2.75	2.00/2.66	1.97/2.59	1.93/2.53	1.88/2.43	1.83/2.35	1.76/2.23	1.72/2.15	1.67/2.06	1.61/1.96	1.58/1.90	1.52/1.82	1.50/1.78	1.46/1.71	1.43/1.66	1.41/1.64
60	4.00/7.08	3.15/4.98	2.76/4.13	2.52/3.65	2.37/3.34	2.25/3.12	2.17/2.95	2.10/2.82	2.04/2.72	1.99/2.63	1.95/2.56	1.92/2.50	1.86/2.40	1.81/2.32	1.75/2.20	1.70/2.12	1.65/2.03	1.59/1.93	1.56/1.87	1.50/1.79	1.48/1.74	1.44/1.68	1.41/1.63	1.39/1.60
65	3.99/7.04	3.14/4.95	2.75/4.10	2.51/3.62	2.36/3.31	2.24/3.09	2.15/2.93	2.08/2.79	2.02/2.70	1.98/2.61	1.94/2.54	1.90/2.47	1.85/2.37	1.80/2.30	1.73/2.18	1.68/2.09	1.63/2.00	1.57/1.90	1.54/1.84	1.49/1.76	1.46/1.71	1.42/1.64	1.39/1.60	1.37/1.56
70	3.98/7.01	3.13/4.92	2.74/4.08	2.50/3.60	2.35/3.29	2.23/3.07	2.14/2.91	2.07/2.77	2.01/2.67	1.97/2.59	1.93/2.51	1.89/2.45	1.84/2.35	1.79/2.28	1.72/2.15	1.67/2.07	1.62/1.98	1.56/1.88	1.53/1.82	1.47/1.74	1.45/1.69	1.40/1.62	1.37/1.56	1.35/1.53
80	3.96/6.96	3.11/4.88	2.72/4.04	2.48/3.56	2.33/3.25	2.21/3.04	2.12/2.87	2.05/2.74	1.99/2.64	1.95/2.55	1.91/2.48	1.88/2.41	1.82/2.32	1.77/2.24	1.70/2.11	1.65/2.03	1.60/1.94	1.54/1.84	1.51/1.78	1.45/1.70	1.42/1.65	1.38/1.57	1.35/1.52	1.32/1.49

DEGREES OF FREEDOM FOR DENOMINATOR

Table D (continued)

DEGREES OF FREEDOM FOR NUMERATOR

df (denom.)	1	2	3	4	5	6	7	8	9	10	11	12	14	16	20	24	30	40	50	75	100	200	500	∞
100	3.94 / 6.90	3.09 / 4.82	2.70 / 3.98	2.46 / 3.51	2.30 / 3.20	2.19 / 2.99	2.10 / 2.82	2.03 / 2.69	1.97 / 2.59	1.92 / 2.51	1.88 / 2.43	1.85 / 2.36	1.79 / 2.26	1.75 / 2.19	1.68 / 2.06	1.63 / 1.98	1.57 / 1.89	1.51 / 1.79	1.48 / 1.73	1.42 / 1.64	1.39 / 1.59	1.34 / 1.51	1.30 / 1.46	1.28 / 1.43
125	3.92 / 6.84	3.07 / 4.78	2.68 / 3.94	2.44 / 3.47	2.29 / 3.17	2.17 / 2.95	2.08 / 2.79	2.01 / 2.65	1.95 / 2.56	1.90 / 2.47	1.86 / 2.40	1.83 / 2.33	1.77 / 2.23	1.72 / 2.15	1.65 / 2.03	1.60 / 1.94	1.55 / 1.85	1.49 / 1.75	1.45 / 1.68	1.39 / 1.59	1.36 / 1.54	1.31 / 1.46	1.27 / 1.40	1.25 / 1.37
150	3.91 / 6.81	3.06 / 4.75	2.67 / 3.91	2.43 / 3.44	2.27 / 3.13	2.16 / 2.92	2.07 / 2.76	2.00 / 2.62	1.94 / 2.53	1.89 / 2.44	1.85 / 2.37	1.82 / 2.30	1.76 / 2.20	1.71 / 2.12	1.64 / 2.00	1.59 / 1.91	1.54 / 1.83	1.47 / 1.72	1.44 / 1.66	1.37 / 1.56	1.34 / 1.51	1.29 / 1.43	1.25 / 1.37	1.22 / 1.33
200	3.89 / 6.76	3.04 / 4.71	2.65 / 3.88	2.41 / 3.41	2.26 / 3.11	2.14 / 2.90	2.05 / 2.73	1.98 / 2.60	1.92 / 2.50	1.87 / 2.41	1.83 / 2.34	1.80 / 2.28	1.74 / 2.17	1.69 / 2.09	1.62 / 1.97	1.57 / 1.88	1.52 / 1.79	1.45 / 1.69	1.42 / 1.62	1.35 / 1.53	1.32 / 1.48	1.26 / 1.39	1.22 / 1.33	1.19 / 1.28
400	3.86 / 6.70	3.02 / 4.66	2.62 / 3.83	2.39 / 3.36	2.23 / 3.06	2.12 / 2.85	2.03 / 2.69	1.96 / 2.55	1.90 / 2.46	1.85 / 2.37	1.81 / 2.29	1.78 / 2.23	1.72 / 2.12	1.67 / 2.04	1.60 / 1.92	1.54 / 1.84	1.49 / 1.74	1.42 / 1.64	1.38 / 1.57	1.32 / 1.47	1.28 / 1.42	1.22 / 1.32	1.16 / 1.24	1.13 / 1.19
1000	3.85 / 6.66	3.00 / 4.62	2.61 / 3.80	2.38 / 3.34	2.22 / 3.04	2.10 / 2.82	2.02 / 2.66	1.95 / 2.53	1.89 / 2.43	1.84 / 2.34	1.80 / 2.26	1.76 / 2.20	1.70 / 2.09	1.65 / 2.01	1.58 / 1.89	1.53 / 1.81	1.47 / 1.71	1.41 / 1.61	1.36 / 1.54	1.30 / 1.44	1.26 / 1.38	1.19 / 1.28	1.13 / 1.19	1.08 / 1.11
∞	3.84 / 6.64	2.99 / 4.60	2.60 / 3.78	2.37 / 3.32	2.21 / 3.02	2.09 / 2.80	2.01 / 2.64	1.94 / 2.51	1.88 / 2.41	1.83 / 2.32	1.79 / 2.24	1.75 / 2.18	1.69 / 2.07	1.64 / 1.99	1.57 / 1.87	1.52 / 1.79	1.46 / 1.69	1.40 / 1.59	1.35 / 1.52	1.28 / 1.41	1.24 / 1.36	1.17 / 1.25	1.11 / 1.15	1.00 / 1.00

DEGREES OF FREEDOM FOR DENOMINATOR

A difference in variances is significant at $\alpha = 0.05$ with df_1 and df_2 if it *equals* or *exceeds* the upper value in each cell or is *less than or equal to* the lower value in that cell; for example, if $\hat{s}_1^2 = 8.69$, df = 12 and $\hat{s}_2^2 = 2.63$, df = 9, $F = 3.30$. The critical values at df_{12} and df_9 are 0.291 and 3.87. Since obtained F is between these values, we fail to reject the hypothesis of homogeneity of variances (*Note:* Had we calculated F with $\hat{s}_1^2$ in the denominator, the critical lower and upper values at 9 and 12 df would have been 0.259 and 3.44.

Table D₁　Critical values of F that cut off the upper and lower 2.5% of the F distributions

df	1	2	3	4	5	6	7	8	9	10	11	12	15	20	24	30	40	50	60	100	200	500
1	.002	.026	.057	.082	.100	.113	.124	.132	.139	.144	.149	.153	.161	.170	.175	.180	.184	.187	.189	.193	.196	.198
	648	800	864	900	922	937	948	957	963	969	973	977	985	993	997	1000	1010	1010	1010	1010	1020	1020
2	.001	.026	.062	.094	.119	.138	.153	.165	.175	.183	.190	.196	.210	.224	.232	.239	.247	.251	.255	.261	.266	.269
	38.5	39.0	39.2	39.2	39.3	39.3	39.4	39.4	39.4	39.4	39.4	39.4	39.4	39.4	39.5	39.5	39.5	39.5	39.5	39.5	39.5	39.5
3	.001	.026	.065	.100	.129	.152	.170	.185	.197	.207	.216	.224	.241	.259	.269	.279	.289	.295	.299	.308	.314	.318
	17.4	16.0	15.4	15.1	14.9	14.7	14.6	14.5	14.5	14.4	14.4	14.3	14.3	14.2	14.1	14.1	14.0	14.0	14.0	14.0	13.9	13.9
4	.001	.026	.066	.104	.135	.161	.181	.198	.212	.224	.234	.243	.263	.284	.296	.308	.320	.327	.332	.342	.351	.356
	12.2	10.6	9.98	9.60	9.36	9.20	9.07	8.98	8.90	8.84	8.79	8.75	8.66	8.56	8.51	8.46	8.41	8.38	8.36	8.32	8.29	8.27
5	.001	.025	.067	.107	.140	.167	.189	.208	.223	.236	.248	.257	.280	.304	.317	.330	.344	.353	.359	.370	.380	.386
	10.0	8.43	7.76	7.39	7.15	6.98	6.85	6.76	6.68	6.62	6.57	6.52	6.43	6.33	6.28	6.23	6.18	6.14	6.12	6.08	6.05	6.03
6	.001	.025	.068	.109	.143	.172	.195	.215	.231	.246	.258	.268	.293	.320	.334	.349	.364	.375	.381	.394	.405	.415
	8.81	7.26	6.60	6.23	5.99	5.82	5.70	5.60	5.52	5.46	5.41	5.37	5.27	5.17	5.12	5.07	5.01	4.98	4.96	4.92	4.88	4.86
7	.001	.025	.068	.110	.146	.176	.200	.221	.238	.253	.266	.277	.304	.333	.348	.364	.381	.392	.399	.413	.426	.433
	8.07	6.54	5.89	5.52	5.29	5.12	4.99	4.90	4.82	4.76	4.71	4.67	4.57	4.47	4.42	4.36	4.31	4.28	4.25	4.21	4.18	4.16
8	.001	.025	.069	.111	.148	.179	.204	.226	.244	.259	.273	.285	.313	.343	.360	.377	.395	.407	.415	.431	.442	.450
	7.57	6.06	5.42	5.05	4.82	4.65	4.53	4.43	4.36	4.30	4.24	4.20	4.10	4.00	3.95	3.89	3.84	3.81	3.78	3.74	3.70	3.68
9	.001	.025	.069	.112	.150	.181	.207	.230	.248	.265	.279	.291	.320	.352	.370	.388	.408	.420	.428	.446	.459	.467
	7.21	5.71	5.08	4.72	4.48	4.32	4.20	4.10	4.03	3.96	3.91	3.87	3.77	3.67	3.61	3.56	3.51	3.47	3.45	3.40	3.37	3.35
10	.001	.025	.069	.113	.151	.183	.210	.233	.252	.269	.283	.296	.327	.360	.379	.398	.419	.431	.441	.459	.474	.483
	6.94	5.46	4.83	4.47	4.24	4.07	3.95	3.85	3.78	3.72	3.66	3.62	3.52	3.42	3.37	3.31	3.26	3.22	3.20	3.15	3.12	3.09

Table D₁ (continued)

11	.001	.025	.069	.114	.152	.185	.212	.236	.256	.273	.288	.301	.332	.368	.386	.407	.429	.442	.450	.472	.485	.495
	6.72	5.26	4.63	4.28	4.04	3.88	3.76	3.66	3.59	3.53	3.47	3.43	3.33	3.23	3.17	3.12	3.06	3.03	3.00	2.96	2.92	2.90
12	.001	.025	.070	.114	.153	.186	.214	.238	.259	.276	.292	.305	.337	.374	.394	.416	.437	.450	.461	.481	.498	.508
	6.55	5.10	4.47	4.12	3.89	3.73	3.61	3.51	3.44	3.37	3.32	3.28	3.18	3.07	3.02	2.96	2.91	2.87	2.85	2.80	2.76	2.74
15	.001	.025	.070	.116	.156	.190	.219	.244	.265	.284	.300	.315	.349	.389	.410	.433	.458	.474	.485	.508	.526	.538
	6.20	4.76	4.15	3.80	3.58	3.41	3.29	3.20	3.12	3.06	3.01	2.96	2.86	2.76	2.70	2.64	2.59	2.55	2.52	2.47	2.44	2.41
20	.001	.025	.071	.117	.158	.193	.224	.250	.273	.292	.310	.325	.363	.406	.430	.456	.484	.503	.514	.541	.562	.575
	5.87	4.46	3.86	3.51	3.29	3.13	3.01	2.91	2.84	2.77	2.72	2.68	2.57	2.46	2.41	2.35	2.29	2.25	2.22	2.17	2.13	2.10
24	.001	.025	.071	.117	.159	.195	.227	.253	.277	.297	.315	.331	.370	.415	.441	.468	.498	.518	.531	.562	.585	.599
	5.72	4.32	3.72	3.38	3.15	2.99	2.87	2.78	2.70	2.64	2.59	2.54	2.44	2.33	2.27	2.21	2.15	2.11	2.08	2.02	1.98	1.95
30	.001	.025	.071	.118	.161	.197	.229	.257	.281	.302	.321	.337	.378	.426	.453	.482	.515	.535	.551	.585	.610	.625
	5.57	4.18	3.59	3.25	3.03	2.87	2.75	2.65	2.57	2.51	2.46	2.41	2.31	2.20	2.14	2.07	2.01	1.97	1.94	1.88	1.84	1.81
40	.001	.025	.071	.119	.162	.199	.232	.260	.285	.307	.327	.344	.387	.437	.466	.498	.533	.556	.573	.610	.641	.662
	5.42	4.05	3.46	3.13	2.90	2.74	2.62	2.53	2.45	2.39	2.33	2.29	2.18	2.07	2.01	1.94	1.88	1.83	1.80	1.74	1.69	1.66
60	.001	.025	.071	.120	.163	.202	.235	.264	.290	.313	.333	.351	.396	.450	.481	.515	.555	.581	.600	.641	.680	.704
	5.29	3.93	3.34	3.01	2.79	2.63	2.51	2.41	2.33	2.27	2.22	2.17	2.06	1.94	1.88	1.82	1.74	1.70	1.67	1.60	1.54	1.51
120	.001	.025	.072	.120	.165	.204	.238	.268	.295	.318	.340	.359	.406	.464	.498	.536	.580	.611	.633	.684	.729	.762
	5.15	3.80	3.23	2.89	2.67	2.52	2.39	2.30	2.22	2.16	2.10	2.05	1.95	1.82	1.76	1.69	1.61	1.56	1.53	1.45	1.39	1.34

Table E   Transformation of r to z_r

r	z_r	r	z_r	r	z_r
.01	.010	.34	.354	.67	.811
.02	.020	.35	.366	.68	.829
.03	.030	.36	.377	.69	.848
.04	.040	.37	.389	.70	.867
.05	.050	.38	.400	.71	.887
.06	.060	.39	.412	.72	.908
.07	.070	.40	.424	.73	.929
.08	.080	.41	.436	.74	.950
.09	.090	.42	.448	.75	.973
.10	.100	.43	.460	.76	.996
.11	.110	.44	.472	.77	1.020
.12	.121	.45	.485	.78	1.045
.13	.131	.46	.497	.79	1.071
.14	.141	.47	.510	.80	1.099
.15	.151	.48	.523	.81	1.127
.16	.161	.49	.536	.82	1.157
.17	.172	.50	.549	.83	1.188
.18	.181	.51	.563	.84	1.221
.19	.192	.52	.577	.85	1.256
.20	.203	.53	.590	.86	1.293
.21	.214	.54	.604	.87	1.333
.22	.224	.55	.618	.88	1.376
.23	.234	.56	.633	.89	1.422
.24	.245	.57	.648	.90	1.472
.25	.256	.58	.663	.91	1.528
.26	.266	.59	.678	.92	1.589
.27	.277	.60	.693	.93	1.658
.28	.288	.61	.709	.94	1.738
.29	.299	.62	.725	.95	1.832
.30	.309	.63	.741	.96	1.946
.31	.321	.64	.758	.97	2.092
.32	.332	.65	.775	.98	2.298
.33	.343	.66	.793	.99	2.647

A given value of r_r is statistically significant if it equals or exceeds the tabled value at the designated α level at a given N. To interpolate, sum the critical values above and below the N of interest and divide by 2. Thus the critical value at $\alpha = 0.05$, two-tailed test, when $N = 21$, is $(0.450 + 0.428)/2 = 0.439$.

Table F   Critical values of r_s

N*	LEVEL OF SIGNIFICANCE FOR ONE-TAILED TEST			
	.05	.025	.01	.005
	LEVEL OF SIGNIFICANCE FOR TWO-TAILED TEST			
	.10	.05	.02	.01
5	.900	1.000	1.000	--
6	.829	.886	.943	1.000
7	.714	.786	.893	.929
8	.643	.738	.833	.881
9	.600	.683	.783	.833
10	.564	.648	.746	.794
12	.506	.591	.712	.777
14	.456	.544	.645	.715
16	.425	.506	.601	.665
18	.399	.475	.564	.625
20	.377	.450	.534	.591
22	.359	.428	.508	.562
24	.343	.409	.485	.537
26	.329	.392	.465	.515
28	.317	.377	.448	.496
30	.306	.364	.432	.478

*N = number of pairs

Table G   Functions of r

r	$\sqrt{r}$	r^2	$\sqrt{r - r^2}$	$\sqrt{1 - r}$	$1 - r^2$	$\sqrt{1 - r^2}$	$100(1 - k)$	r
						k	**% EFF.**	
1.00	1.0000	1.0000	0.0000	0.0000	0.0000	0.0000	100.00	1.00
.99	.9950	.9801	.0995	.1000	.0199	.1411	85.89	.99
.98	.9899	.9604	.1400	.1414	.0396	.1990	80.10	.98
.97	.9849	.9409	.1706	.1732	.0591	.2431	75.69	.97
.96	.9798	.9216	.1960	.2000	.0784	.2800	72.00	.96
.95	.9747	.9025	.2179	.2236	.0975	.3122	68.78	.95
.94	.9695	.8836	.2375	.2449	.1164	.3412	65.88	.94
.93	.9644	.8649	.2551	.2646	.1351	.3676	63.24	.93
.92	.9592	.8464	.2713	.2828	.1536	.3919	60.81	.92
.91	.9539	.8281	.2862	.3000	.1719	.4146	58.54	.91
.90	.9487	.8100	.3000	.3162	.1900	.4359	56.41	.90
.89	.9434	.7921	.3129	.3317	.2079	.4560	54.40	.89
.88	.9381	.7744	.3250	.3464	.2256	.4750	52.50	.88
.87	.9327	.7569	.3363	.3606	.2431	.4931	50.69	.87
.86	.9274	.7396	.3470	.3742	.2604	.5103	48.97	.86
.85	.9220	.7225	.3571	.3873	.2775	.5268	47.32	.85
.84	.9165	.7056	.3666	.4000	.2944	.5426	45.74	.84
.83	.9110	.6889	.3756	.4123	.3111	.5578	44.22	.83
.82	.9055	.6724	.3842	.4243	.3276	.5724	42.76	.82
.81	.9000	.6561	.3923	.4359	.3439	.5864	41.36	.81
.80	.8944	.6400	.4000	.4472	.3600	.6000	40.00	.80
.79	.8888	.6241	.4073	.4583	.3759	.6131	38.69	.79
.78	.8832	.6084	.4142	.4690	.3916	.6258	37.42	.78
.77	.8775	.5929	.4208	.4796	.4071	.6380	36.20	.77
.76	.8718	.5776	.4271	.4899	.4224	.6499	35.01	.76
.75	.8660	.5625	.4330	.5000	.4375	.6614	33.86	.75
.74	.8602	.5476	.4386	.5099	.4524	.6726	32.74	.74
.73	.8544	.5329	.4440	.5196	.4671	.6834	31.66	.73
.72	.8485	.5184	.4490	.5292	.4816	.6940	30.60	.72
.71	.8426	.5041	.4538	.5385	.4959	.7042	29.58	.71
.70	.8367	.4900	.4583	.5477	.5100	.7141	28.59	.70
.69	.8307	.4761	.4625	.5568	.5239	.7238	27.62	.69
.68	.8246	.4624	.4665	.5657	.5376	.7332	26.68	.68
.67	.8185	.4489	.4702	.5745	.5511	.7424	25.76	.67
.66	.8124	.4356	.4737	.5831	.5644	.7513	24.87	.66
.65	.8062	.4225	.4770	.5916	.5775	.7599	24.01	.65
.64	.8000	.4096	.4800	.6000	.5904	.7684	23.16	.64
.63	.7937	.3969	.4828	.6083	.6031	.7766	22.34	.63
.62	.7874	.3844	.4854	.6164	.6156	.7846	21.54	.62
.61	.7810	.3721	.4877	.6245	.6279	.7924	20.76	.61
.60	.7746	.3600	.4899	.6325	.6400	.8000	20.00	.60
.59	.7681	.3481	.4918	.6403	.6519	.8074	19.26	.59
.58	.7616	.3364	.4936	.6481	.6636	.8146	18.54	.58
.57	.7550	.3249	.4951	.6557	.6751	.8216	17.84	.57
.56	.7483	.3136	.4964	.6633	.6864	.8285	17.15	.56
.55	.7416	.3025	.4975	.6708	.6975	.8352	16.48	.55
.54	.7348	.2916	.4984	.6782	.7084	.8417	15.83	.54
.53	.7280	.2809	.4991	.6856	.7191	.8480	15.20	.53
.52	.7211	.2704	.4996	.6928	.7296	.8542	14.58	.52
.51	.7141	.2601	.4999	.7000	.7399	.8602	13.98	.51
.50	.7071	.2500	.5000	.7071	.7500	.8660	13.40	.50

Table G (continued)

r	$\sqrt{r}$	r^2	$\sqrt{r - r^2}$	$\sqrt{1 - r}$	$1 - r^2$	$\sqrt{1 - r^2}$	100(1 - k)	r
						k	% EFF.	
.50	.7071	.2500	.5000	.7071	.7500	.8660	13.40	.50
.49	.7000	.2401	.4999	.7141	.7599	.8717	12.83	.49
.48	.6928	.2304	.4996	.7211	.7696	.8773	12.27	.48
.47	.6856	.2209	.4991	.7280	.7791	.8827	11.73	.47
.46	.6782	.2116	.4984	.7348	.7884	.8879	11.21	.46
.45	.6708	.2025	.4975	.7416	.7975	.8930	10.70	.45
.44	.6633	.1936	.4964	.7483	.8064	.8980	10.20	.44
.43	.6557	.1849	.4951	.7550	.8151	.9028	9.72	.43
.42	.6481	.1764	.4936	.7616	.8236	.9075	9.25	.42
.41	.6403	.1681	.4918	.7681	.8319	.9121	8.79	.41
.40	.6325	.1600	.4899	.7746	.8400	.9165	8.35	.40
.39	.6245	.1521	.4877	.7810	.8479	.9208	7.92	.39
.38	.6164	.1444	.4854	.7874	.8556	.9250	7.50	.38
.37	.6083	.1369	.4828	.7937	.8631	.9290	7.10	.37
.36	.6000	.1296	.4800	.8000	.8704	.9330	6.70	.36
.35	.5916	.1225	.4770	.8062	.8775	.9367	6.33	.35
.34	.5831	.1156	.4737	.8124	.8844	.9404	5.96	.34
.33	.5745	.1089	.4702	.8185	.8911	.9440	5.60	.33
.32	.5657	.1024	.4665	.8246	.8976	.9474	5.25	.32
.31	.5568	.0961	.4625	.8307	.9039	.9507	4.93	.31
.30	.5477	.0900	.4583	.8367	.9100	.9539	4.61	.30
.29	.5385	.0841	.4538	.8426	.9159	.9570	4.30	.29
.28	.5292	.0784	.4490	.8485	.9216	.9600	4.00	.28
.27	.5196	.0729	.4440	.8544	.9271	.9629	3.71	.27
.26	.5099	.0676	.4386	.8602	.9324	.9656	3.44	.26
.25	.5000	.0625	.4330	.8660	.9375	.9682	3.18	.25
.24	.4899	.0576	.4271	.8718	.9424	.9708	2.92	.24
.23	.4796	.0529	.4208	.8775	.9471	.9732	2.68	.23
.22	.4690	.0484	.4142	.8832	.9516	.9755	2.45	.22
.21	.4583	.0441	.4073	.8888	.9559	.9777	2.23	.21
.20	.4472	.0400	.4000	.8944	.9600	.9798	2.02	.20
.19	.4359	.0361	.3923	.9000	.9639	.9818	1.82	.19
.18	.4243	.0324	.3842	.9055	.9676	.9837	1.63	.18
.17	.4123	.0289	.3756	.9110	.9711	.9854	1.46	.17
.16	.4000	.0256	.3666	.9165	.9744	.9871	1.29	.16
.15	.3873	.0225	.3571	.9220	.9775	.9887	1.13	.15
.14	.3742	.0196	.3470	.9274	.9804	.9902	.98	.14
.13	.3606	.0169	.3363	.9327	.9831	.9915	.85	.13
.12	.3464	.0144	.3250	.9381	.9856	.9928	.72	.12
.11	.3317	.0121	.3129	.9434	.9879	.9939	.61	.11
.10	.3162	.0100	.3000	.9487	.9900	.9950	.50	.10
.09	.3000	.0081	.2862	.9539	.9919	.9959	.41	.09
.08	.2828	.0064	.2713	.9592	.9936	.9968	.32	.08
.07	.2646	.0049	.2551	.9644	.9951	.9975	.25	.07
.06	.2449	.0036	.2375	.9695	.9964	.9982	.18	.06
.05	.2236	.0025	.2179	.9747	.9975	.9987	.13	.05
.04	.2000	.0016	.1960	.9798	.9984	.9992	.08	.04
.03	.1732	.0009	.1706	.9849	.9991	.9995	.05	.03
.02	.1414	.0004	.1400	.9899	.9996	.9998	.02	.02
.01	.1000	.0001	.0995	.9950	.9999	.9999	.01	.01
.00	.0000	.0000	.0000	1.0000	1.0000	1.0000	.00	.00

Table H Binomial coefficients

N	$\binom{N}{0}$	$\binom{N}{1}$	$\binom{N}{2}$	$\binom{N}{3}$	$\binom{N}{4}$	$\binom{N}{5}$	$\binom{N}{6}$	$\binom{N}{7}$	$\binom{N}{8}$	$\binom{N}{9}$	$\binom{N}{10}$
0	1										
1	1	1									
2	1	2	1								
3	1	3	3	1							
4	1	4	6	4	1						
5	1	5	10	10	5	1					
6	1	6	15	20	15	6	1				
7	1	7	21	35	35	21	7	1			
8	1	8	28	56	70	56	28	8	1		
9	1	9	36	84	126	126	84	36	9	1	
10	1	10	45	120	210	252	210	120	45	10	1
11	1	11	55	165	330	462	462	330	165	55	11
12	1	12	66	220	495	792	924	792	495	220	66
13	1	13	78	286	715	1287	1716	1716	1287	715	286
14	1	14	91	364	1001	2002	3003	3432	3003	2002	1001
15	1	15	105	455	1365	3003	5005	6435	6435	5005	3003
16	1	16	120	560	1820	4368	8008	11440	12870	11440	8008
17	1	17	136	680	2380	6188	12376	19448	24310	24310	19448
18	1	18	153	816	3060	8568	18564	31824	43758	48620	43758
19	1	19	171	969	3876	11628	27132	50388	75582	92378	92378
20	1	20	190	1140	4845	15504	38760	77520	125970	167960	184756

In Table I, x is the frequency in the P category, and $N - x$ is the frequency in the Q category. The obtained x or $N - x$ must be *equal to* or *greater than* the value shown for significance at the chosen level. Dashes indicate that no decision is possible for N at the given α-level.

Table I   Critical values of x or $N - x$ (whichever is larger) at 0.05 and 0.01 levels when $P = Q = 1/2$

N	ONE-TAILED TEST		TWO-TAILED TEST	
	0.05	*0.01*	*0.05*	*0.01*
5	5	—	—	—
6	6	—	6	—
7	7	7	7	—
8	7	8	8	—
9	8	9	8	9
10	9	10	9	10
11	9	10	10	11
12	10	11	10	11
13	10	12	11	12
14	11	12	12	13
15	12	13	12	13
16	12	14	13	14
17	13	14	13	15
18	13	15	14	15
19	14	15	15	16
20	15	16	15	17
21	15	17	16	17
22	16	17	17	18
23	16	18	17	19
24	17	19	18	19
25	18	19	18	20
26	18	20	19	20
27	19	20	20	21
28	19	21	20	22
29	20	22	21	22
30	20	22	21	23
31	21	23	22	24
32	22	24	23	24
33	22	24	23	25
34	23	25	24	25
35	23	25	24	26
36	24	26	25	27
37	24	27	25	27
38	25	27	26	28
39	26	28	27	28
40	26	28	27	29
41	27	29	28	30
42	27	29	28	30
43	28	30	29	31
44	28	31	29	31
45	29	31	30	32
46	30	32	31	33
47	30	32	31	33
48	31	33	32	34
49	31	34	32	35
50	32	34	33	35

Table J was prepared to expedite decision-making when dealing with binomial populations in which $P \neq Q$. *Example:* A researcher has conducted twelve independent repetitions of the same study, using $\alpha = 0.01$. Four of these studies achieved statistical significance. Is this result (four out of twelve statistically significant outcomes) itself statistically significant, or is it within chance expectations? Looking in the column headed .01 opposite $N = 12$, we find that two or more differences significant at $\alpha = .01$ is in itself significant at $\alpha = .01$. Thus the researcher may conclude that the overall results of his or her investigations justify rejecting H_0.

A given value of x is significant at a given α-level if it equals or exceeds the critical value shown in table. All values shown are one-tailed. Since the binomial is not symmetrical when $P \neq Q \neq 1/2$, there is no straightforward way to obtain two-tailed values.

Table J   Critical values of x at $\alpha = 0.05$ (lightface) and $\alpha = 0.01$ (boldface) at varying values of P and Q for N's equal to 2 through 49

N	.01	.02	.03	.04	.05	.06	.07	.08	.09	.10	.11	.12	.13	.14	.15	.16	.17	.18	.19	.20	.21	.22	.23	.24	.25
Q →	.99	.98	.97	.96	.95	.94	.93	.92	.91	.90	.89	.88	.87	.86	.85	.84	.83	.82	.81	.80	.79	.78	.77	.76	.75
2	1	1	2	2	2	2	2	2	2	2	2	2	2	2	2	2	2	2	2	2	2	2	—	—	—
2	**1**	**2**	**2**	**2**	**2**	**2**	**2**	**2**	**2**	**2**	—	—	—	—	—	—	—	—	—	—	—	—	—	—	—
3	1	2	2	2	2	2	2	2	2	2	2	2	3	3	3	3	3	3	3	3	3	3	3	3	3
3	**2**	**2**	**2**	**2**	**2**	**3**	**3**	**3**	**3**	**3**	**3**	**3**	**3**	**3**	**3**	**3**	**3**	**3**	**3**	**3**	**3**	—	—	—	—
4	1	2	2	2	2	2	2	3	3	3	3	3	3	3	3	3	3	3	3	3	3	3	3	3	4
4	**2**	**2**	**2**	**3**	**3**	**3**	**3**	**3**	**3**	**3**	**3**	**3**	**3**	**4**	**4**	**4**	**4**	**4**	**4**	**4**	**4**	**4**	**4**	**4**	**4**
5	1	2	2	2	2	2	2	3	3	3	3	3	3	3	3	3	3	3	4	4	4	4	4	4	4
5	**2**	**2**	**2**	**3**	**3**	**3**	**3**	**3**	**3**	**3**	**4**	**4**	**4**	**4**	**4**	**4**	**4**	**4**	**4**	**4**	**4**	**5**	**5**	**5**	**5**
6	2	2	2	2	2	2	3	3	3	3	3	3	3	3	3	4	4	4	4	4	4	4	4	5	5
6	**2**	**2**	**3**	**3**	**3**	**3**	**3**	**3**	**4**	**4**	**4**	**4**	**4**	**4**	**4**	**4**	**5**	**5**	**5**	**5**	**5**	**5**	**5**	**5**	**5**
7	2	2	2	2	2	3	3	3	3	4	4	4	4	4	4	5	5	5	5	5	5	6	6	6	6
7	**2**	**2**	**3**	**3**	**3**	**3**	**3**	**4**	**4**	**4**	**4**	**4**	**4**	**5**	**5**	**5**	**5**	**5**	**5**	**6**	**6**	**6**	**6**	**6**	**6**
8	2	2	2	2	3	3	3	3	3	3	3	4	4	4	4	4	4	4	5	5	5	5	5	5	5
8	**2**	**3**	**3**	**3**	**3**	**3**	**4**	**4**	**4**	**4**	**4**	**5**	**5**	**5**	**5**	**5**	**5**	**5**	**6**	**6**	**6**	**6**	**6**	**6**	**6**
9	2	2	2	2	3	3	3	3	3	4	4	4	4	4	4	5	5	5	5	5	5	5	5	5	5
9	**2**	**3**	**3**	**3**	**3**	**4**	**4**	**4**	**4**	**4**	**5**	**5**	**5**	**5**	**5**	**5**	**5**	**6**	**6**	**6**	**6**	**6**	**6**	**6**	**6**
10	2	2	2	3	3	3	3	4	4	4	4	4	4	5	5	5	5	5	5	6	6	7	7	7	7
10	**2**	**3**	**3**	**3**	**4**	**4**	**4**	**4**	**5**	**5**	**5**	**5**	**5**	**5**	**6**	**6**	**6**	**6**	**6**	**6**	**7**	**7**	**7**	**7**	**7**
11	2	2	2	3	3	3	3	4	4	4	4	4	5	5	5	5	5	5	6	6	6	6	6	6	6
11	**2**	**3**	**3**	**3**	**4**	**4**	**4**	**4**	**5**	**5**	**5**	**5**	**5**	**6**	**6**	**6**	**6**	**6**	**6**	**7**	**7**	**7**	**7**	**7**	**7**
12	2	2	2	3	3	3	3	4	4	4	4	4	5	5	5	5	5	5	6	6	6	6	6	6	7
12	**2**	**3**	**3**	**4**	**4**	**4**	**4**	**5**	**5**	**5**	**5**	**5**	**6**	**6**	**6**	**6**	**6**	**7**	**7**	**7**	**7**	**7**	**7**	**8**	**8**
13	2	2	3	3	3	3	4	4	4	4	4	5	5	5	5	5	6	6	6	6	6	6	7	7	7
13	**2**	**3**	**3**	**4**	**4**	**4**	**5**	**5**	**5**	**5**	**6**	**6**	**6**	**6**	**6**	**7**	**7**	**7**	**7**	**7**	**8**	**8**	**8**	**8**	**8**
14	2	2	3	3	3	3	4	4	4	4	5	5	5	5	5	6	6	6	6	6	7	7	7	7	7
14	**2**	**3**	**3**	**4**	**4**	**4**	**5**	**5**	**5**	**5**	**6**	**6**	**6**	**6**	**7**	**7**	**7**	**7**	**7**	**8**	**8**	**8**	**8**	**8**	**9**
15	2	2	3	3	3	4	4	4	4	5	5	5	5	5	6	6	6	6	6	7	7	7	7	7	8
15	**2**	**3**	**3**	**4**	**4**	**5**	**5**	**5**	**5**	**6**	**6**	**6**	**6**	**7**	**7**	**7**	**7**	**8**	**8**	**8**	**8**	**8**	**9**	**9**	**9**

Table J (continued)

| .26 | .27 | .28 | .29 | .30 | .31 | .32 | .33 | .34 | .35 | .36 | .37 | .38 | .39 | .40 | .41 | .42 | .43 | .44 | .45 | .46 | .47 | .48 | .49 | .50 |
.74	.73	.72	.71	.70	.69	.68	.67	.66	.65	.64	.63	.62	.61	.60	.59	.58	.57	.56	.55	.54	.53	.52	.51	.50
—	—	—	—	—	—	—	—	—	—	—	—	—	—	—	—	—	—	—	—	—	—	—	—	—
3	3	3	3	3	3	3	3	3	3	3	—	—	—	—	—	—	—	—	—	—	—	—	—	—
—	—	—	—	—	—	—	—	—	—	—	—	—	—	—	—	—	—	—	—	—	—	—	—	—
4	4	4	4	4	4	4	4	4	4	4	4	4	4	4	4	4	4	4	4	4	—	—	—	—
4	4	4	4	4	4	—	—	—	—	—	—	—	—	—	—	—	—	—	—	—	—	—	—	—
4	4	4	4	4	4	4	4	4	4	5	5	5	5	5	5	5	5	5	5	5	5	5	5	5
5	5	5	5	5	5	5	5	5	5	5	5	5	5	5	—	—	—	—	—	—	—	—	—	—
5	5	5	5	5	5	5	5	5	5	5	5	5	5	5	6	6	6	6	6	6	6	6	6	6
5	5	5	5	6	6	6	6	6	6	6	6	6	6	6	6	6	6	6	6	6	—	—	—	—
5	5	5	5	5	5	5	5	5	6	6	6	6	6	6	6	6	6	6	6	6	7	7	7	7
6	6	6	6	6	6	6	6	6	6	7	7	7	7	7	7	7	7	7	7	7	7	7	7	7
5	5	5	6	6	6	6	6	6	6	6	6	6	6	6	7	7	7	7	7	7	7	7	7	7
6	6	6	6	7	7	7	7	7	7	7	7	7	7	7	8	8	8	8	8	8	8	8	8	8
6	6	6	6	6	6	6	6	6	7	7	7	7	7	7	7	7	7	7	8	8	8	8	8	8
7	7	7	7	7	7	7	7	7	8	8	8	8	8	8	8	8	8	8	8	9	9	9	9	9
6	6	6	6	6	7	7	7	7	7	7	7	7	7	8	8	8	8	8	8	8	8	8	8	9
7	7	7	7	8	8	8	8	8	8	8	8	8	9	9	9	9	9	9	9	9	9	9	9	10
6	6	7	7	7	7	7	7	7	8	8	8	8	8	8	8	8	9	9	9	9	9	9	9	9
7	8	8	8	8	8	8	8	9	9	9	9	9	9	9	9	9	10	10	10	10	10	10	10	10
7	7	7	7	7	7	8	8	8	8	8	8	8	8	9	9	9	9	9	9	9	9	10	10	10
8	8	8	8	8	9	9	9	9	9	9	9	10	10	10	10	10	10	10	10	10	11	11	11	11
7	7	7	8	8	8	8	8	8	8	9	9	9	9	9	9	9	10	10	10	10	10	10	10	10
8	8	9	9	9	9	9	9	10	10	10	10	10	10	10	10	11	11	11	11	11	11	11	11	12
7	8	8	8	8	8	8	9	9	9	9	9	9	9	10	10	10	10	10	10	11	11	11	11	11
9	9	9	9	9	10	10	10	10	10	10	10	11	11	11	11	11	11	12	12	12	12	12	12	12
8	8	8	8	9	9	9	9	9	9	10	10	10	10	10	10	10	11	11	11	11	11	11	12	12
9	9	9	10	10	10	10	10	10	11	11	11	11	11	11	12	12	12	12	12	12	12	13	13	13

Table J   (continued)

N	P / Q	.01 / .99	.02 / .98	.03 / .97	.04 / .96	.05 / .95	.06 / .94	.07 / .93	.08 / .92	.09 / .91	.10 / .90	.11 / .89	.12 / .88	.13 / .87	.14 / .86	.15 / .85	.16 / .84	.17 / .83	.18 / .82	.19 / .81	.20 / .80	.21 / .79	.22 / .78	.23 / .77	.24 / .76	.25 / .75
16	P	2	2	3	3	3	4	4	4	4	5	5	5	5	6	6	6	6	7	7	7	7	7	8	8	8
	Q	3	3	4	4	4	5	5	5	6	6	6	6	7	7	7	7	8	8	8	8	8	9	9	9	9
17	P	2	2	3	3	4	4	4	4	5	5	5	5	6	6	6	6	7	7	7	7	7	8	8	8	8
	Q	3	3	4	4	4	5	5	5	6	6	6	7	7	7	7	8	8	8	8	9	9	9	9	9	10
18	P	2	2	3	3	4	4	4	5	5	5	5	6	6	6	6	7	7	7	7	8	8	8	8	8	9
	Q	3	3	4	4	5	5	5	6	6	6	7	7	7	7	8	8	8	8	9	9	9	9	10	10	10
19	P	2	3	3	3	4	4	4	5	5	5	6	6	6	6	7	7	7	7	8	8	8	8	9	9	9
	Q	3	3	4	4	5	5	5	6	6	6	7	7	7	8	8	8	8	9	9	9	9	10	10	10	10
20	P	2	3	3	3	4	4	4	5	5	5	6	6	6	7	7	7	7	8	8	8	8	9	9	9	9
	Q	3	3	4	4	5	5	6	6	6	7	7	7	8	8	8	8	9	9	9	9	10	10	10	10	11
21	P	2	3	3	3	4	4	5	5	5	6	6	6	6	7	7	7	8	8	8	8	9	9	9	9	10
	Q	3	3	4	4	5	5	6	6	6	7	7	7	8	8	8	9	9	9	10	10	10	10	11	11	11
22	P	2	3	3	4	4	4	5	5	5	6	6	6	7	7	7	8	8	8	9	9	9	9	10	10	10
	Q	3	3	4	5	5	5	6	6	7	7	7	8	8	8	9	9	9	10	10	10	10	11	11	11	11
23	P	2	3	3	4	4	4	5	5	6	6	6	6	7	7	7	8	8	8	9	9	9	9	10	10	10
	Q	3	4	4	5	5	6	6	6	7	7	7	8	8 ·	9	9	9	9	10	10	10	11	11	11	12	12
24	P	2	3	3	4	4	5	5	5	6	6	6	7	7	7	8	8	8	9	9	9	9	10	10	10	11
	Q	3	4	4	5	5	6	6	7	7	7	8	8	8	9	9	9	10	10	10	11	11	11	12	12	12
25	P	2	3	3	4	4	5	5	5	6	6	6	7	7	8	8	8	9	9	9	9	10	10	10	11	11
	Q	3	4	4	5	5	6	6	7	7	7	8	8	9	9	9	10	10	10	11	11	11	12	12	12	13
26	P	2	3	4	4	4	5	5	6	6	6	7	7	7	8	8	8	9	9	9	10	10	10	11	11	11
	Q	3	4	4	5	5	6	6	7	7	8	8	8	9	9	10	10	10	11	11	11	12	12	12	13	13
27	P	2	3	3	4	4	5	5	6	6	6	7	7	8	8	8	9	9	9	10	10	10	11	11	11	12
	Q	3	4	4	5	6	6	6	7	7	8	8	9	9	9	10	10	11	11	11	12	12	12	13	13	13
28	P	2	3	4	4	4	5	5	6	6	7	7	7	8	8	8	9	9	10	10	10	11	11	11	12	12
	Q	3	4	4	5	6	6	7	7	8	8	8	9	9	10	10	10	11	11	12	12	12	13	13	13	14
29	P	2	3	4	4	5	5	5	6	6	7	7	8	8	8	9	9	9	10	10	10	11	11	12	12	12
	Q	3	4	5	5	6	6	7	7	8	8	9	9	9	10	10	11	11	11	12	12	13	13	13	14	14
30	P	2	3	4	4	5	5	6	6	6	7	7	8	8	8	9	9	10	10	10	11	11	11	12	12	13
	Q	3	4	5	5	6	6	7	7	8	8	9	9	10	10	10	11	11	12	12	12	13	13	14	14	14
31	P	2	3	4	4	5	5	6	6	7	7	7	8	8	9	9	9	10	10	11	11	11	12	12	12	13
	Q	3	4	5	5	6	6	7	7	8	8	9	9	10	10	11	11	12	12	12	13	13	14	14	14	15
32	P	2	3	4	4	5	5	6	6	7	7	8	8	9	9	9	10	10	10	11	11	12	12	12	13	13
	Q	3	4	5	5	6	7	7	8	8	9	9	10	10	10	11	11	12	12	13	13	13	14	14	15	15
33	P	2	3	4	4	5	5	6	6	7	7	8	8	9	9	9	10	10	11	11	12	12	12	13	13	13
	Q	3	4	5	5	6	7	7	8	8	9	9	10	10	11	11	12	12	12	13	13	14	14	15	15	15
34	P	2	3	4	4	5	6	6	7	7	7	8	8	9	9	10	10	11	11	11	12	12	12	13	13	14
	Q	3	4	5	6	6	7	7	8	8	9	9	10	10	11	11	12	12	13	13	14	14	14	15	15	16
35	P	2	3	4	5	5	6	6	7	7	8	8	9	9	9	10	10	11	11	12	12	12	13	13	14	14
	Q	3	4	5	6	6	7	7	8	9	9	10	10	11	11	12	12	13	13	13	14	14	15	15	16	16
36	P	3	3	4	5	5	6	6	7	7	8	8	9	9	10	10	11	11	11	12	12	13	13	14	14	14
	Q	3	4	5	6	6	7	8	8	9	9	10	10	11	11	12	12	13	13	14	14	15	15	15	16	16
37	P	3	3	4	5	5	6	6	7	7	8	8	9	9	10	10	11	11	12	12	13	13	13	14	14	15
	Q	3	4	5	6	6	7	8	8	9	9	10	11	11	12	12	13	13	13	14	14	15	15	16	16	17

Table J   (*continued*)

.26	.27	.28	.29	.30	.31	.32	.33	.34	.35	.36	.37	.38	.39	.40	.41	.42	.43	.44	.45	.46	.47	.48	.49	.50
.74	.73	.72	.71	.70	.69	.68	.67	.66	.65	.64	.63	.62	.61	.60	.59	.58	.57	.56	.55	.54	.53	.52	.51	.50
8	8	9	9	9	9	9	9	10	10	10	10	10	10	11	11	11	11	11	11	12	12	12	12	12
9	**10**	**10**	**10**	**10**	**10**	**11**	**11**	**11**	**11**	**11**	**11**	**12**	**12**	**12**	**12**	**12**	**12**	**13**	**13**	**13**	**14**	**14**	**14**	**14**
8	9	9	9	9	9	10	10	10	10	10	11	11	11	11	11	12	12	12	12	12	12	13	13	13
10	**10**	**10**	**10**	**11**	**11**	**11**	**11**	**11**	**12**	**12**	**12**	**12**	**12**	**13**	**13**	**13**	**13**	**13**	**13**	**14**	**14**	**14**	**14**	**14**
9	9	9	9	10	10	10	10	10	11	11	11	11	11	12	12	12	12	12	13	13	13	13	13	13
10	**10**	**11**	**11**	**11**	**11**	**12**	**12**	**12**	**12**	**12**	**13**	**13**	**13**	**13**	**13**	**13**	**14**	**14**	**14**	**14**	**14**	**14**	**15**	**15**
9	9	10	10	10	10	10	11	11	11	11	12	12	12	12	12	13	13	13	13	13	13	14	14	14
11	**11**	**11**	**11**	**12**	**12**	**12**	**12**	**13**	**13**	**13**	**13**	**13**	**14**	**14**	**14**	**14**	**15**	**15**	**15**	**15**	**15**	**15**	**15**	**15**
10	10	10	11	11	11	11	12	12	12	12	12	13	13	13	13	14	14	14	14	14	15	15	15	15
11	**11**	**11**	**12**	**12**	**12**	**12**	**13**	**13**	**13**	**13**	**14**	**14**	**14**	**14**	**14**	**15**	**15**	**15**	**15**	**15**	**16**	**16**	**16**	**16**
10	10	10	11	11	11	11	12	12	12	12	12	13	13	13	13	14	14	14	14	14	15	15	15	15
11	**12**	**12**	**12**	**12**	**13**	**13**	**13**	**13**	**14**	**14**	**14**	**14**	**14**	**15**	**15**	**15**	**15**	**16**	**16**	**16**	**16**	**16**	**17**	**17**
10	10	11	11	11	11	12	12	12	12	13	13	13	13	14	14	14	14	15	15	15	15	15	16	16
12	**12**	**12**	**13**	**13**	**13**	**13**	**14**	**14**	**14**	**14**	**15**	**15**	**15**	**15**	**15**	**16**	**16**	**16**	**16**	**17**	**17**	**17**	**17**	**17**
11	11	11	11	12	12	12	12	13	13	13	13	14	14	14	14	15	15	15	15	16	16	16	16	16
12	**12**	**13**	**13**	**13**	**13**	**14**	**14**	**14**	**15**	**15**	**15**	**15**	**16**	**16**	**16**	**16**	**17**	**17**	**17**	**17**	**18**	**18**	**18**	**18**
11	11	11	12	12	12	13	13	13	13	14	14	14	14	15	15	15	15	16	16	16	16	17	17	17
13	**13**	**13**	**13**	**14**	**14**	**14**	**14**	**15**	**15**	**15**	**15**	**16**	**16**	**16**	**16**	**17**	**17**	**17**	**17**	**18**	**18**	**18**	**18**	**19**
11	12	12	12	12	13	13	13	13	14	14	14	15	15	15	15	16	16	16	16	17	17	17	17	18
13	**13**	**13**	**14**	**14**	**14**	**15**	**15**	**15**	**15**	**16**	**16**	**16**	**17**	**17**	**17**	**17**	**18**	**18**	**18**	**18**	**19**	**19**	**19**	**19**
12	12	12	12	13	13	13	14	14	14	14	15	15	15	15	16	16	16	17	17	17	17	18	18	18
13	**14**	**14**	**14**	**14**	**15**	**15**	**15**	**16**	**16**	**16**	**16**	**17**	**17**	**17**	**18**	**18**	**18**	**18**	**19**	**19**	**19**	**19**	**20**	**20**
12	12	12	13	13	13	14	14	14	15	15	15	15	16	16	16	17	17	17	17	18	18	18	18	19
14	**14**	**14**	**15**	**15**	**15**	**15**	**16**	**16**	**16**	**17**	**17**	**17**	**18**	**18**	**18**	**18**	**19**	**19**	**19**	**19**	**20**	**20**	**20**	**20**
12	13	13	13	13	14	14	14	15	15	15	16	16	16	16	17	17	17	18	18	18	18	19	19	19
14	**14**	**15**	**15**	**15**	**15**	**16**	**16**	**17**	**17**	**17**	**17**	**18**	**18**	**18**	**19**	**19**	**19**	**20**	**20**	**20**	**21**	**21**	**21**	**21**
13	13	13	14	14	14	14	15	15	15	16	16	16	16	17	17	17	18	18	18	19	19	19	19	20
14	**15**	**15**	**15**	**16**	**16**	**16**	**17**	**17**	**17**	**18**	**18**	**18**	**19**	**19**	**19**	**19**	**20**	**20**	**20**	**21**	**21**	**21**	**21**	**22**
13	13	14	14	14	15	15	15	16	16	16	17	17	17	17	18	18	18	19	19	19	20	20	20	20
15	**15**	**15**	**16**	**16**	**16**	**17**	**17**	**17**	**18**	**18**	**19**	**19**	**19**	**20**	**20**	**20**	**21**	**21**	**21**	**21**	**22**	**22**	**22**	**22**
13	14	14	14	15	15	15	16	16	16	17	17	17	18	18	18	19	19	19	20	20	20	20	21	21
15	**15**	**16**	**16**	**16**	**17**	**17**	**17**	**18**	**18**	**19**	**19**	**19**	**20**	**20**	**20**	**21**	**21**	**21**	**22**	**22**	**22**	**23**	**23**	**23**
14	14	14	15	15	15	16	16	16	17	17	17	18	18	18	19	19	19	20	20	20	21	21	21	22
16	**16**	**16**	**17**	**17**	**18**	**18**	**18**	**19**	**19**	**19**	**19**	**20**	**20**	**20**	**21**	**21**	**21**	**22**	**22**	**22**	**23**	**23**	**23**	**24**
14	14	15	15	15	16	16	16	17	17	17	18	18	19	19	19	20	20	20	21	21	21	22	22	22
16	**16**	**16**	**17**	**17**	**18**	**18**	**18**	**19**	**19**	**19**	**20**	**20**	**20**	**21**	**21**	**21**	**22**	**22**	**22**	**23**	**23**	**23**	**24**	**24**
14	15	15	15	16	16	16	17	17	18	18	18	19	19	19	20	20	20	21	21	21	22	22	22	23
16	**16**	**17**	**17**	**17**	**18**	**18**	**19**	**19**	**20**	**20**	**20**	**21**	**21**	**21**	**22**	**22**	**22**	**23**	**23**	**24**	**24**	**24**	**24**	**25**
14	15	15	15	16	16	17	17	18	18	18	19	19	19	20	20	21	21	21	22	22	22	23	23	23
16	**17**	**17**	**18**	**18**	**18**	**19**	**19**	**20**	**20**	**20**	**21**	**21**	**21**	**22**	**22**	**23**	**23**	**23**	**24**	**24**	**24**	**25**	**25**	**25**
15	15	16	16	16	17	17	18	18	18	19	19	20	20	20	21	21	21	22	22	22	23	23	24	24
17	**17**	**18**	**18**	**18**	**19**	**19**	**20**	**20**	**20**	**21**	**21**	**22**	**22**	**22**	**23**	**23**	**23**	**24**	**24**	**25**	**25**	**25**	**26**	**26**
15	16	16	16	17	17	18	18	18	19	19	20	20	20	21	21	22	22	22	23	23	23	24	24	24
17	**18**	**18**	**18**	**19**	**19**	**20**	**20**	**20**	**21**	**21**	**22**	**22**	**22**	**23**	**23**	**24**	**24**	**24**	**25**	**25**	**25**	**26**	**26**	**27**

Table J   (continued)

N	P Q	.01 .99	.02 .98	.03 .97	.04 .96	.05 .95	.06 .94	.07 .93	.08 .92	.09 .91	.10 .90	.11 .89	.12 .88	.13 .87	.14 .86	.15 .85	.16 .84	.17 .83	.18 .82	.19 .81	.20 .80	.21 .79	.22 .78	.23 .77	.24 .76	.25 .75
38		3	3	4	5	5	6	6	7	8	8	9	9	10	10	10	11	11	12	12	13	13	14	14	15	15
		3	4	5	6	7	7	8	8	9	10	10	11	11	12	12	13	13	14	14	15	15	16	16	17	17
39		3	3	4	5	5	6	6	7	8	8	9	9	10	10	11	11	12	12	13	13	14	14	14	15	15
		3	4	5	6	7	7	8	9	9	10	10	11	11	12	12	13	13	14	14	15	15	16	16	17	17
40		3	3	4	5	5	6	7	7	8	8	9	9	10	10	11	11	12	12	13	13	14	14	15	15	16
		3	4	5	6	7	7	8	9	9	10	10	11	12	12	13	13	14	14	15	15	16	16	17	17	18
41		3	3	4	5	6	6	7	7	8	8	9	10	10	11	11	12	12	13	13	14	14	15	15	15	16
		3	4	5	6	7	8	8	9	9	10	11	11	12	12	13	13	14	14	15	16	16	17	17	18	18
42		3	4	4	5	6	6	7	7	8	9	9	10	10	11	11	12	12	13	13	14	14	15	15	16	16
		3	4	5	6	7	8	8	9	10	10	11	11	12	12	13	14	14	15	15	16	16	17	17	18	18
43		3	4	4	5	6	6	7	8	8	9	9	10	10	11	11	12	13	13	14	14	15	15	16	16	17
		3	5	5	6	7	8	8	9	10	10	11	12	12	13	13	14	14	15	16	16	17	17	18	18	19
44		3	4	4	5	6	6	7	8	8	9	9	10	11	11	12	12	13	13	14	14	15	15	16	16	17
		3	5	5	6	7	8	9	9	10	11	11	12	12	13	14	14	15	15	16	16	17	17	18	18	19
45		3	4	4	5	6	7	7	8	8	9	10	10	11	11	12	12	13	14	14	15	15	16	16	17	17
		4	5	6	6	7	8	9	9	10	11	11	12	13	13	14	14	15	15	16	17	17	18	18	19	19
46		3	4	4	5	6	7	7	8	9	9	10	10	11	11	12	13	13	14	14	15	15	16	16	17	17
		4	5	6	6	7	8	9	9	10	11	11	12	13	13	14	15	15	16	16	17	17	18	19	19	20
47		3	4	5	5	6	7	7	8	9	9	10	10	11	12	12	13	13	14	15	15	16	16	17	17	18
		4	5	6	7	7	8	9	10	10	11	12	12	13	14	14	15	15	16	17	17	18	18	19	19	20
48		3	4	5	5	6	7	7	8	9	9	10	11	11	12	13	13	14	14	15	15	16	16	17	18	18
		4	5	6	7	7	8	9	10	10	11	12	12	13	14	14	15	16	16	17	17	18	19	19	20	20
49		3	4	5	5	6	7	8	8	9	10	10	11	11	12	13	13	14	14	15	16	16	17	17	18	18
		4	5	6	7	8	8	9	10	11	11	12	13	13	14	15	15	16	16	17	18	18	19	19	20	21

.26	.27	.28	.29	.30	.31	.32	.33	.34	.35	.36	.37	.38	.39	.40	.41	.42	.43	.44	.45	.46	.47	.48	.49	.50
.74	.73	.72	.71	.70	.69	.68	.67	.66	.65	.64	.63	.62	.61	.60	.59	.58	.57	.56	.55	.54	.53	.52	.51	.50
15	16	16	17	17	18	18	18	19	19	20	20	20	21	21	22	22	22	23	23	24	24	24	25	25
17	**18**	**18**	**19**	**19**	**20**	**20**	**20**	**21**	**21**	**22**	**22**	**23**	**23**	**23**	**24**	**24**	**24**	**25**	**25**	**26**	**26**	**26**	**27**	**27**
16	16	17	17	17	18	18	19	19	20	20	20	21	21	22	22	22	23	23	24	24	24	25	25	26
18	**18**	**19**	**19**	**20**	**20**	**20**	**21**	**21**	**22**	**22**	**23**	**23**	**23**	**24**	**24**	**25**	**25**	**25**	**26**	**26**	**27**	**27**	**27**	**28**
16	17	17	17	18	18	19	19	20	20	20	21	21	22	22	23	23	23	24	24	25	25	25	26	26
18	**19**	**19**	**20**	**20**	**20**	**21**	**21**	**22**	**22**	**23**	**23**	**23**	**24**	**24**	**25**	**25**	**26**	**26**	**26**	**27**	**27**	**28**	**28**	**28**
16	17	17	18	18	19	19	20	20	20	21	21	22	22	23	23	23	24	24	25	25	26	26	26	27
18	**19**	**19**	**20**	**20**	**21**	**21**	**22**	**22**	**23**	**23**	**23**	**24**	**24**	**25**	**25**	**26**	**26**	**26**	**27**	**27**	**28**	**28**	**28**	**29**
17	17	18	18	19	19	19	20	20	21	21	22	22	23	23	23	24	24	25	25	26	26	26	27	27
19	**19**	**20**	**20**	**21**	**21**	**22**	**22**	**23**	**23**	**23**	**24**	**24**	**25**	**25**	**26**	**26**	**27**	**27**	**27**	**28**	**28**	**29**	**29**	**29**
17	18	18	18	19	19	20	20	21	21	22	22	23	23	24	24	24	25	25	26	26	27	27	27	28
19	**20**	**20**	**21**	**21**	**22**	**22**	**23**	**23**	**23**	**24**	**24**	**25**	**25**	**26**	**26**	**27**	**27**	**28**	**28**	**28**	**29**	**29**	**30**	**30**
17	18	18	19	19	20	20	21	21	22	22	23	23	24	24	24	25	25	26	26	27	27	28	28	28
19	**20**	**21**	**21**	**21**	**22**	**22**	**23**	**23**	**24**	**24**	**25**	**25**	**26**	**26**	**27**	**27**	**28**	**28**	**28**	**29**	**29**	**30**	**30**	**31**
18	18	19	19	20	20	21	21	22	22	23	23	24	24	24	25	25	26	26	27	27	28	28	29	29
20	**20**	**21**	**21**	**22**	**22**	**23**	**23**	**24**	**24**	**25**	**25**	**26**	**26**	**27**	**27**	**28**	**28**	**28**	**29**	**29**	**30**	**30**	**31**	**31**
18	18	19	20	20	21	21	22	22	22	23	23	24	24	25	25	26	26	27	27	28	28	29	29	30
20	**21**	**21**	**22**	**22**	**23**	**23**	**24**	**24**	**25**	**25**	**26**	**26**	**27**	**27**	**28**	**28**	**29**	**29**	**30**	**30**	**30**	**31**	**31**	**32**
18	19	19	20	20	21	21	22	22	23	23	24	24	25	25	26	26	27	27	28	28	29	29	30	30
21	**21**	**22**	**22**	**23**	**23**	**24**	**24**	**25**	**25**	**26**	**26**	**27**	**27**	**28**	**28**	**29**	**29**	**30**	**30**	**31**	**31**	**31**	**32**	**32**
19	19	20	20	21	21	22	22	23	23	24	24	25	25	26	26	27	27	28	28	29	29	30	30	31
21	**21**	**22**	**22**	**23**	**24**	**24**	**25**	**25**	**26**	**26**	**27**	**27**	**28**	**28**	**29**	**29**	**30**	**30**	**31**	**31**	**32**	**32**	**33**	**33**
19	19	20	21	21	22	22	23	23	24	24	25	25	26	26	27	27	28	28	29	29	30	30	31	31
21	**22**	**22**	**23**	**23**	**24**	**24**	**25**	**26**	**26**	**27**	**27**	**28**	**28**	**29**	**29**	**30**	**30**	**31**	**31**	**32**	**32**	**33**	**33**	**34**

Table K   Percentage points of the Studentized range

Error df	a	K = NUMBER OF MEANS OR NUMBER OF STEPS BETWEEN ORDERED MEANS									
		2	3	4	5	6	7	8	9	10	11
5	.05	3.64	4.60	5.22	5.67	6.03	6.33	6.58	6.80	6.99	7.17
	.01	5.70	6.98	7.80	8.42	8.91	9.32	9.67	9.97	10.24	10.48
6	.05	3.46	4.34	4.90	5.30	5.63	5.90	6.12	6.32	6.49	6.65
	.01	5.24	6.33	7.03	7.56	7.97	8.32	8.61	8.87	9.10	9.30
7	.05	3.34	4.16	4.68	5.06	5.36	5.61	5.82	6.00	6.16	6.30
	.01	4.95	5.92	6.54	7.01	7.37	7.68	7.94	8.17	8.37	8.55
8	.05	3.26	4.04	4.53	4.89	5.17	5.40	5.60	5.77	5.92	6.05
	.01	4.75	5.64	6.20	6.62	6.96	7.24	7.47	7.68	7.86	8.03
9	.05	3.20	3.95	4.41	4.76	5.02	5.24	5.43	5.59	5.74	5.87
	.01	4.60	5.43	5.96	6.35	6.66	6.91	7.13	7.33	7.49	7.65
10	.05	3.15	3.88	4.33	4.65	4.91	5.12	5.30	5.46	5.60	5.72
	.01	4.48	5.27	5.77	6.14	6.43	6.67	6.87	7.05	7.21	7.36
11	.05	3.11	3.82	4.26	4.57	4.82	5.03	5.20	5.35	5.49	5.61
	.01	4.39	5.15	5.62	5.97	6.25	6.48	6.67	6.84	6.99	7.13
12	.05	3.08	3.77	4.20	4.51	4.75	4.95	5.12	5.27	5.39	5.51
	.01	4.32	5.05	5.50	5.84	6.10	6.32	6.51	6.67	6.81	6.94
13	.05	3.06	3.73	4.15	4.45	4.69	4.88	5.05	5.19	5.32	5.43
	.01	4.26	4.96	5.40	5.73	5.98	6.19	6.37	6.53	6.67	6.79
14	.05	3.03	3.70	4.11	4.41	4.64	4.83	4.99	5.13	5.25	5.36
	.01	4.21	4.89	5.32	5.63	5.88	6.08	6.26	6.41	6.54	6.66
15	.05	3.01	3.67	4.08	4.37	4.59	4.78	4.94	5.08	5.20	5.31
	.01	4.17	4.84	5.25	5.56	5.80	5.99	6.16	6.31	6.44	6.55
16	.05	3.00	3.65	4.05	4.33	4.56	4.74	4.90	5.03	5.15	5.26
	.01	4.13	4.79	5.19	5.49	5.72	5.92	6.08	6.22	6.35	6.46
17	.05	2.98	3.63	4.02	4.30	4.52	4.70	4.86	4.99	5.11	5.21
	.01	4.10	4.74	5.14	5.43	5.66	5.85	6.01	6.15	6.27	6.38
18	.05	2.97	3.61	4.00	4.28	4.49	4.67	4.82	4.96	5.07	5.17
	.01	4.07	4.70	5.09	5.38	5.60	5.79	5.94	6.08	6.20	6.31
19	.05	2.96	3.59	3.98	4.25	4.47	4.65	4.79	4.92	5.04	5.14
	.01	4.05	4.67	5.05	5.33	5.55	5.73	5.89	6.02	6.14	6.25
20	.05	2.95	3.58	3.96	4.23	4.45	4.62	4.77	4.90	5.01	5.11
	.01	4.02	4.64	5.02	5.29	5.51	5.69	5.84	5.97	6.09	6.19
24	.05	2.92	3.53	3.90	4.17	4.37	4.54	4.68	4.81	4.92	5.01
	.01	3.96	4.55	4.91	5.17	5.37	5.54	5.69	5.81	5.92	6.02
30	.05	2.89	3.49	3.85	4.10	4.30	4.46	4.60	4.72	4.82	4.92
	.01	3.89	4.45	4.80	5.05	5.24	5.40	5.54	5.65	5.76	5.85
40	.05	2.86	3.44	3.79	4.04	4.23	4.39	4.52	4.63	4.73	4.82
	.01	3.82	4.37	4.70	4.93	5.11	5.26	5.39	5.50	5.60	5.69
60	.05	2.83	3.40	3.74	3.98	4.16	4.31	4.44	4.55	4.65	4.73
	.01	3.76	4.28	4.59	4.82	4.99	5.13	5.25	5.36	5.45	5.53
120	.05	2.80	3.36	3.68	3.92	4.10	4.24	4.36	4.47	4.56	4.64
	.01	3.70	4.20	4.50	4.71	4.87	5.01	5.12	5.21	5.30	5.37
∞	.05	2.77	3.31	3.63	3.86	4.03	4.17	4.29	4.39	4.47	4.55
	.01	3.64	4.12	4.40	4.60	4.76	4.88	4.99	5.08	5.16	5.23

Table L   Squares, square roots, and reciprocals of numbers from 1 to 1000

N	N²	√N	1/N	N	N²	√N	1/N	N	N²	√N	1/N
1	1	1.0000	1.000000	61	3721	7.8102	.016393	121	14641	11.0000	.00826446
2	4	1.4142	.500000	62	3844	7.8740	.016129	122	14884	11.0454	.00819672
3	9	1.7321	.333333	63	3969	7.9373	.015873	123	15129	11.0905	.00813008
4	16	2.0000	.250000	64	4096	8.0000	.015625	124	15376	11.1355	.00800452
5	25	2.2361	.200000	65	4225	8.0623	.015385	125	15625	11.1803	.00800000
6	36	2.4495	.166667	66	4356	8.1240	.015152	126	15876	11.2250	.00793651
7	49	2.6458	.142857	67	4489	8.1854	.014925	127	16129	11.2694	.00787402
8	64	2.8284	.125000	68	4624	8.2462	.014706	128	16384	11.3137	.00781250
9	81	3.0000	.111111	69	4761	8.3066	.014493	129	16641	11.3578	.00775194
10	100	3.1623	.100000	70	4900	8.3666	.014286	130	16900	11.4018	.00769231
11	121	3.3166	.090909	71	5041	8.4261	.014085	131	17161	11.4455	.00763359
12	144	3.4641	.083333	72	5184	8.4853	.013889	132	17424	11.4891	.00757576
13	169	3.6056	.076923	73	5329	8.5440	.013699	133	17689	11.5326	.00751880
14	196	3.7417	.071429	74	5476	8.6023	.013514	134	17956	11.5758	.00746269
15	225	3.8730	.066667	75	5625	8.6603	.013333	135	18225	11.6190	.00740741
16	256	4.0000	.062500	76	5776	8.7178	.013158	136	18496	11.6619	.00735294
17	289	4.1231	.058824	77	5929	8.7750	.012987	137	18769	11.7047	.00729927
18	324	4.2426	.055556	78	6084	8.8318	.012821	138	19044	11.7473	.00724638
19	361	4.3589	.052632	79	6241	8.8882	.012658	139	19321	11.7898	.00719424
20	400	4.4721	.050000	80	6400	8.9443	.012500	140	19600	11.8322	.00714286
21	441	4.5826	.047619	81	6561	9.0000	.012346	141	19881	11.8743	.00709220
22	484	4.6904	.045455	82	6724	9.0554	.012195	142	20164	11.9164	.00704225
23	529	4.7958	.043478	83	6889	9.1104	.012048	143	20449	11.9583	.00699301
24	576	4.8990	.041667	84	7056	9.1652	.011905	144	20736	12.0000	.00694444
25	625	5.0000	.040000	85	7225	9.2195	.011765	145	21025	12.0416	.00689655
26	676	5.0990	.038462	86	7396	9.2736	.011628	146	21316	12.0830	.00684932
27	729	5.1962	.037037	87	7569	9.3274	.011494	147	21609	12.1244	.00680272
28	784	5.2915	.035714	88	7744	9.3808	.011364	148	21904	12.1655	.00675676
29	841	5.3852	.034483	89	7921	9.4340	.011236	149	22201	12.2066	.00671141
30	900	5.4772	.033333	90	8100	9.4868	.011111	150	22500	12.2474	.00666667
31	961	5.5678	.032258	91	8281	9.5394	.010989	151	22801	12.2882	.00662252
32	1024	5.6569	.031250	92	8464	9.5917	.010870	152	23104	12.3288	.00657895
33	1089	5.7446	.030303	93	8649	9.6437	.010753	153	23409	12.3693	.00653595
34	1156	5.8310	.029412	94	8836	9.6954	.010638	154	23716	12.4097	.00649351
35	1225	5.9161	.028571	95	9025	9.7468	.010526	155	24025	12.4499	.00645161
36	1296	6.0000	.027778	96	9216	9.7980	.010417	156	24336	12.4900	.00641026
37	1369	6.0828	.027027	97	9409	9.8489	.010309	157	24649	12.5300	.00636943
38	1444	6.1644	.026316	98	9604	9.8995	.010204	158	24964	12.5698	.00632911
39	1521	6.2450	.025641	99	9801	9.9499	.010101	159	25281	12.6095	.00628931
40	1600	6.3246	.025000	100	10000	10.0000	.010000	160	25600	12.6491	.00625000
41	1681	6.4031	.024390	101	10201	10.0499	.00990099	161	25921	12.6886	.00621118
42	1764	6.4807	.023810	102	10404	10.0995	.00980392	162	26244	12.7279	.00617284
43	1849	6.5574	.023256	103	10609	10.1489	.00970874	163	26569	12.7671	.00613497
44	1936	6.6332	.022727	104	10816	10.1980	.00961538	164	26896	12.8062	.00609756
45	2025	6.7082	.022222	105	11025	10.2470	.00952381	165	27225	12.8452	.00606061
46	2116	6.7823	.021739	106	11236	10.2956	.00943396	166	27556	12.8841	.00602410
47	2209	6.8557	.021277	107	11449	10.3441	.00934579	167	27889	12.9228	.00598802
48	2304	6.9282	.020833	108	11664	10.3923	.00925926	168	28224	12.9615	.00595238
49	2401	7.0000	.020408	109	11881	10.4403	.00917431	169	28561	13.0000	.00591716
50	2500	7.0711	.020000	110	12100	10.4881	.00909091	170	28900	13.0384	.00588235
51	2601	7.1414	.019608	111	12321	10.5357	.00900901	171	29241	13.0767	.00584795
52	2704	7.2111	.019231	112	12544	10.5830	.00892857	172	29584	13.1149	.00581395
53	2809	7.2801	.018868	113	12769	10.6301	.00884956	173	29929	13.1529	.00578035
54	2916	7.3485	.018519	114	12996	10.6771	.00877193	174	30276	13.1909	.00574713
55	3025	7.4162	.018182	115	13225	10.7238	.00869565	175	30625	13.2288	.00571429
56	3136	7.4833	.017857	116	13456	10.7703	.00862069	176	30976	13.2665	.00568182
57	3249	7.5498	.017544	117	13689	10.8167	.00854701	177	31329	13.3041	.00564972
58	3364	7.6158	.017241	118	13924	10.8628	.00847458	178	31684	13.3417	.00561798
59	3481	7.6811	.016949	119	14161	10.9087	.00840336	179	32041	13.3791	.00558659
60	3600	7.7460	.016667	120	14400	10.9545	.00833333	180	32400	13.4164	.00555556

Table L   (continued)

N	N²	√N̄	1/N	N	N²	√N̄	1/N	N	N²	√N̄	1/N
181	32761	13.4536	.00552486	241	58081	15.5242	.00414938	301	90601	17.3494	.00332226
182	33124	13.4907	.00549451	242	58564	15.5563	.00413223	302	91204	17.3781	.00331126
183	33489	13.5277	.00546448	243	59049	15.5885	.00411523	303	91809	17.4069	.00330033
184	33856	13.5647	.00543478	244	59536	15.6205	.00409836	304	92416	17.4356	.00328047
185	34225	13.6015	.00540541	245	60025	15.6525	.00408163	305	93025	17.4642	.00328947
186	34596	13.6382	.00537634	246	60516	15.6844	.00406504	306	93636	17.4929	.00326797
187	34969	13.6748	.00534759	247	61009	15.7162	.00404858	307	94249	17.5214	.00325733
188	35344	13.7113	.00531915	248	61504	15.7480	.00403226	308	94864	17.5499	.00321675
189	35721	13.7477	.00529101	249	62001	15.7797	.00401606	309	95481	17.5784	.00323625
190	36100	13.7840	.00526316	250	62500	15.8114	.00400000	310	96100	17.6068	.00322581
191	36481	13.8203	.00523560	251	63001	15.8430	.00398406	311	96721	17.6352	.00321543
192	36864	13.8564	.00520833	252	63504	15.8745	.00396825	312	97344	17.6635	.00320513
193	37249	13.8924	.00518135	253	64009	15.9060	.00395257	313	97969	17.6918	.00319489
194	37636	13.9284	.00515464	254	64516	15.9374	.00393701	314	98596	17.7200	.00318471
195	38025	13.9642	.00512821	255	65025	15.9687	.00392157	315	99225	17.7482	.00317460
196	38416	14.0000	.00510204	256	65536	16.0000	.00390625	316	99856	17.7764	.00316456
197	38809	14.0357	.00507614	257	66049	16.0312	.00389105	317	100489	17.8045	.00315457
198	39204	14.0712	.00505051	258	66564	16.0624	.00387597	318	101124	17.8326	.00314465
199	39601	14.1067	.00502513	259	67081	16.0935	.00386100	319	101761	17.8606	.00313480
200	40000	14.1421	.00500000	260	67600	16.1245	.00384615	320	102400	17.8885	.00312500
201	40401	14.1774	.00497512	261	68121	16.1555	.00383142	321	103041	17.9165	.00311526
202	40804	14.2127	.00495050	262	68644	16.1864	.00381679	322	103684	17.9444	.00310559
203	41209	14.2478	.00492611	263	69169	16.2173	.00380228	323	104329	17.9722	.00309598
204	41616	14.2829	.00490196	264	69696	16.2481	.00378788	324	104976	18.0000	.00308642
205	42025	14.3178	.00487805	265	70225	16.2788	.00377358	325	105625	18.0278	.00307692
206	42436	14.3527	.00485437	266	70756	16.3095	.00375940	326	106276	18.0555	.00306748
207	42849	14.3875	.00483092	267	71289	16.3401	.00374532	327	106929	18.0831	.00305810
208	43264	14.4222	.00480769	268	·71824	16.3707	.00373134	328	107584	18.1108	.00304878
209	43681	14.4568	.00478469	269	72361	16.4012	.00371747	329	108241	18.1384	.00303951
210	44100	14.4914	.00476190	270	72900	16.4317	.00370370	330	108900	18.1659	.00303030
211	44521	14.5258	.00473934	271	73441	16.4621	.00369004	331	109561	18.1934	.00302115
212	44944	14.5602	.00471698	272	73984	16.4924	.00367647	332	110224	18.2209	.00301205
213	45369	14.5945	.00469484	273	74529	16.5227	.00366300	333	110889	18.2483	.00300300
214	45796	14.6287	.00467290	274	75076	16.5529	.00364964	334	111556	18.2757	.00299401
215	46225	14.6629	.00465116	275	75625	16.5831	.00363636	335	112225	18.3030	.00298507
216	46656	14.6969	.00462963	276	76176	16.6132	.00362319	336	112896	18.3303	.00297619
217	47089	14.7309	.00460829	277	76729	16.6433	.00361011	337	113569	18.3576	.00296736
218	47524	14.7648	.00458716	278	77284	16.6733	.00359712	338	114244	18.3848	.00295858
219	47961	14.7986	.00456621	279	77841	16.7033	.00358423	339	114921	18.4120	.00294985
220	48400	14.8324	.00454545	280	78400	16.7332	.00357143	340	115600	18.4391	.00294118
221	48841	14.8661	.00452489	281	78961	16.7631	.00355872	341	116281	18.4662	.00293255
222	49284	14.8997	.00450450	282	79524	16.7929	.00354610	342	116964	18.4932	.00292398
223	49729	14.9332	.00448430	283	80089	16.8226	.00353357	343	117649	18.5203	.00291545
224	50176	14.9666	.00446429	284	80656	16.8523	.00352113	344	118336	18.5472	.00290698
225	50625	15.0000	.00444444	285	81225	16.8819	.00350877	345	119025	18.5742	.00289855
226	51076	15.0333	.00442478	286	81796	16.9115	.00349650	346	119716	18.6011	.00289017
227	51529	15.0665	.00440529	287	82369	16.9411	.00348432	347	120409	18.6279	.00288184
228	51984	15.0997	.00438596	288	82944	16.9706	.00347222	348	121104	18.6548	.00287356
229	52441	15.1327	.00436681	289	83521	17.0000	.00346021	349	121801	18.6815	.00286533
230	52900	15.1658	.00434783	290	84100	17.0294	.00344828	350	122500	18.7083	.00285714
231	53361	15.1987	.00432900	291	84681	17.0587	.00343643	351	123201	18.7350	.00284900
232	53824	15.2315	.00431034	292	85264	17.0880	.00342466	352	123904	18.7617	.00284091
233	54289	15.2643	.00429185	293	85849	17.1172	.00341297	353	124609	18.7883	.00283286
234	54756	15.2971	.00427350	294	86436	17.1464	.00340136	354	125316	18.8149	.00282486
235	55225	15.3297	.00425532	295	87025	17.1756	.00338983	355	126025	18.8414	.00281690
236	55696	15.3623	.00423729	296	87616	17.2047	.00337838	356	126736	18.8680	.00280899
237	56169	15.3948	.00421941	297	88209	17.2337	.00336700	357	127449	18.8944	.00280112
238	56644	15.4272	.00420168	298	88804	17.2627	.00335570	358	128164	18.9209	.00279330
239	57121	15.4596	.00418410	299	89401	17.2916	.00334448	359	128881	18.9473	.00278552
240	57600	15.4919	.00416667	300	90000	17.3205	.00333333	360	129600	18.9737	.00277778

Table L   (*continued*)

N	N²	√N	1/N	N	N²	√N	1/N	N	N²	√N	1/N
361	130321	19.0000	.00277008	421	177241	20.5183	.00237530	481	231361	21.9317	.00207900
362	131044	19.0263	.00276243	422	178084	20.5426	.00236967	482	232324	21.9545	.00207469
363	131769	19.0526	.00275482	423	178929	20.5670	.00236407	483	233289	21.9773	.00207039
364	132496	19.0788	.00274725	424	179776	20.5913	.00235849	484	234256	22.0000	.00206612
365	133225	19.1050	.00273973	425	180625	20.6155	.00235294	485	235225	22.0227	.00206186
366	133956	19.1311	.00273224	426	181476	20.6398	.00234742	486	236196	22.0454	.00205761
367	134689	19.1572	.00272480	427	182329	20.6640	.00234192	487	237169	22.0681	.00205339
368	135424	19.1833	.00271739	428	183184	20.6882	.00233645	488	238144	22.0907	.00204918
369	136161	19.2094	.00271003	429	184041	20.7123	.00233100	489	239121	22.1133	.00204499
370	136900	19.2354	.00270270	430	184900	20.7364	.00232558	490	240100	22.1359	.00204082
371	137641	19.2614	.00269542	431	185761	20.7605	.00232019	491	241081	22.1585	.00203666
372	138384	19.2873	.00268817	432	186624	20.7846	.00231481	492	242064	22.1811	.00203252
373	139129	19.3132	.00268097	433	187489	20.8087	.00230947	493	243049	22.2036	.00202840
374	139876	19.3391	.00267380	434	188356	20.8327	.00230415	494	244036	22.2261	.00202429
375	140625	19.3649	.00266667	435	189225	20.8567	.00229885	495	245025	22.2486	.00202020
376	141376	19.3907	.00265957	436	190096	20.8806	.00229358	496	246016	22.2711	.00201613
377	142129	19.4165	.00265252	437	190969	20.9045	.00228833	497	247009	22.2935	.00201207
378	142884	19.4422	.00264550	438	191844	20.9284	.00228311	498	248004	22.3159	.00200803
379	143641	19.4679	.00263852	439	192721	20.9523	.00227790	499	249001	22.3383	.00200401
380	144400	19.4936	.00263158	440	193600	20.9762	.00227273	500	250000	22.3607	.00200000
381	145161	19.5192	.00262467	441	194481	21.0000	.00226757	501	251001	22.3830	.00199601
382	145924	19.5448	.00261780	442	195364	21.0238	.00226244	502	252004	22.4054	.00199203
383	146689	19.5704	.00261097	443	196249	21.0476	.00225734	503	253009	22.4277	.00198807
384	147456	19.5959	.00260417	444	197136	21.0713	.00225225	504	254016	22.4499	.00198413
385	148225	19.6214	.00259740	445	198025	21.0950	.00224719	505	255025	22.4722	.00198020
386	148996	19.6469	.00259067	446	198916	21.1187	.00224215	506	256036	22.4944	.00197628
387	149769	19.6723	.00258398	447	199809	21.1424	.00223714	507	257049	22.5167	.00197239
388	150544	19.6977	.00257732	448	200704	21.1660	.00223214	508	258064	22.5389	.00196850
389	151321	19.7231	.00257069	449	201601	21.1896	.00222717	509	259081	22.5610	.00196464
390	152100	19.7484	.00256410	450	202500	21.2132	.00222222	510	260100	22.5832	.00196078
391	152881	19.7737	.00255754	451	203401	21.2368	.00221729	511	261121	22.6053	.00195695
392	153664	19.7990	.00255102	452	204304	21.2603	.00221239	512	262144	22.6274	.00195312
393	154449	19.8242	.00254453	453	205209	21.2838	.00220751	513	263169	22.6495	.00194932
394	155236	19.8494	.00253807	454	206116	21.3073	.00220264	514	264196	22.6716	.00194553
395	156025	19.8746	.00253165	455	207025	21.3307	.00219870	515	265225	22.6936	.00194175
396	156816	19.8997	.00252525	456	207936	21.3542	.00219298	516	266256	22.7156	.00193798
397	157609	19.9249	.00251889	457	208849	21.3776	.00218818	517	267289	22.7376	.00193424
398	158404	19.9499	.00251256	458	209764	21.4009	.00218341	518	268324	22.7596	.00193050
399	159201	19.9750	.00250627	459	210681	21.4243	.00217865	519	269361	22.7816	.00192678
400	160000	20.0000	.00250000	460	211600	21.4476	.00217391	520	270400	22.8035	.00192308
401	160801	20.0250	.00249377	461	212521	21.4709	.00216920	521	271441	22.8254	.00191939
402	161604	20.0499	.00248756	462	213444	21.4942	.00216450	522	272484	22.8473	.00191571
403	162409	20.0749	.00248139	463	214369	21.5174	.00215983	523	273529	22.8692	.00191205
404	163216	20.0998	.00247525	464	215296	21.5407	.00215517	524	274576	22.8910	.00190840
405	164025	20.1246	.00246914	465	216225	21.5639	.00215054	525	275625	22.9129	.00190476
406	164836	20.1494	.00246305	466	217156	21.5870	.00214592	526	276676	22.9347	.00190114
407	165649	20.1742	.00245700	467	218089	21.6102	.00214133	527	277729	22.9565	.00189753
408	166464	20.1990	.00245098	468	219024	21.6333	.00213675	528	278784	22.9783	.00189394
409	167281	20.2237	.00244499	469	219961	21.6564	.00213220	529	279841	23.0000	.00189036
410	168100	20.2485	.00243902	470	220900	21.6795	.00212766	530	280900	23.0217	.00188679
411	168921	20.2731	.00243309	471	221841	21.7025	.00212314	531	281961	23.0434	.00188324
412	169744	20.2978	.00242718	472	222784	21.7256	.00211864	532	283024	23.0651	.00187970
413	170569	20.3224	.00242131	473	223729	21.7486	.00211416	533	284089	23.0868	.00187617
414	171396	20.3470	.00241546	474	224676	21.7715	.00210970	534	285156	23.1084	.00187266
415	172225	20.3715	.00240964	475	225625	21.7945	.00210526	535	286225	23.1301	.00186916
416	173056	20.3961	.00240385	476	226576	21.8174	.00210084	536	287296	23.1517	.00186567
417	173889	20.4206	.00239808	477	227529	21.8403	.00209644	537	288369	23.1733	.00186220
418	174724	20.4450	.00239234	478	228484	21.8632	.00209205	538	289444	23.1948	.00185874
419	175561	20.4695	.00238663	479	229441	21.8861	.00208768	539	290521	23.2164	.00185529
420	176400	20.4939	.00238095	480	230400	21.9089	.00208333	540	291600	23.2379	.00185185

Table L   (continued)

N	N²	√N	1/N	N	N²	√N	1/N	N	N²	√N	1/N
541	292681	23.2594	.00184843	601	361201	24.5153	.00166389	661	436921	25.7099	.00151286
542	293764	23.2809	.00184502	602	302404	24.5357	.00166113	662	438244	25.7294	.00151057
543	294849	23.3024	.00184162	603	363609	24.5561	.00165837	663	439569	25.7488	.00150830
544	295936	23.3238	.00183824	604	364816	24.5764	.00165563	664	440896	25.7682	.00150602
545	297025	23.3452	.00183486	605	366025	24.5967	.00165289	665	442225	25.7876	.00150376
546	298116	23.3666	.00183150	606	367236	24.6171	.00165017	666	443556	25.8070	.00150150
547	299209	23.3880	.00182815	607	368449	24.6374	.00164745	667	444889	25.8263	.00149925
548	300304	23.4094	.00182482	608	369664	24.6577	.00164474	668	446224	25.8457	.00149701
549	301401	23.4307	.00182149	609	370881	24.6779	.00164204	669	447561	25.8650	.00149477
550	302500	23.4521	.00181818	610	372100	24.6982	.00163934	670	448900	25.8844	.00149254
551	303601	23.4734	.00181488	611	373321	24.7184	.00163666	671	450241	25.9037	.00149031
552	304704	23.4947	.00181159	612	374544	24.7386	.00163399	672	451584	25.9230	.00148810
553	305809	23.5160	.00180832	613	375769	24.7588	.00163132	673	452929	25.9422	.00148588
554	306916	23.5372	.00180505	614	376996	24.7790	.00162866	674	454276	25.9615	.00148368
555	308025	23.5584	.00180180	615	378225	24.7992	.00162602	675	455625	25.9808	.00148148
556	309136	23.5797	.00179856	616	379456	24.8193	.00162338	676	456976	26.0000	.00147929
557	310249	23.6008	.00179533	617	380689	24.8395	.00162075	677	458329	26.0192	.00147710
558	311364	23.6220	.00179211	618	381924	24.8596	.00161812	678	459684	26.0384	.00147493
559	312481	23.6432	.00178891	619	383161	24.8797	.00161551	679	461041	26.0576	.00147275
560	313600	23.6643	.00178571	620	384400	24.8998	.00161290	680	462400	26.0768	.00147059
561	314721	23.6854	.00178253	621	385641	24.9199	.00161031	681	463761	26.0960	.00146843
562	315844	23.7065	.00177936	622	386884	24.9399	.00160772	682	465124	26.1151	.00146628
563	316969	23.7276	.00177620	623	388129	24.9600	.00160514	683	466489	26.1343	.00146413
564	318096	23.7487	.00177305	624	389376	24.9800	.00160256	684	467856	26.1534	.00146199
565	319225	23.7697	.00176991	625	390625	25.0000	.00160000	685	469225	26.1725	.00145985
566	320356	23.7908	.00176678	626	391876	25.0200	.00159744	686	470596	26.1916	.00145773
567	321489	23.8118	.00176367	627	393129	25.0400	.00159490	687	471969	26.2107	.00145560
568	322624	23.8328	.00176056	628	394384	25.0599	.00159236	688	473344	26.2298	.00145349
569	323761	23.8537	.00175747	629	395641	25.0799	.00158983	689	474721	26.2488	.00145138
570	324900	23.8747	.00175439	630	396900	25.0998	.00158730	690	476100	26.2679	.00144928
571	326041	23.8956	.00175131	631	398161	25.1197	.00158479	691	477481	26.2869	.00144718
572	327184	23.9165	.00164825	632	399424	25.1396	.00158228	692	478864	26.3059	.00144509
573	328329	23.9374	.00174520	633	400689	25.1595	.00157978	693	480249	26.3249	.00144300
574	329476	23.9583	.00174216	634	401956	25.1794	.00157729	694	481636	26.3439	.00144092
575	330625	23.9792	.00173913	635	403225	25.1992	.00157480	695	483025	26.3629	.00143885
576	331776	24.0000	.00173611	636	404496	25.2190	.00157233	696	484416	26.3818	.00143678
577	332929	24.0208	.00173310	637	405769	25.2389	.00156986	697	485809	26.4008	.00143472
578	334084	24.0416	.00173010	638	407044	25.2587	.00156740	698	487204	26.4197	.00143266
579	335241	24.0624	.00172712	639	408321	25.2784	.00156495	699	488601	26.4386	.00143062
580	336400	24.0832	.00172414	640	409600	25.2982	.00156250	700	490000	26.4575	.00142857
581	337561	24.1039	.00172117	641	410881	25.3180	.00156006	701	491401	26.4764	.00142653
582	338724	24.1247	.00171821	642	412164	25.3377	.00155763	702	492804	26.4953	.00142450
583	339889	24.1454	.00171527	643	413449	25.3574	.00155521	703	494209	26.5141	.00142248
584	341056	24.1661	.00171233	644	414736	25.3772	.00155280	704	495616	26.5330	.00142045
585	342225	24.1868	.00170940	645	416025	25.3969	.00155039	705	497025	26.5518	.00141844
586	343396	24.2074	.00170648	646	417316	25.4165	.00154799	706	498436	26.5707	.00141643
587	344569	24.2281	.00170358	647	418609	25.4362	.00154560	707	499849	26.5895	.00141443
588	345744	24.2487	.00170068	648	419904	25.4558	.00154321	708	501264	26.6083	.00141243
589	346921	24.2693	.00169779	649	421201	25.4755	.00154083	709	502681	26.6271	.00141044
590	348100	24.2899	.00169492	650	422500	25.4951	.00153846	710	504100	26.6458	.00140845
591	349281	24.3105	.00169205	651	423801	25.5147	.00153610	711	505521	26.6646	.00140647
592	350464	24.3311	.00168919	652	425104	25.5343	.00153374	712	506944	26.6833	.00140449
593	351649	24.3516	.00168634	653	426409	25.5539	.00153139	713	508369	26.7021	.00140252
594	352836	24.3721	.00168350	654	427716	25.5734	.00152905	714	509796	26.7208	.00140056
595	354025	24.3926	.00168067	655	429025	25.5930	.00152672	715	511225	26.7395	.00139860
596	355216	24.4131	.00167785	656	430336	25.6125	.00152439	716	512656	26.7582	.00139665
597	356409	24.4336	.00167504	657	431649	25.6320	.00152207	717	514089	26.7769	.00139470
598	357604	24.4540	.00167224	658	432964	25.6515	.00151976	718	515524	26.7955	.00139276
599	358801	24.4745	.00166945	659	434281	25.6710	.00151745	719	516961	26.8142	.00139082
600	360000	24.4949	.00166667	660	435600	25.6905	.00151515	720	518400	26.8328	.00138889

Table L   (continued)

N	N²	√N	1/N	N	N²	√N	1/N	N	N²	√N	1/N
721	519841	26.8514	.00138696	781	609961	27.9464	.00128041	841	707281	29.0000	.00118906
722	521284	26.8701	.00138504	782	611524	27.9643	.00127877	842	708964	29.0172	.00118765
723	522729	26.8887	.00138313	783	613089	27.9821	.00127714	843	710649	29.0345	.00118624
724	524176	26.9072	.00138122	784	614656	28.0000	.00127551	844	712336	29.0517	.00118483
725	525625	26.9258	.00137931	785	616225	28.0179	.00127389	845	714025	29.0689	.00118343
726	527076	26.9444	.00137741	786	617796	28.0357	.00127226	846	715716	29.0861	.00118203
727	528529	26.9629	.00137552	787	619369	28.0535	.00127065	847	717409	29.1033	.00118064
728	529984	26.9815	.00137363	788	620944	28.0713	.00126904	848	719104	29.1204	.00117925
729	531441	27.0000	.00137174	789	622521	28.0891	.00126743	849	720801	29.1376	.00117786
730	532900	27.0185	.00136986	790	624100	28.1069	.00126582	850	722500	29.1548	.00117647
731	534361	27.0370	.00136799	791	625681	28.1247	.00126422	851	724201	29.1719	.00117509
732	535824	27.0555	.00136612	792	627264	28.1425	.00126263	852	725904	29.1890	.00117371
733	537289	27.0740	.00136426	793	628849	28.1603	.00126103	853	727609	29.2062	.00117233
734	538756	27.0924	.00136240	794	630436	28.1780	.00125945	854	729316	29.2233	.00117096
735	540225	27.1109	.00136054	795	632025	28.1957	.00125786	855	731025	29.2404	.00116959
736	541696	27.1293	.00135870	796	633616	28.2135	.00125628	856	732736	29.2575	.00116822
737	543169	27.1477	.00135685	797	635209	28.2312	.00125471	857	734449	29.2746	.00116686
738	544644	27.1662	.00135501	798	636804	28.2489	.00125313	858	736164	29.2916	.00116550
739	546121	27.1846	.00135318	799	638401	28.2666	.00125156	859	737881	29.3087	.00116414
740	547600	27.2029	.00135135	800	640000	28.2843	.00125000	860	739600	29.3258	.00116279
741	549081	27.2213	.00134953	801	641601	28.3019	.00124844	861	741321	29.3428	.00116144
742	550564	27.2397	.00134771	802	643204	28.3196	.00124688	862	743044	29.3598	.00116009
743	552049	27.2580	.00134590	803	644809	28.3373	.00124533	863	744769	29.3769	.00115875
744	553536	27.2764	.00134409	804	646416	28.3549	.00124378	864	746496	29.3939	.00115741
745	555025	27.2947	.00134228	805	648025	28.3725	.00124224	865	748225	29.4109	.00115607
746	556516	27.3130	.00134048	806	649636	28.3901	.00124069	866	749956	29.4279	.00115473
747	558009	27.3313	.00133869	807	651249	28.4077	.00123916	867	751689	29.4449	.00115340
748	559504	27.3496	.00133690	808	625864	28.4253	.00123762	868	753424	29.4618	.00115207
749	561001	27.3679	.00133511	809	654481	28.4429	.00123609	869	755161	29.4788	.00115075
750	562500	27.3861	.00133333	810	656100	28.4605	.00123457	870	756900	29.4958	.00114943
751	564001	27.4044	.00133156	811	657721	28.4781	.00123305	871	758641	29.5127	.00114811
752	565504	27.4226	.00132979	812	659344	28.4956	.00123153	872	760384	29.5296	.00114679
753	567009	27.4408	.00132802	813	660969	28.5132	.00123001	873	762129	29.5466	.00114548
754	568516	27.4591	.00132626	814	662596	28.5307	.00122850	874	763876	29.5635	.00114416
755	570025	27.4773	.00132450	815	664225	28.5482	.00122699	875	765625	29.5804	.00114286
756	571536	27.4955	.00132275	816	665856	28.5657	.00122549	876	767376	29.5973	.00114155
757	573049	27.5136	.00132100	817	667489	28.5832	.00122399	877	769129	29.6142	.00114025
758	574564	27.5318	.00131926	818	669124	28.6007	.00122249	878	770884	29.6311	.00113895
759	576081	27.5500	.00131752	819	670761	28.6182	.00122100	879	772641	29.6479	.00113766
760	577600	27.5681	.00131579	820	672400	28.6356	.00121951	880	774400	29.6648	.00113636
761	579121	27.5862	.00131406	821	674041	28.6531	.00121803	881	776161	29.6816	.00113507
762	580644	27.6043	.00131234	822	675684	28.6705	.00121655	882	777924	29.6985	.00113379
763	582169	27.6225	.00131062	823	677329	28.6880	.00121507	883	779689	29.7153	.00113250
764	583696	27.6405	.00130890	824	678976	28.7054	.00121359	884	781456	29.7321	.00113122
765	585225	27.6586	.00130719	825	680625	28.7228	.00121212	885	783225	29.7489	.00112994
766	586756	27.6767	.00130548	826	682276	28.7402	.00121065	886	784996	29.7658	.00112867
767	588289	27.6948	.00130378	827	683929	28.7576	.00120919	887	786769	29.7825	.00112740
768	589824	27.7128	.00130208	828	685584	28.7750	.00120773	888	788544	29.7993	.00112613
769	591361	27.7308	.00130039	829	687241	28.7924	.00120627	889	790321	29.8161	.00112486
770	592900	27.7489	.00129870	830	688900	28.8097	.00120482	890	792100	29.8329	.00112360
771	594441	27.7669	.00129702	831	690561	28.8271	.00120337	891	793881	29.8496	.00112233
772	595984	27.7849	.00129534	832	692224	28.8444	.00120192	892	795664	29.8664	.00112108
773	597529	27.8029	.00129366	833	693889	28.8617	.00120048	893	797449	29.8831	.00111982
774	599076	27.8209	.00129199	834	695556	28.8791	.00119904	894	799236	29.8998	.00111857
775	600625	27.8388	.00129032	835	697225	28.8964	.00119760	895	801025	29.9166	.00111732
776	602176	27.8568	.00128866	836	698896	28.9137	.00119617	896	802816	29.9333	.00111607
777	603729	27.8747	.00128700	837	700569	28.9310	.00119474	897	804609	29.9500	.00111483
778	605284	27.8927	.00128535	838	702244	28.9482	.00119332	898	806404	29.9666	.00111359
779	606841	27.9106	.00128370	839	703921	28.9655	.00119190	899	808201	29.9833	.00111235
780	608400	27.9285	.00128205	840	705600	28.9828	.00119048	900	810000	30.0000	.00111111

Table L   (continued)

N	N²	√N	1/N	N	N²	√N	1/N	N	N²	√N	1/N
901	811801	30.0167	.00110988	936	876096	30.5941	.00106838	971	942841	31.1609	.00102987
902	813604	30.0333	.00110865	937	877969	30.6105	.00106724	972	944784	31.1769	.00102881
903	815409	30.0500	.00110742	938	879844	30.6268	.00106610	973	946729	31.1929	.00102775
904	817216	30.0666	.00110619	939	881721	30.6431	.00106496	974	948676	31.2090	.00102669
905	819025	30.0832	.00110497	940	883600	30.6594	.00106383	975	950625	31.2250	.00102564
906	820836	30.0998	.00110375	941	885481	30.6757	.00106270	976	952576	31.2410	.00102459
907	822649	30.1164	.00110254	942	887364	30.6920	.00106157	977	954529	31.2570	.00102354
908	824464	30.1330	.00110132	943	889249	30.7083	.00106045	978	956484	31.2730	.00102249
909	826281	30.1496	.00110011	944	891136	30.7246	.00105932	979	958441	31.2890	.00102145
910	828100	30.1662	.00109890	945	893025	30.7409	.00105820	980	960400	31.3050	.00102041
911	829921	30.1828	.00109769	946	894916	30.7571	.00105708	981	962361	31.3209	.00101937
912	831744	30.1993	.00109649	947	896809	30.7734	.00105597	982	964324	31.3369	.00101833
913	833569	30.2159	.00109529	948	898704	30.7896	.00105485	983	966289	31.3528	.00101729
914	835396	30.2324	.00109409	949	900601	30.8058	.00105374	984	968256	31.3688	.00101626
915	837225	30.2490	.00109290	950	902500	30.8221	.00105263	985	970225	31.3847	.00101523
916	839056	30.2655	.00109170	951	904401	30.8383	.00105152	986	972196	31.4006	.00101420
917	840889	30.2820	.00109051	952	906304	30.8545	.00105042	987	974169	31.4166	.00101317
918	842724	30.2985	.00108932	953	908209	30.8707	.00104932	988	976144	31.4325	.00101215
919	844561	30.3150	.00108814	954	910116	30.8869	.00104822	989	978121	31.4484	.00101112
920	846400	30.3315	.00108696	955	912025	30.9031	.00104712	990	980100	31.4643	.00101010
921	848241	30.3480	.00108578	956	913936	30.9192	.00104603	991	982081	31.4802	.00100908
922	850084	30.3645	.00108460	957	915849	30.9354	.00104493	992	984064	31.4960	.00100806
923	851929	30.3809	.00108342	958	917764	30.9516	.00104384	993	986049	31.5119	.00100705
924	853776	30.3974	.00108225	959	919681	30.9677	.00104275	994	988036	31.5278	.00100604
925	855625	30.4138	.00108108	960	921600	30.9839	.00104167	995	990025	31.5436	.00100503
926	857476	30.4302	.00107991	961	923521	31.0000	.00104058	996	992016	31.5595	.00100402
927	859329	30.4467	.00107875	962	925444	31.0161	.00103950	997	994009	31.5753	.00103842
928	861184	30.4631	.00107759	963	927369	31.0322	.00103842	998	996004	31.5911	.00100200
929	863041	30.4795	.00107643	964	929296	31.0483	.00103734	999	998001	31.6070	.00100100
930	864900	30.4959	.00107527	965	931225	31.0644	.00103627	1000	1000000	31.6228	.00100000
931	866761	30.5123	.00107411	966	933156	31.0805	.00103520				
932	868624	30.5287	.00107296	967	935089	31.0966	.00103413				
933	870489	30.5450	.00107181	968	937024	31.1127	.00103306				
934	872356	30.5614	.00107066	969	938961	31.1288	.00103199				
935	874225	30.5778	.00106952	970	940900	31.1448	.00103093				

Table M   Random digits

ROW NUMBER										
00000	10097	32533	76520	13586	34673	54876	80959	09117	39292	74945
00001	37542	04805	64894	74296	24805	24037	20636	10402	00822	91665
00002	08422	68953	19645	09303	23209	02560	15953	34764	35080	33606
00003	99019	02529	09376	70715	38311	31165	88676	74397	04436	27659
00004	12807	99970	80157	36147	64032	36653	98951	16877	12171	76833
00005	66065	74717	34072	76850	36697	36170	65813	39885	11199	29170
00006	31060	10805	45571	82406	35303	42614	86799	07439	23403	09732
00007	85269	77602	02051	65692	68665	74818	73053	85247	18623	88579
00008	63573	32135	05325	47048	90553	57548	28468	28709	83491	25624
00009	73796	45753	03529	64778	35808	34282	60935	20344	35273	88435
00010	98520	17767	14905	68607	22109	40558	60970	93433	50500	73998
00011	11805	05431	39808	27732	50725	68248	29405	24201	52775	67851
00012	83452	99634	06288	98033	13746	70078	18475	40610	68711	77817
00013	88685	40200	86507	58401	36766	67951	90364	76493	29609	11062
00014	99594	67348	87517	64969	91826	08928	93785	61368	23478	34113
00015	65481	17674	17468	50950	58047	76974	73039	57186	40218	16544
00016	80124	35635	17727	08015	45318	22374	21115	78253	14385	53763
00017	74350	99817	77402	77214	43236	00210	45521	64237	96286	02655
00018	69916	26803	66252	29148	36936	87203	76621	13990	94400	56418
00019	09893	20505	14225	68514	46427	56788	96297	78822	54382	14598
00020	91499	14523	68479	27686	46162	83554	94750	89923	37089	20048
00021	80336	94598	26940	36858	70297	34135	53140	33340	42050	82341
00022	44104	81949	85157	47954	32979	26575	57600	40881	22222	06413
00023	12550	73742	11100	02040	12860	74697	96644	89439	28707	25815
00024	63606	49329	16505	34484	40219	52563	43651	77082	07207	31790
00025	61196	90446	26457	47774	51924	33729	65394	59593	42582	60527
00026	15474	45266	95270	79953	59367	83848	82396	10118	33211	59466
00027	94557	28573	67897	54387	54622	44431	91190	42592	92927	45973
00028	42481	16213	97344	08721	16868	48767	03071	12059	25701	46670
00029	23523	78317	73208	89837	68935	91416	26252	29663	05522	82562
00030	04493	52494	75246	33824	45862	51025	61962	79335	65337	12472
00031	00549	97654	64051	88159	96119	63896	54692	82391	23287	29529
00032	35963	15307	26898	09354	33351	35462	77974	50024	90103	39333
00033	59808	08391	45427	26842	83609	49700	13021	24892	78565	20106
00034	46058	85236	01390	92286	77281	44077	93910	83647	70617	42941
00035	32179	00597	87379	25241	05567	07007	86743	17157	85394	11838
00036	69234	61406	20117	45204	15956	60000	18743	92423	97118	96338
00037	19565	41430	01758	75379	40419	21585	66674	36806	84962	85207
00038	45155	14938	19476	07246	43667	94543	59047	90033	20826	69541
00039	94864	31994	36168	10851	34888	81553	01540	35456	05014	51176
00040	98086	24826	45240	28404	44999	08896	39094	73407	35441	31880
00041	33185	16232	41941	50949	89435	48581	88695	41994	37548	73043
00042	80951	00406	96382	70774	20151	23387	25016	25298	94624	61171
00043	79752	49140	71961	28296	69861	02591	74852	20539	00387	59579
00044	18633	32537	98145	06571	31010	24674	05455	61427	77938	91936
00045	74029	43902	77557	32270	97790	17119	52527	58021	80814	51748
00046	54178	45611	80993	37143	05335	12969	56127	19255	36040	90324
00047	11664	49883	52079	84827	59381	71539	09973	33440	88461	23356
00048	48324	77928	31249	64710	02295	36870	32307	57546	15020	09994
00049	69074	94138	87637	91976	35584	04401	10518	21615	01848	76938
00050	09188	20097	32825	39527	04220	86304	83389	87374	64278	58044
00051	90045	85497	51981	50654	94938	81997	91870	76150	68476	64659
00052	73189	50207	47677	26269	62290	64464	27124	67018	41361	82760
00053	75768	76490	20971	87749	90429	12272	95375	05871	93823	43178
00054	54016	44056	66281	31003	00682	27398	20714	53295	07706	17813
00055	08358	69910	78542	42785	13661	58873	04618	97553	31223	08420
00056	28306	03264	81333	10591	40510	07893	32604	60475	94119	01840
00057	53840	86233	81594	13628	51215	90290	28466	68795	77762	20791
00058	91757	53741	61613	62669	50263	90212	55781	76514	83483	47055
00059	89415	92694	00397	58391	12607	17646	48949	72306	94541	37408

Table M   (continued)

ROW NUMBER										
00060	77513	03820	86864	29901	68414	82774	51908	13980	72893	55507
00061	19502	37174	69979	20288	55210	29773	74287	75251	65344	67415
00062	21818	59313	93278	81757	05686	73156	07082	85046	31853	38452
00063	51474	66499	68107	23621	94049	91345	42836	09191	08007	45449
00064	99559	68331	62535	24170	69777	12830	74819	78142	43860	72834
00065	33713	48007	93584	72869	51926	64721	58303	29822	93174	93972
00066	85274	86893	11303	22970	28834	34137	73515	90400	71148	43643
00067	84133	89640	44035	52166	73852	70091	61222	60561	62327	18423
00068	56732	16234	17395	96131	10123	91622	85496	57560	81604	18880
00069	65138	56806	87648	85261	34313	65861	45875	21069	85644	47277
00070	38001	02176	81719	11711	71602	92937	74219	64049	65584	49698
00071	37402	96397	01304	77586	56271	10086	47324	62605	40030	37438
00072	97125	40348	87083	31417	21815	39250	75237	62047	15501	29578
00073	21826	41134	47143	34072	64638	85902	49139	06441	03856	54552
00074	73135	42742	95719	09035	85794	74296	08789	88156	64691	19202
00075	07638	77929	03061	18072	96207	44156	23821	99538	04713	66994
00076	60528	83441	07954	19814	59175	20695	05533	52139	61212	06455
00077	83596	35655	06958	92983	05128	09719	77433	53783	92301	50498
00078	10850	62746	99599	10507	13499	06319	53075	71839	06410	19362
00079	39820	98952	43622	63147	64421	80814	43800	09351	31024	73167
00080	59580	06478	75569	78800	88835	54486	23768	06156	04111	08408
00081	38508	07341	23793	48763	90822	97022	17719	04207	95954	49953
00082	30692	70668	94688	16127	56196	80091	82067	63400	05462	69200
00083	65443	95659	18238	27437	49632	24041	08337	65676	96299	90836
00084	27267	50264	13192	72294	07477	44606	17985	48911	97341	30358
00085	91307	06991	19072	24210	36699	53728	28825	35793	28976	66252
00086	68434	94688	84473	13622	62126	98408	12843	82590	09815	93146
00087	48908	15877	54745	24591	35700	04754	83824	52692	54130	55160
00088	06913	45197	42672	78601	11883	09528	63011	98901	14974	40344
00089	10455	16019	14210	33712	91342	37821	88325	80851	43667	70883
00090	12883	97343	65027	61184	04285	01392	17974	15077	90712	26769
00091	21778	30976	38807	36961	31649	42096	63281	02023	08816	47449
00092	19523	59515	65122	59659	86283	68258	69572	13798	16435	91529
00093	67245	52670	35583	16563	79246	86686	76463	34222	26655	90802
00094	60584	47377	07500	37992	45134	26529	26760	83637	41326	44344
00095	53853	41377	36066	94850	58838	73859	49364	73331	96240	43642
00096	24637	38736	74384	89342	52623	07992	12369	18601	03742	83873
00097	83080	12451	38992	22815	07759	51777	97377	27585	51972	37867
00098	16444	24334	36151	99073	27493	70939	85130	32552	54846	54759
00099	60790	18157	57178	65762	11161	78576	45819	52979	65130	04860
00100	03991	10461	93716	16894	66083	24653	84609	58232	88618	19161
00101	38555	95554	32886	59780	08355	60860	29735	47762	71299	23853
00102	17546	73704	92052	46215	55121	29281	59076	07936	27954	58909
00103	32643	52861	95819	06831	00911	98936	76355	93779	80863	00514
00104	69572	68777	39510	35905	14060	40619	29549	69616	33564	60780
00105	24122	66591	27699	06494	14845	46672	61958	77100	90899	75754
00106	61196	30231	92962	61773	41839	55382	17267	70943	78038	70267
00107	30532	21704	10274	12202	39685	23309	10061	68829	55986	66485
00108	03788	97599	75867	20717	74416	53166	35208	33374	87539	08823
00109	48228	63379	85783	47619	53152	67433	35663	52972	16818	60311
00110	60365	94653	35075	33949	42614	29297	01918	28316	98953	73231
00111	83799	42402	56623	34442	34994	41374	70071	14736	09958	18065
00112	32960	07405	36409	83232	99385	41600	11133	07586	15917	06253
00113	19322	53845	57620	52606	66497	68646	78138	66559	19640	99413
00114	11220	94747	07399	37408	48509	23929	27482	45476	85244	35159
00115	31751	57260	68980	05339	15470	48355	88651	22596	03152	19121
00116	88492	99382	14454	04504	20094	98977	74843	93413	22109	78508
00117	30934	47744	07481	83828	73788	06533	28597	20405	94205	20380
00118	22888	48893	27499	98748	60530	45128	74022	84617	82037	10268
00119	78212	16993	35902	91386	44372	15486	65741	14014	87481	37220

Table M (*continued*)

ROW NUMBER										
00120	41849	84547	46850	52326	34677	58300	74910	64345	19325	81549
00121	46352	33049	69248	93460	45305	07521	61318	31855	14413	70951
00122	11087	96294	14013	31792	59747	67277	76503	34513	39663	77544
00123	52701	08337	56303	97315	16520	69676	11654	99893	02181	68161
00124	57275	36898	81304	48595	68652	27376	92852	55866	88448	03584
00125	20857	73156	70284	24326	79375	95220	01159	63267	10622	48391
00126	15633	84924	90415	93614	33521	26665	55823	47641	86225	31704
00127	92694	48297	39904	02115	59589	49067	66821	41575	49767	04037
00128	77613	19019	88152	00080	20554	91409	96277	48257	50816	97616
00129	38688	32486	45134	63545	59404	72059	43947	51680	43852	59693
00130	25163	01889	70014	15021	41290	67312	71857	15957	68971	11403
00131	65251	07629	37239	33295	05870	01119	92784	26340	18477	65622
00132	36815	43625	18637	37509	82444	99005	04921	73701	14707	93997
00133	64397	11692	05327	82162	20247	81759	45197	25332	83745	22567
00134	04515	25624	95096	67946	48460	85558	15191	18782	16930	33361
00135	83761	60873	43253	84145	60833	25983	01291	41349	20368	07126
00136	14387	06345	80854	09279	43529	06318	38384	74761	41196	37480
00137	51321	92246	80088	77074	88722	56736	66164	49431	66919	31678
00138	72472	00008	80890	18002	94813	31900	54155	83436	35352	54131
00139	05466	55306	93128	18464	74457	90561	72848	11834	79982	68416
00140	39528	72484	82474	25593	48545	35247	18619	13674	18611	19241
00141	81616	18711	53342	44276	75122	11724	74627	73707	58319	15997
00142	07586	16120	82641	22820	92904	13141	32392	19763	61199	67940
00143	90767	04235	13574	17200	69902	63742	78464	22501	18627	90872
00144	40188	28193	29593	88627	94972	11598	62095	36787	00441	58997
00145	34414	82157	86887	55087	19152	00023	12302	80783	32624	68691
00146	63439	75363	44989	16822	36024	00867	76378	41605	65961	73488
00147	67049	09070	93399	45547	94458	74284	05041	49807	20288	34060
00148	79495	04146	52162	90286	54158	34243	46978	35482	59362	95938
00149	91704	30552	04737	21031	75051	93029	47665	64382	99782	93478
00150	94015	46874	32444	48277	59820	96163	64654	25843	41145	42820
00151	74108	88222	88570	74015	25704	91035	01755	14750	48968	38603
00152	62880	87873	95160	59221	22304	90314	72877	17334	39283	04149
00153	11748	12102	80580	41867	17710	59621	06554	07850	73950	79552
00154	17944	05600	60478	03343	25852	58905	57216	39618	49856	99326
00155	66067	42792	95043	52680	46780	56487	09971	59481	37006	22186
00156	54244	91030	45547	70818	59849	96169	61459	21647	87417	17198
00157	30945	57589	31732	57260	47670	07654	46376	25366	94746	49580
00158	69170	37403	86995	90307	94304	71803	26825	05511	12459	91314
00159	08345	88975	35841	85771	08105	59987	87112	21476	14713	71181
00160	27767	43584	85301	88977	29490	69714	73035	41207	74699	09310
00161	13025	14338	54066	15243	47724	66733	47431	43905	31048	56699
00162	80217	36292	98525	24335	24432	24896	43277	58874	11466	16082
00163	10875	62004	90391	61105	57411	06368	53856	30743	08670	84741
00164	54127	57326	26629	19087	24472	88779	30540	27886	61732	75454
00165	60311	42824	37301	42678	45990	43242	17374	52003	70707	70214
00166	49739	71484	92003	98086	76668	73209	59202	11973	02902	33250
00167	78626	51594	16453	94614	39014	97066	83012	09832	25571	77628
00168	66692	13986	99837	00582	81232	44987	09504	96412	90193	79568
00169	44071	28091	07362	97703	76447	42537	98524	97831	65704	09514
00170	41468	85149	49554	17994	14924	39650	95294	00556	70481	06905
00171	94559	37559	49678	53119	70312	05682	66986	34099	74474	20740
00172	41615	70360	64114	58660	90850	64618	80620	51790	11436	38072
00173	50273	93113	41794	86861	24781	89683	55411	85667	77535	99892
00174	41396	80504	90670	08289	40902	05069	95083	06783	28102	57816
00175	25807	24260	71529	78920	72682	07385	90726	57166	98884	08583
00176	06170	97965	88302	98041	21443	41808	68984	83620	89747	98882
00177	60808	54444	74412	81105	01176	28838	36421	16489	18059	51061
00178	80940	44893	10408	36222	80582	71944	92638	40333	67054	16067
00179	19516	90120	46759	71643	13177	55292	21036	82808	77501	97427

Table M (*continued*)

ROW NUMBER									
00180	49386	54480	23604	23554	21785	41101	91178	10174	29420	90438
00181	06312	88940	15995	69321	47458	64809	98189	81851	29651	84215
00182	60942	00307	11897	92674	40405	68032	96717	54244	10701	41393
00183	92329	98932	78284	46347	71209	92061	39448	93136	25722	08564
00184	77936	63574	31384	51924	85561	29671	58137	17820	22751	36518
00185	38101	77756	11657	13897	95889	57067	47648	13885	70669	93406
00186	39641	69457	91339	22502	92613	89719	11947	56203	19324	20504
00187	84054	40455	99396	63680	67667	60631	69181	96845	38525	11600
00188	47468	03577	57649	63266	24700	71594	14004	23153	69249	05747
00189	43321	31370	28977	23896	76479	68562	62342	07589	08899	05985
00190	64281	61826	18555	64937	13173	33365	78851	16499	87064	13075
00191	66847	70495	32350	02985	86716	38746	26313	77463	55387	72681
00192	72461	33230	21529	53424	92581	02262	78438	66276	18396	73538
00193	21032	91050	13058	16218	12470	56500	15292	76139	59526	52113
00194	95362	67011	06651	16136	01016	00857	55018	56374	35824	71708
00195	49712	97380	10404	55452	34030	60726	75211	10271	36633	68424
00196	58275	61764	97586	54716	50259	46345	87195	46092	26787	60939
00197	89514	11788	68224	23417	73959	76145	30342	40277	11049	72049
00198	15472	50669	48139	36732	46874	37088	63465	09819	58869	35220
00199	12120	86124	51247	44302	60883	52109	21437	36786	49226	77837

Acknowledgments

We are grateful to the following authors and publishers for permission to adapt from the following tables.

Table B   R. A. Fisher, *Statistical Methods for Research Workers* (14th ed.). Reprinted with permission of Macmillan Publishing Company, Inc. Copyright © 1970, University of Adelaide.

Table C   Table III of R. A. Fisher and F. Yates. *Statistical Tables for Biological, Agricultural, and Medical Research.* Edinburgh: Oliver and Boyd, Ltd., Second ed., 1974.

Table D and Table D₁   G. W. Snedecor and William G. Cochran, *Statistical Methods,* 7th edition, Ames, Iowa: Iowa State University Press©1980. Reprinted by permission.

Table E   Q. McNemar, Table B of *Psychological Statistics.* New York: John Wiley and Sons, Inc., 1962.

Table F   E. G. Olds, The 5 percent significance levels of sums of squares of rank differences and a correction. *Ann. Math. Statist.* **20**, 117–118, 1949. E. G. Olds, Distribution of sums of squares of rank differences for small numbers of individuals. *Ann. Math. Statist.* **9**, 133–148, 1938.

Table G   W. V. Bingham, Table XVII of *Aptitudes and Aptitude Testing.* New York: Harper and Row, 1937.

Table H   S. Siegel, *Nonparametric Statistics.* New York: McGraw-Hill, 1956.

Table I   R. P. Runyon, Table A of *Nonparametric Statistics,* Reading, Mass.: Addison-Wesley Publishing Co., 1977.

Table J   R. P. Runyon, Table B of *Nonparametric Statistics,* Reading, Mass.: Addison-Wesley Publishing Co., 1977.

Table K   E. S. Pearson and H. O. Hartley, *Biometrika Tables for Statisticians,* Vol. 1, 2nd ed. New York: Cambridge, 1958.

Table L   A. L. Edwards, *Statistical Analysis,* 3rd ed. New York: Holt Rinehart and Winston, Inc., 1969.
J. W. Dunlap and A. K. Kurtz, *Handbook of Statistical Nomographs, Tables, and Formulas.* New York: World Book Company, 1932.

Table M   RAND Corporation, *A Million Random Digits,* Glencoe, Ill.: Free Press of Glencoe, 1955.

Glossary of Terms

D

Abscissa (*X-axis*)   Horizontal axis of a graph.

Absolute value of a number   The value of a number without regard to sign.

Addition rule   If A and B are mutually exclusive events, the probability of obtaining *either of them* is equal to the probability of A plus the probability of B. Symbolically,

$$p(A \text{ or } B) = p(A) + p(B).$$

Alpha (α) *level*   The level of significance set by the researcher for inferring the operation of nonchance factors.

Alternative hypothesis (H_1)   A statement specifying that the population parameter is some value other than the one specified under the null hypothesis.

Analysis of variance   A method, described initially by R. A. Fisher, for partitioning the sum of squares for experimental or survey data into known components of variation.

A posteriori comparisons   Comparisons not planned in advance to investigate specific hypotheses concerning population parameters.

A priori or planned comparisons   Comparisons planned in advance to investigate specific hypotheses concerning population parameters.

Array   Arrangement of data according to their magnitude from the smallest to the largest value.

Asymmetric measure of association   A measure of the one-way effect of one variable upon another.

Bar graph   A form of graph that uses bars to indicate the frequency of occurrence of observations within each nominal or ordinal category.

Before-after design   A correlated-samples design in which each individual is measured on the dependent variable both before and after the introduction of the experimental conditions.

Between-group variance   Estimate of variance based upon variability between groups.

Bias   In sampling, when selections favor certain events or certain collections of events.

Binomial distribution   A model with known mathematical properties used to describe the distributions of discrete random variables.

Cardinal numbers   Numbers used to represent quantity.

Centile   A percentage rank that divides a distribution into 100 equal parts. Same as a percentile.

Central limit theorem   If random samples of a fixed N are drawn from *any* population (regardless of the form of the population distribution), as N becomes larger, the distribution of sample means approaches normality, with the overall mean approaching μ, the variance of the sample means $\sigma_{\bar{x}}^2$ being equal to σ^2/N, and a standard error of $\sigma_{\bar{x}}$ of $\sigma/\sqrt{N}$.

Coefficient of determination (r^2)   The ratio of the explained variation to the total variation.

Coefficient of multiple determination (R^2)   A measure of the proportion of total variation in a dependent variable that is explained jointly by two or more independent variables.

Concordant pair   Two cases are ranked similarly on two variables.

Conditional distribution   The distribution of the categories of one variable under the differing conditions or categories of another.

Conditional probability   The probability of an event given that another event has occurred. Represented symbolically as $p(A\,|\,B)$, the probability of A given that B has occurred.

Conditional table   A contingency table that summarizes the relationship between two variables for categories or conditions of one or the other.

Confidence interval   A confidence interval for a parameter specifies a range of values bounded by two endpoints called confidence limits. Common confidence intervals are 95% and 99%.

Confidence limits   The two endpoints of a confidence interval.

Contingency coefficient   Nominal measure of association based on χ^2 and used primarily for square tables.

Contingency table   A table showing the joint distribution of two variables.

Continuous scales   Scales in which the variables can assume an unlimited number of intermediate values.

Control variable   A variable whose subgroups or categories are used in a conditional table to examine the relationship between two variables. Same as a test variable.

Correlation coefficient   A measure that expresses the extent to which two variables are related.

Correlation matrix   Summary of the statistical relationships between all possible pairs of variables.

Covariation   Extent to which two variables vary together.

Cramer's V   Nominal measure of association based on χ^2 suitable for tables of any dimension.

Critical region　That portion of the area under a curve that includes those values of a statistic that lead to rejection of the null hypothesis.

Critical values of t　Those values that bound the critical rejection regions corresponding to varying levels of significance.

Cumulative frequency　The number of cases (frequencies) at and below a given point.

Cumulative frequency curve (Ogive)　A curve that shows the number of cases below the upper real limit of an interval.

Cumulative frequency distribution　A distribution that shows the cumulative frequency at and below the upper real limit of the corresponding class interval.

Cumulative percentage　The percentage of cases (frequencies) at and below a given point.

Cumulative percentage distribution　A distribution that shows the cumulative percentage at and below the upper real limit of the corresponding class interval.

Cumulative proportion　The proportion of cases (frequencies) at and below a given point.

Cumulative proportion distribution　A distribution that shows the cumulative proportion at and below the upper real limit of the corresponding class interval.

Data　Numbers or measurements that are collected as a result of observations, interviews, etc.

Decile　A percentage rank that divides a distribution into 10 equal parts.

Degrees of freedom (df)　The number of values that are free to vary after we have placed certain restrictions upon our data.

Dependence　The condition that exists when the occurrence of a given event affects the probability of the occurrence of another event.

Dependent variable   The variable that is being predicted. Also referred to as the criterion variable.

Descriptive statistics   Procedures used to organize and present data in a convenient, summary form.

Deviation   The distance and direction of a score from a reference point.

Directional hypothesis   An alternative hypothesis that states the direction in which the population parameter differs from the one specified under H_0.

Discrete scales (Discontinuous scales)   Scales in which the variables have equality of counting units.

Dispersion   The spread or variability of scores about the measure of central tendency.

Element   A single member of a population.

Epsilon (ε)   The percentage difference within a category of the dependent variable between the two extreme categories of the independent variable.

Exhaustive   Two or more events are said to be exhaustive if they exhaust all possible outcomes. Symbolically,

$$p(A \text{ or } B \text{ or } \ldots) = 1.00.$$

Explained variation   Variation of predicted scores about the mean of the distribution. Same as the regression sum of squares.

F-ratio   The between-group variance estimate divided by the within-group variance estimate.

Finite population   A population whose elements or members can be listed.

Frequency curve (Frequency polygon)   A form of graph, representing a frequency distribution, in which a continuous line is used to indicate the frequency of the corresponding scores.

Frequency distribution   A frequency distribution shows the number of times each score occurs when the values of a variable are arranged in order according to their magnitudes.

Gamma (γ)   A symmetric measure of association for ordinal-level data based upon pair-by-pair comparison with a PRE interpretation.

Goodman's and Kruskal's tau (τ)An asymmetric measure of association for nominal-level data with a PRE interpretation.

Grouped frequency distribution   A frequency distribution in which the values of the variable have been grouped into class intervals.

Histogram   A form of bar graph used with interval- or ratio-scaled frequency distribution.

Homogeneity of variance   The condition that exists when two or more sample variances have been drawn from populations with equal variances.

Homoscedasticity   Condition that exists when the distribution of Y scores for each value of X has the same variability.

Independence   The condition that exists when the occurrence of a given event will not affect the probability of the occurrence of another event. Symbolically,

$$p(A\,|\,B) = p(A) \quad \text{and} \quad p(B\,|\,A) = p(B).$$

Independent variable   The predictor variable. Referred to as the experimental variable in an experiment.

Inferential or inductive statistics   Procedures used to arrive at broader generalizations or inferences from sample data to populations.

Infinite population   A population whose elements or members cannot be listed.

Inflated N An error produced whenever several observations are made on the same individual and treated as though they were independent observations.

Interquartile range   A measure of variability obtained by subtracting the score at the 1st quartile from the score at the 3rd quartile.

Interval estimation   The determination of an interval within which the population parameter is presumed to fall.

Interval scale   Scale in which exact distances can be known between categories. The zero point in this scale is arbitrary, and arithmetic operations are permitted.

Intervening variable   A variable that causally links two other variables.

Joint occurrence   The occurrence of two events simultaneously. Such events cannot be mutually exclusive.

Lambda (λ)   An asymmetric or symmetric measure of association for nominal-level data.

Law of large numbers   As the sample size increases, the standard error of the mean decreases.

Leptokurtic distribution   Bell-shaped distribution characterized by a piling up of scores in the center of the distribution.

Marginals   The column and row totals of a contingency table.

Matched-group design   A correlated-samples design in which pairs of individuals are matched on a variable correlated with the criterion measure. Each member of a pair receives different experimental conditions.

Mean   Sum of the scores or values of a variable divided by their number.

Mean deviation (average deviation)   Sum of the deviation of each score from the mean, without regard to sign, divided by the number of scores.

Measurement   The assignment of numbers to objects or events according to sets of predetermined (or arbitrary) rules.

Measure of central tendency   Index of central location used in the description of frequency distributions.

Median　Score in a distribution of scores, above and below which one-half of the frequencies fall.

Mesokurtic distribution　Bell-shaped distribution; "ideal" form of normal curve.

Mode　Score that occurs with the greatest frequency.

Multiple correlation coefficient (R)　A measure of the linear relationship between a dependent variable and the combined effects of two or more independent variables.

Multiple regression coefficient　A measure of the influence of an independent variable on a dependent variable when the effects of all other independent variables in the multiple regression equation have been held constant.

Multiplication rule　Given two events A and B, the probability of obtaining both A and B jointly is the product of the probability of obtaining one of these events times the conditional probability of obtaining one event, given that the other event has occurred. Symbolically,

$$p(A \text{ and } B) = p(A)p(B|A) = p(B)p(A|B).$$

Multivariate statistical technique　A statistical technique used to determine the nature of the relationships between three or more variables.

Mutually exclusive　Events A and B are said to be mutually exclusive if both cannot occur simultaneously. Symbolically, for mutually exclusive events,

$$p(A \text{ and } B) = 0.00.$$

Negatively skewed distribution　Distribution that has relatively fewer frequencies at the low end of the horizontal axis.

Negative Relationship　Variables are said to be negatively related when high scores on one variable tend to be associated with low scores on the other variable. Conversely, low scores on one variable tend to be associated with high scores on the other.

Nominal numbers　Numbers used to name.

Nominal scale   Scales in which the categories are homogeneous, mutually exclusive, and unordered.

Nondirectional hypothesis   An alternative hypothesis (H_1) that states only that the population parameter is *different* from the one specified under H_0.

Normal curve   A frequency curve with a characteristic bell-shaped form.

Null hypothesis (H_0) A statement that specifies hypothesized values for one or more of the population parameters. Commonly, although not necessarily, involves the hypothesis of "no difference."

Ogive   A cumulative frequency curve that shows the number of cases below the upper real limit of an interval.

One-tailed probability value   Probability values obtained by examining only one tail of the distribution.

One-variable test ("goodness-of-fit" technique)   Test of whether or not a significant difference exists between the *observed* number of cases falling into each category and the *expected* number of cases, based on the null hypothesis.

One-way analysis of variance   Statistical analysis of various categories, groups, or levels of one independent (experimental) variable.

Order of a conditional or partial table   The order of a table is the number of control or test variables that are included.

Order of a partial correlation coefficient   The order of a partial correlation coefficient indicates the number of control or test variables.

Ordinal numbers   Numbers used to represent position or order in a series.

Ordinal scale   Scale in which the classes can be rank ordered, that is, expressed in terms of the algebra of inequalities (for example, $a < b$ or $a > b$).

Ordinate (Y-axis)   Vertical axis of a graph.

Pair-by-pair comparison An approach to prediction for ordinal-level data in contingency form.

Panel design A research design in which a sample is drawn and the same respondents are interviewed and then reinterviewed at a later time.

Parameter Any characteristic of a finite population that can be estimated and is measurable or of an infinite population that can be estimated.

Partial correlation coefficient A measure of the linear relationship between two interval-level variables controlling for one or more other interval-level variables.

Partial table A contingency table that summarizes the relationship between two variables for categories or conditions of one or more other variables. Same as a conditional table.

Pearson's r (product-moment correlation coefficient) Correlation coefficient used with interval- or ratio-scaled variables.

Percentage A proportion that has been multiplied by 100.

Percentage difference Differences in percentages normally measured between the two extreme categories of the independent variable within a category of the dependent variable. Also referred to as epsilon (ϵ).

Percentile rank Number that represents the percentage of cases in a distribution that had scores at or lower than the one cited.

Percentiles Numbers that divide a distribution into 100 equal parts. Same as a centile.

Perfect relationship A relationship in which knowledge of the independent variable allows a perfect prediction of the dependent variable, or vice versa.

Phi coefficient Nominal measure of association based on χ^2 used with 2×2 tables.

Platykurtic distribution Frequency distribution characterized by a flattening in the central position.

Point estimation   An estimate of a population parameter that involves a single sample value, selected by the criterion of "best estimate."

Population   A complete set of individuals, objects, or measurements having some common observable characteristic.

Positively skewed distribution   Distribution that has relatively fewer frequencies at the high end of the horizontal axis.

Positive relationship   High scores on one variable tend to be associated with high scores on the other variable and, conversely, low scores on one variable tend to be associated with low scores on the other variable.

Post hoc fallacy   Faulty causal inferences from correlational data.

Power efficiency   The increase in sample size required to make one test as powerful as a competing test.

Power of a test   The probability of rejecting the null hypothesis when it is, in fact, false.

Probability   A theory concerned with possible outcomes of studies. Symbolically,

$$p(A) = \frac{\text{number of outcomes in which } A \text{ occurs}}{\text{total number of outcomes}}.$$

Proportion   A value calculated by dividing the quantity in one category by the total of all the components.

Proportional reduction in error (*PRE*)   A ratio of the prediction errors without information about the independent variable to the prediction errors having information about the independent variable.

Quartile   A percentage rank that divides a distribution into 4 equal parts.

Random sampling   Samples are selected in such a way that each sample of a given size has precisely the same probability of being selected.

Range   Measure of dispersion; the scale distance between the largest and the smallest score.

Rate   A ratio of the occurrences in a group category to the total number of elements in the group with which we are concerned.

Ratio scale   Same as interval scale, except that there is a true zero point.

Regression line (line of "best fit")   Straight line that makes the squared deviations around it minimal.

Regression sum of squares   Variation of predicted scores about the mean of the distribution. Same as the explained variation.

Residual   Size and direction of the error in prediction or the vertical distance from the observed data coordinate to the regression line.

Residual sum of squares   Variation of scores around the regression line. Same as the unexplained variation.

Residual variance   Variance around the regression line.

Restricted range   Truncated range on one or both variables, resulting in a deceptively low correlation between these variables.

Robust test   A statistical test from which statistical inferences are likely to be valid even when there are fairly large departures from normality in the population distribution.

Sample   A subset or part of a population.

Sampling distribution   A theoretical probability distribution of a statistic that would result from drawing all possible samples of a given size from some population.

Scatter diagram   Graphic device used to summarize visually the relationship between two variables.

Significance level   A probability value that is considered so rare in the sampling distribution specified under the null hypothesis that one is willing to assert the operation of nonchance factors. Common significance levels are 0.05 and 0.01, and 0.001.

Simple random sampling   A method of selecting samples so that each sample of a given size in a population has an equal chance of being selected.

Skewed distribution   Distribution that departs from symmetry and tails off at one end.

Somer's d   An asymmetrical measure of association for ordinal-level data that incorporates tied ranks.

Spearman's rho (r_s) A measure of association for ordinal-level data that provides a measure of the extent to which two sets of ranks are in agreement or disagreement.

Spurious relationship   Exists when the zero-order relationship between an independent and dependent variable disappears or becomes significantly weaker with the introduction of a control variable.

Standard deviation   Measure of dispersion defined as the square root of the sum of the squared deviations from the mean, divided by N. Also can be defined as the square root of the variance.

Standard error of the difference between means   Standard deviation of the sampling distribution of the difference between means,

$$\sqrt{\sigma_{\bar{X}_1}^2 + \sigma_{\bar{X}_2}^2}.$$

Standard error of estimate   Standard deviation of scores around the regression line.

Standard error of the mean   A theoretical standard deviation of sample means, of a given sample size, drawn from some specified population. When based upon a known population standard deviation, $\sigma_{\bar{X}} = \sigma/\sqrt{N}$; when estimated from a single sample, $s_{\bar{X}} = s/\sqrt{N - 1}$.

Standard error of the mean difference   Standard error based upon a correlated samples design. Used in the Student's *t*-ratio for correlated samples $(s_{\bar{D}})$.

Standard error of the proportion   An estimate of the standard deviation of the sampling distribution of proportions.

Standard normal distribution   A frequency distribution that has a mean of 0, a standard deviation of 1, and a total area equal to 1.00.

Standard score (*z*)   A score that represents the deviation of a specific score from the mean, expressed in standard deviation units.

Statistic   A number that describes a characteristic of a sample.

Statistical elaboration   The process of introducing a control variable to determine whether a bivariate relationship remains the same or varies under the different categories or conditions of the control variable.

Statistics   Collection of numerical facts expressed in summarizing statements; method of dealing with data: a tool for collecting, organizing, and analyzing numerical facts or observations.

Sum of squares   Deviations from the mean, squared and then summed.

Symmetric measure of association   A measure of the mutual association between two variables.

t-distributions   Theoretical symmetrical sampling distributions with a mean of zero and a standard deviation that becomes smaller as degrees of freedom (df) increase. Used in relation to the Student *t*-ratio.

t-ratio   A test statistic for determining the significance of a difference between means (two-sample case) or for testing the hypothesis that a given sample mean was drawn from a population with the mean specified under the null hypothesis (one-sample case). Used when population standard deviation (or standard deviations) is not known.

Test variable   A variable whose subgroups or categories are used in a conditional table to examine the relationship between two variables. Same as a control variable.

Tied pair   Two cases are ranked similarly on one or both of two variables.

Total sum of squares   Variation of scores around the sample mean. Same as the total variation.

Total variation   Variation of scores around the sample mean. Same as the total sum of squares.

True limits of a number   The true limits of a value of a continuous variable are equal to that number plus or minus one-half of the unit of measurement.

Two-tailed probability value   Probability values that take into account both tails of the distribution.

Type I error (type α error)   The rejection of H_0 when it is actually true. The probability of a type I error is given by the α level.

Type II error (type β error)   The probability of accepting H_0 when it is actually false. The probability of a type II error is given by β.

Unbiased estimate of a parameter   An estimate that equals, on the average, the value of the parameter.

Unexplained variation   Variation of scores around the regression line. Same as the residual sum of squares.

Variable   Any characteristic of a person, group, or environment that can vary or denotes a difference.

Variance   Sum of the squared deviations from the mean, divided by N.

Variance estimate   Sum of the squared deviations from the mean divided by degrees of freedom.

Weighted mean   Sum of the mean of each group multiplied by its respective weight (the N in each group), divided by the sum of the weights (total N).

Within-group variance   Estimate of variance based upon the variability within groups.

References

E

Bentler, P. M., and M. O. Newcomb, (1978) Longitudinal study of marital success and failure. *Journal of Consulting and Clinical Psychology* 46:1053–1070.

Berenson, W. M., K. W. Elifson, and T. Tollerson, III, (1976) Preachers in politics: A study of political activism among the black ministry. *J. of Black Studies* 6:373–383.

Bingham, W. V., (1937) *Aptitudes and Aptitude Testing*. New York: Harper and Bros.

Blalock, H. M., (1979) *Social Statistics* (Revised 2nd ed.). New York: McGraw-Hill Book Co.

Dunlap, J. W., and A. K. Kurtz (1932) *Handbook of Statistical Nomographs, Tables, and Formulas*. New York: World Book Company.

Edwards, A. L., (1969) *Statistical Analysis* (3rd ed.). New York: Holt, Rinehart and Winston, Inc.

Fisher, R. A., (1950) *Statistical Methods for Research Workers*. Edinburgh: Oliver and Boyd, Ltd.

Fisher, R. A., and F. Yates, (1948) *Statistical Tables for Biological, Agricultural, and Medical Research*. Edinburgh: Oliver and Boyd, Ltd.

Haber, A., R. P. Runyon, and P. Badia, (1970) *Readings in Statistics*. Reading, Mass.: Addison-Wesley Publishing Co., Inc.

Henkel, R. E., (1975) Part-whole correlations and the treatment of ordinal and quasi-interval data as interval data. *Pacific Soc. Rev.* 18:3–26.

Hess, E. H., A. L. Seltzer, and J. M. Shlien, (1965) Pupil response of hetero- and homo-sexual males to pictures of men and women: A pilot study, *J. Abn. Soc. Psych.* 70:165–168.

Huff, D., (1954) *How to Lie with Statistics.* New York: W. W. Norton and Co., Inc.

Kirk, R. E., (1968) *Experimental Design: Procedures for the Behavioral Sciences.* Belmont, California: Brooks/Cole.

Labovitz, S., (1970) The assignment of numbers to rank order categories. *Amer. Soc. Rev.* 35:515–524.

McNemar, Q., (1962) *Psychological Statistics.* New York: John Wiley and Sons, Inc.

Olds, E. G., (1949) The 5 percent significance levels of sums of squares of rank differences and a correction. *Ann. Math. Statist.* 20:117–118.

Pearson, E. S., and H. O. Hartley, (1958) *Biometrika Tables for Statisticians* (Vol. 1, 2nd ed.). New York: Cambridge University Press.

Rand Corporation, (1955) *A Million Random Digits*. Glencoe, Ill.: The Free Press of Glencoe.

Runyon, R. P., (1977) *Non-Parametric Statistics*. Reading, Mass.: Addison-Wesley Publishing Co.

Siegel, S., (1956) *Non-Parametric Statistics*. New York: McGraw-Hill.

Snedecor, G. W., and W. G. Cochran, (1956) *Statistical Methods* (6th ed.). Ames, Iowa: Iowa State University Press.

Tukey, J. W., (1953) The Problem of Multiple Comparisons. Ditto: Princeton University, 396 pp.

Answers to
Selected
Exercises

Answers to Selected Exercises

In problems involving many steps, you may occasionally find a discrepancy between the answers you obtained and those shown here. Where the discrepancies are small, they are probably due to rounding errors. These disparities are more common today because of the wide differences in methods used; that is, varying degrees of sophistication among calculators, adding machines, and hand calculations.

Chapter 1

2 **a)** constant   **b)** variable   **c)** constant   **d)** variable

    e) constant   **f)** constant   **g)** both   **h)** constant

    i) constant

6 U.S. Senators, astronauts

7 population

10 **a)** data   **b)** data

Chapter 2

1 a)

	PROPORTION	PERCENTAGE
Too little	0.71	71
About right	0.24	24
Too much	0.03	3
No opinion	0.02	2

b)

	PROPORTION*	PERCENTAGE*
Too little	0.77	77
About right	0.21	21
Too much	0.02	2
No opinion	0.01	1

c)

	PROPORTION	PERCENTAGE
Too little	0.83	83
About right	0.13	13
Too much	0.02	2
No opinion	0.02	2

d)

	PROPORTION*	PERCENTAGE*
Too little	0.25	25
About right	0.28	28
Too much	0.43	43
No opinion	0.03	3

*The slight disparity from 1.00 and 100 is due to rounding error.

The majority of the respondents appear to want more information about earthquakes and how to prepare for them. Thus, the results appear to contradict the view that Californians would prefer not thinking about earthquakes.

2 $a = 5$

3 $y = 22$

4 $\Sigma X = 1200$

5 $N = 4$

6 $N = 41$

7 $s^2 = 20$

8 a) 100.00 **b)** 46.41 **c)** 2.96 **d)** 0.01

e) 16.46 **f)** 1.05 **g)** 86.21 **h)** 10.00

10 6 to 7

12   **(a)**   **(b)**

YEAR	PERCENTAGE OF MALE* HOMICIDE VICTIMS	PERCENTAGE OF FEMALE* HOMICIDE VICTIMS
1968	13.85	13.85
1969	14.14	14.67
1970	15.91	15.79
1971	17.51	17.16
1972	18.50	17.56
1973	20.10	20.98

*The slight disparity from 100.00 is due to rounding error.

c) 74.21%   **d)** 24.98%

13 a) 25   **b)** 60   **c)** 37   **d)** 31   **e)** 60   **f)** 44

14 a) $\displaystyle\sum_{i=1}^{3} X_i$ **b)** $\displaystyle\sum_{i=1}^{N} X_i$ **c)** $\displaystyle\sum_{i=3}^{6} X_i^2$ **d)** $\displaystyle\sum_{i=4}^{N} X_i^2$

17 a) 12.65   **b)** 4.00   **c)** 1.26   **d)** 0.40   **e)** 0.13

18 a) -0.5 to $+0.5$ **b)** 0.45 to 0.55   **c)** 0.95 to 1.05

   d) 0.485 to 0.495   **e)** -4.5 to -5.5 **f)** -4.45 to -4.55

19 $\displaystyle\sum_{i=1}^{N} X_i^2 = 4^2 + 5^2 + 7^2 + 9^2 + 10^2 + 11^2 + 14^2 = 588$

$$\left(\sum_{i=1}^{N} X_i\right)^2 = (4 + 5 + 7 + 9 + 10 + 11 + 14)^2$$

$$= 60^2 = 3600,$$

   $588 \neq 3600$

23 91.53

26 a)

	AGGRESSIVE MALES (PERCENT)	NONAGGRESSIVE MALES (PERCENT)
Attempted to get girl intoxicated	9.1	37.9
Falsely promised marriage	7.5	8.0
Falsely professed love	14.6	44.8
Threatened to terminate relationship	3.5	9.2

b) The results do not support the hypothesis. In fact, they are in a direction opposite to the hypothesis.

Chapter 3

1

	TRUE LIMITS	WIDTH
a)	7.5–12.5	5
b)	5.5–7.5	2
c)	(−0.5)–(+2.5)	3
d)	4.5–14.5	10
e)	(−1.5)–(−8.5)	7
f)	2.45–3.55	1.1
g)	1.495–1.755	0.26
h)	(−3.5)–(+3.5)	7

2

	WIDTH	APPARENT LIMITS	TRUE LIMITS
i)	7	0–6	0.5–6.5
ii)	1	29	28.5–29.5
iii)	2	18–19	17.5–19.5
iv)	4	(−30)–(−27)	(−30.5)–(−26.5)
v)	0.01	0.30	0.295–0.305
vi)	0.006	0.206–0.211	0.2055–0.2115

3

CLASS INTERVALS	TRUE LIMITS	f	CUMULATIVE f	CUMULATIVE %
40–44	39.5–44.5	1	1	2.5
45–49	44.5–49.5	1	2	5.0
50–54	49.5–54.5	1	3	7.5
55–59	54.5–59.5	0	3	7.5
60–64	59.5–64.5	3	6	15.0
65–69	64.5–69.5	3	9	22.5
70–74	69.5–74.5	4	13	32.5
75–79	74.5–79.5	11	24	60.0
80–84	79.5–84.5	8	32	80.0
85–89	84.5–89.5	4	36	90.0
90–94	89.5–94.5	3	39	97.5
95–99	94.5–99.5	1	40	100.0

4 b) $i = 3$

CLASS INTERVALS	f	CLASS INTERVALS	f
40–42	1	70–72	3
43–45	0	73–75	2
46–48	1	76–78	8
49–51	0	79–81	6
52–54	1	82–84	4
55–57	0	85–87	3
58–60	1	88–90	1
61–63	2	91–93	2
64–66	1	94–96	1
67–69	2	97–99	1

c) $i = 10$

CLASS INTERVALS	f
40–49	2
50–59	1
60–69	6
70–79	15
80–89	12
90–99	4

d) $i = 20$

CLASS INTERVALS	f
40–59	3
60–79	21
80–99	16

8

CLASS INTERVALS	f
5–9	1
10–14	2
15–19	3
20–24	4
25–29	5
30–34	6

CLASS INTERVALS	f
35–39	8
40–44	6
45–49	5
50–54	4
55–59	3
60–64	2
65–69	1

9

CLASS INTERVALS	f
3–7	1
8–12	0
13–17	5
18–22	0
23–27	9
28–32	0

CLASS INTERVALS	f
33–37	14
38–42	0
43–47	11
48–52	0
53–57	7
58–62	0
63–67	3

10

CLASS INTERVALS	f
4–5	1
6–7	0
8–9	0
10–11	0
12–13	1
14–15	2
16–17	2
18–19	0
20–21	0
22–23	2
24–25	3
26–27	4
28–29	0
30–31	0
32–33	3
34–35	4

CLASS INTERVALS	f
36–37	7
38–39	0
40–41	0
42–43	2
44–45	7
46–47	2
48–49	0
50–51	0
52–53	2
54–55	3
56–57	2
58–59	0
60–61	0
62–63	1
64–65	1
66–67	1

11

CLASS INTERVALS	f
0–9	1
10–19	5
20–29	9
30–39	14
40–49	11
50–59	7
60–69	3

14

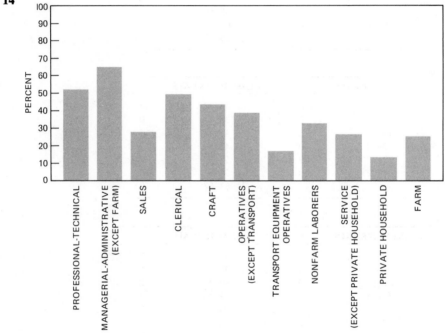

16 a) positively skewed
d) normal

19

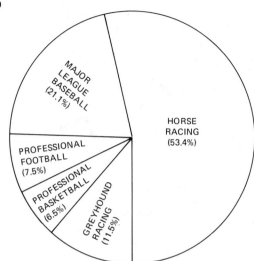

Chapter 4

2 a) 2.45	**b)** 18.85	**c)** 55.65
4 a) 16.13	**b)** 30.64	**c)** 43.48
6 a) 39.0	**b)** first quartile (4.7 times)	
8 b) 520,560	**c)** 540	
10 b) 41.0	**c)** 45.7	

Chapter 5

1 a) $\overline{X}$ = 5.0; median = 5.5; mode = 8   **2** (c)

b) $\overline{X}$ = 5.0; median = 5.0; mode = 5.0

c) $\overline{X}$ = 17.5; median = 3.83; mode = 4

4 All measures of central tendency will be multiplied by 100.

6 $\overline{X}$ = 5.0; median unchanged, mode unchanged.

7 a) negative skew   **b)** positive skew

c) no evidence of skew   **d)** no evidence of skew

9 $\overline{X}$ = 5.0   **a)** $\overline{X}$ = 7.0   **b)** $\overline{X}$ = 3.0   **c)** $\overline{X}$ = 26.7

d) $\overline{X}$ = 10   **e)** $\overline{X}$ = 2.5

11 The distribution is symmetrical.

13 a) 71.9   **b)**

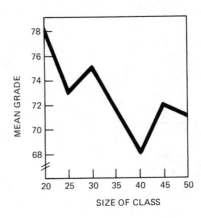

14 raise median: (c)
raise mean: (b)

17 3.04

Chapter 6

1 $s^2 = 1.66$; $s = 1.29$ **a)** no change   **b)** increase

2 s for sentence $= 46.7$; s for time served $= 16.2$

4 mean $= 746.7$; median $= 600$; mode $= 1000, 600,$ and 400;
standard deviation $= 703.9$; variance $= 495,488.9$

6 a) 3.52   **b)** 2.31   **c)** 5.89   **d)** 0

7 The standard deviation is very skewed to the right. Extreme deviations increase
the size of the standard deviation.

8 a) 10   **b)** 8   **c)** 20   **d)** 0

9 s for males $= 0.78$
s for females $= 0.73$

11 a) 1.64   **b)** 2.69

Chapter 7

1 a) 0.94   **b)** -0.40 **c)** 0.00   **d)** -1.32 **e)** 2.23   **f)** -2.53

2 a) 0.4798   **b)** 0.4713   **c)** 0.0987

d) 0.1554   **e)** 0.4505   **f)** 0.4750

g) 0.4901   **h)** 0.4951   **i)** 0.4990

3 a) i) 0.3413; 341   ii) 0.4772; 477

iii) 0.1915; 192   iv) 0.4938; 494

b) i) 0.1587; 159   ii) 0.0228; 23   iii) 0.6915; 692

iv) 0.9938; 994   v) 0.5000; 500

c) i) 0.1359; 136   ii) 0.8351; 835

iii) 0.6687; 669   iv) 0.3023; 302

5 a) 40.13; 57.93   **b)** 60.26

7 a) $z = -0.67$ **b)** $z = 0.67$

$X = 63.96$ $X = 80.04$

c) $z = 1.28$ **d)** $z = 0.67$

$X = 87.36$ 25.14% score above

e) $z = -0.5$ **f)** $z = -0.67$ and $z = 0.67$

30.85% score below   63.96 − 80.04

g) $z = \pm 1.64$ **h)** $z = \pm 2.58$

below 52.32,   below 41.04,

above 91.68   above 102.96

8 $\mu = 72,\ \sigma = 8$:

a) 66.64   **b)** 77.36

c) 82.24   **d)** $z = 1.00$

e) $z = -0.75$ 15.87% score above

22.66% score below   **f)** 66.64 − 77.36

g) below 58.88,   **h)** below 51.36,

above 85.12   above 92.64

8 $\mu = 72,\ \sigma = 4$:

a) 69.32   **b)** 74.68

c) 77.12   **d)** $z = 2.00$

e) $z = -1.5$ 2.28% score above

6.68% score below   **f)** 69.32 − 74.68

g) below 65.44,   **h)** below 61.68

above 78.56   above 82.32

8 $\mu = 72,\ \sigma = 2$:

a) 70.66   **b)** 73.34

c) 74.56   **d)** $z = 4.00$

e) $z = -3.00$ 0.003% score above

0.13% score below   **f)** 70.66 − 73.34

g) below 68.72,
   above 75.28

h) below 66.84,
   above 77.16

9 test 2, test 1

10 No, z scores merely reflect the form of the distributions from which they were derived. Normally distributed variables will yield normally distributed z scores.

Chapter 8

1   Percentaging down

	LOW	MODERATE	HIGH
Excellent	0.0	25.0	50.0
Average	60.0	50.0	33.3
Poor	40.0	25.0	16.7
Total	100.0	100.0	100.0

   Percentaging across

	LOW	MODERATE	HIGH	TOTAL
Excellent	0.0	25.0	75.0	100.0
Average	42.9	28.6	28.6	100.1*
Poor	50.0	25.0	25.0	100.0

*Does not sum to 100.0 because of rounding error.

   Percentaging on the total

	LOW	MODERATE	HIGH	TOTAL
Excellent	0.0	6.7	20.0	
Average	20.0	13.3	13.3	
Poor	13.3	6.7	6.7	
				100.0

4 lambda = 0.125   tau = 0.022

5 gamma = -0.538 Somer's d = -0.375

6

REASON FOR NOT SEEKING WORK	SEX		
	Male	Female	lambda = 0.193
In school	51.1 (693)	24.0 (681)	
Ill or disabled	24.0 (326)	13.9 (394)	
Keeping house	2.4 (32)	43.2 (1226)	
Thinks cannot get a job	22.5 (305)	19.0 (540)	
Total	100.0 (1356)	100.1* (2841)	N = 4197

*Does not add to 100.0 because of rounding.

10

RACIAL COMPOSITION OF MARRIAGE PARTNERS	YEAR	
	1970	1977
Husband black, wife white	13.2 (41)	22.6 (95)
Wife black, husband white	7.7 (24)	7.1 (30)
Husband black, wife other	2.6 (8)	4.8 (20)
Wife black, husband other	1.3 (4)	0.5 (2)
Husband white, wife other	44.8 (139)	42.0 (177)
Wife white, husband other	30.3 (94)	23.0 (97)
Total	99.9* (310)	100.0 (421) N = 731

*Does not add to 100.0 because of rounding error.

12 lambda = 0.125,   tau = 0.123

Chapter 9

1 $r = 0.48$ **3** $r = 0.80$

5 a) nonlinear relationship   **b)** restricted range   **c)** restricted range

6 $r = 0.83$ **7** $r = 0.00; r_s = 0.50$

8 Nonlinear relationships give rise to artificially low Pearson's r correlation coefficients.

9 If the range is markedly restricted (truncated range), Pearson's r will be artificially low.

10 Reversing the scoring of a variable included in a correlation matrix will reverse the sign of its correlation with all other variables in the matrix.

12 $r = 0.76$

14 a)

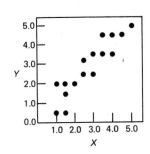

b)

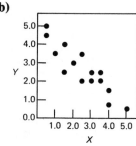

c)

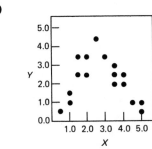

d)

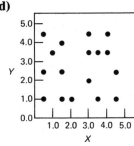

15 a) (c)   **b)** (a)   **c)** (d)   **d)** (b)

17 $r = 0.85$

19 $r = 0.9107$

Chapter 10

1 $Y' = 5.69 - 0.89X$

2 a) 1.72   **b)** 118.48   **c)** $s_{est\ y} = 0.3928$

3 a) 1.59   **b)** $s_{est\ y} = 0.47$ **c)** 0.1296

5 a) $a = -51.081, b_y = 2.442$ **c)** $r = 0.995, r^2 = 0.989$

7 a) 33.35   b) $p = 0.1539$ c) 46.82
  e) 53.19   f) yes; $p = 0.4840$ g) yes; $p = 0.2912$

8 a) If group size equals 15, the predicted group cohesiveness score is 3.61.
  b) 57.75%

9 b) $r = 0.9658$ d) Chile: 429.858   Ireland: 831.364   Belgium: 1390.606

10 a) $r = 0.958$
   b) Y' for 3 = 3.222 Y' for 4 = 4.139
      Y' for 5 = 5.055 Y' for 6 = 5.972
      Y' for 7 = 6.889 Y' for 8 = 7.806
      Y' for 9 = 8.722 Y' for 10 = 9.639
      Y' for 11 = 10.556
   c) 54.889   d) 4.473   e) 50.424   f) 0.919

11 a) 5.56   b) 4.16   c) 0.857

Chapter 11

4 a) The independent variable is level in school; the dependent variable is political orientation; and the control variable is age.

b) The younger undergraduate students tend to be more liberal (and therefore less conservative) in their political orientation than the younger graduate students. While the differences in political orientation are not as marked between the older undergraduates and graduate students, the same pattern is evident.

c)

POLITICAL ORIENTATION	AGE	
	Young	Old*
Liberal	45.0 (45)	55.0 (55)
Conservative	55.0 (55)	45.0 (45)
Total	100.0 (100)	100.0 (100)

*The older students tend to be slightly more liberal than the younger students.

6 $r_{21.3} = -0.278$

8 The partial correlation coefficient is an "average" of the correlation between two variables across all categories of the control variable and would be misleading if the association were not uniform across all categories of the control variable.

9 $r_{xy \cdot z} = 0.752$

11 The equation is in unstandardized form because the Y-intercept (1.1) is shown and not equal to 0.

12

EMPLOYEE	Y'	Residual*
1	4.84	1.16
2	6.71	−1.71
3	4.97	2.03
4	9.53	−0.53
5	19.00	2.00

*The sum of the 5 residuals does not equal 0 because the scores of other employees included in the study are not provided.

13 a) The model predicts the math achievement scores of the white students better than those of the black students because the coefficient of multiple determination is 0.52 for the white students and 0.14 for the black students.

b) Family socioeconomic status (SES) is the most important predictor variable for both black and white students. For white students, cumulative grade point average and teacher's ability are the least important predictors, and for the black students reading test scores and percentage of white students in class are least important.

c) A standardized regression coefficient allows for an assessment of the relative importance of the independent variables as predictors of the dependent variable. The coefficient of 0.17 indicates that if teacher's ability increases one standard deviation unit, the math achievement scores of the white students increase 0.17 standard deviation units when the remaining independent variables are controlled. The coefficient of 0.34 indicates that if teacher's ability increases one standard deviation unit, the math achievement scores of the black students increase 0.34 standard deviation units when the remaining independent variables are controlled. Thus, the teacher's ability is more important for black students than for white students when the other independent variables have been controlled.

Chapter 12

1 a)

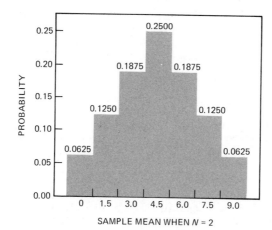

b)

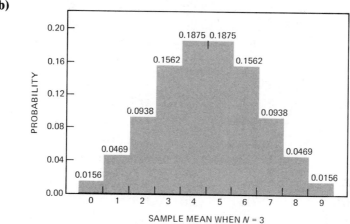

2 An extreme deviation is rarer as sample size increases.

3 a) 0.0625; 0.0156   **b)** 0.1250; 0.0312
   c) 0.3750; 0.3125   **d)** 0.7500; 0.6250

4 a) 0.6250   **b)** 0.1875   **c)** 0.2500

5 a) 0.6874   **b)** 0.3750   **c)** 0.1407

6 (a) and (b)   **c)** Mean = 0.00;
                              Standard deviation
                              = 1.63

X	f	$p(\bar{X})$
4	1	0.0123
3	4	0.0494
2	10	0.1235
1	16	0.1975
0	19	0.2346
−1	16	0.1975
−2	10	0.1235
−3	4	0.0494
−4	1	0.0123
$N_{\bar{x}} = 81$		$\sum p(\bar{X}) = 1.0000$

7 a) 0.2346　　**b)** 0.6296　　**c)** 0.0123　　**d)** 0.0246
　e) 0.0617　　**f)** 0.0617　　**g)** 0.1234　　**h)** 0.8766

8 a) 0.1250　　**b)** 0.1250　　**c)** 0.3750　　**d)** 0.5000

9 a) 0.0192　　**b)** 0.0769　　**c)** 0.3077　　**d)** 0.4231

10 Problem 1:　　**a)** 7 to 1 against,　　**b)** 7 to 1 against
　　Problem 2:　　**a)** 51 to 1 against,　　**b)** 12 to 1 against

11 a) 0.1667　　**b)** 0.1667　　**c)** 0.2778　　**d)** 0.5000

12 a) 0.0046　　　**b)** 0.0278　　　　**c)** 0.0139
　d) 0.0556　　　**e)** 0.5787

13 a) 0.0129　　　**b)** 0.4135　　　**c)** 0.0100

14 a) 0.0156　　　**b)** 0.4219　　　**c)** 0.0123

15 a) 0.0838　　**b)** 126.40　　**c)** < 68.64 and > 131.36　　**d)** 0.4648
　e) i) 0.0070　　**ii)** 0.2160　　**iii)** 2(0.3895) = 0.0779

16 (b)　　**17 a)** $p = 0.20$　　**b)** $p = 0.04$　　**c)** $p = 0.80$　　**d)** $p = 0.64$

18 a) $\dfrac{1}{3}$　　**b)** $\dfrac{1}{2}$　　**c)** $\dfrac{1}{6}$　　**d)** 0

19 a) $\dfrac{1}{11}$　　**b)** $\dfrac{1}{66}$　　**c)** $\dfrac{14}{99}$

20 a) $\dfrac{1}{8}$　　**b)** $\dfrac{1}{36}$　　**c)** $\dfrac{16}{81}$

21 a) 0.40　　**b)** 0.12　　**c)** 0.16　　**d)** 2(0.0336) = 0.0672

22 a) $p = 0.3085$　　**b)** $p = 0.2266$　　**c)** $p = 0.4649$
　d) $p = 0.2564$　　**e)** $p = 0.2934$　　**f)** $p = 0.0947$　　**g)** $p = 0.3296$

23 0.33, 0.33

Chapter 13

2 a) In the event a critical theoretical issue is involved, the commission of a type I error might lead to false conclusion concerning the validity of the theory.

b) In many studies in which the toxic effects of new drugs are being studied, the null hypothesis is that the drug has no adverse effects. In the event of a type II error, a toxic drug might be mistakenly introduced into the market.

3 The null hypothesis cannot be proved. Failure to reject the null hypothesis does not constitute proof that the null hypothesis is correct.

4 All studies involve two statistical hypotheses: the null hypothesis and the alternative hypothesis. One designs the study so that the rejection of the null hypothesis leads to affirmation of the alternative hypothesis.

5 a) $\frac{1}{1024}$ or 0.00098   **b)** 0.055   **c)** 17.28 to 1 against passing

6 a) H_0 **b)** H_1 **c)** H_1 **d)** H_0

7 a) type II   **b)** type I   **c)** no error   **d)** no error

8 a) type II   **b)** type II   **c)** no error

9 It is more likely that he is accepting a false H_0 since the probability that the observed event occurred by chance is still quite low ($p = 0.02$).

10 By chance, one would expect five differences to be statistically significant at the 0.05 level.

11 No

12 Yes

13 No. The only deviations that count are those in the direction specified under H_1.

15 (b)

16 male = 1/2; female = 1/2; 1:1   **17** 0.038   **18** No

19 On the basis of the data, we would reject H_0. However, since H_0 is true, we have made a type I or type α error.

20 On the basis of the data, we would fail to reject H_0. Since H_0 is true, we have made the correct decision.

21 We fail to reject H_0. Since H_0 is false, we have failed to reject a false H_0. Therefore, we have made a type II or type β error.

22 We reject H_0. Since H_0 is false, we have made a correct decision.

23 We reject H_0. Since H_0 is false, we have made a correct decision.

24 The probability of making a type I error is equal to α. In the present case, $\alpha = 0.05$.

25 a) $p = 0.062$ **b)** $p = 0.773$ **c)** $p = 0.124$ **d)** $p = 0.938$

Chapter 14

1 a) As sample size increases, the dispersion of sample means decreases.

b) As you increase the number of samples, you are more likely to obtain extreme values of the sample mean. For example, suppose the probability of obtaining a sample mean of a given value is 0.01. If the number of samples drawn is 10, you probably will not obtain any sample means with that value. However, if you draw as many as 1,000 samples, you would expect to obtain approximately 10 sample means with values so extreme that the probability of their occurrence is 0.01.

2 Suppose we have a population in which the mean = 100 and the standard deviation = 10. The probability of obtaining scores as extreme as 80 or less or 120 or more (± 2 standard deviations) is 0.0456, or less than five in a hundred. The probability of obtaining scores even more extreme (for example, 50, 60, 70, or 130, 140, 150) is even lower. Thus, we would expect very few (if any) of these extreme scores to occur in a given sample. Since the value of the standard deviation is a direct function of the number of extreme scores, the standard deviation of a sample will usually underestimate the standard deviation of the population.

3 a) $21.71 - 26.29$ **b)** $20.82 - 27.18$

4 a) $23.28 - 24.72$ **b)** $23.05 - 24.95$

5 Reject H_0; $z = 2.68$ in which $z_{0.01} = \pm 2.58$

6 Accept H_0; $t = 2.618$ in which $t_{0.01} = \pm 2.831$, df = 21

7 In Exercise 5 the value of the population standard deviation is known; thus, we may employ the z-statistic and determine probability values in terms of areas under the normal curve. In Exercise 6 the value of the population standard deviation is not known and must be estimated from the sample data. Thus, the test statistic is the t-ratio.

Since the t-distributions are more spread out than the normal curve, the proportion of area beyond a specific value of t is greater than the proportion of area beyond the corresponding value of z. Thus, a larger value of t is required to mark off the bounds of the critical region of rejection. In Exercise 5, the absolute value of the obtained z must equal or exceed 2.58. In Exercise 6, the absolute value of the obtained t must equal or exceed 2.83.

To generalize: the probability of making a type II error is less when we know the population standard deviation.

8 Reject H_0; $t = 2.134$ in which $t_{0.05} = 1.833$ (one-tailed test), df = 9

9 a) $z = 2.73$, $p = 0.9968$ **b)** $z = -0.91$, $p = 0.1814$

c) $z = 1.82$, $p = 0.0344$ **d)** $p = 0.6372$

10 a) $36.63 - 43.37$ **b)** $35.43 - 44.57$

11 $t = 2.083$, df = 25

12 27.14 – 32.08

13 $t = 1.75$, df $= 625$

14 263.34 – 273.66

15 a) $\mu = 6.0$, $\sigma = 3.42$

16 a) $p = \dfrac{1}{15}$ **b)** $p = \dfrac{1}{15}$ **c)** $p = \dfrac{8}{15}$ **d)** $p = 0$

17 a) $p = \dfrac{1}{12}$ **b)** $p = \dfrac{1}{12}$ **c)** $p = \dfrac{5}{9}$ **d)** $p = \dfrac{1}{36}$

18 The critical value of r_s for $N = 25$ ($\alpha = 0.05$, two-tailed test) may be obtained by interpolating: 0.409 for $N = 24$, 0.392 for $N = 26$; thus the critical value for $N = 25$ is 0.400. Since the absolute value of the obtained r_s is less than the critical value, we accept H_0.

19 a) $z = 1.01$ **b)** $z = 1.99$

22 The statistics used to describe the distribution of a sample are $\overline{X}$ (the mean) and s (the standard deviation). The statistics used to describe the distribution of a sample statistic are the mean ($\mu_{\overline{x}}$) and the standard error of the mean ($\sigma_{\overline{x}}$).

23 s^2 is not an unbiased estimate of σ^2. When we use N in the denominator of the variance equation, we underestimate the population variance.

24 s^2 is an unbiased estimate of σ^2 since $N - 1$ is used in the denominator of the equation.

26 Using the example in Section 14.7 of the text, we find the 99% confidence interval:

$$\text{upper limit } \mu_0 = 108 + (2.787)(3.0) = 116.36,$$
$$\text{lower limit } \mu_0 = 108 - (2.787)(3.0) = 99.64.$$

Thus, the 99% confidence limits are 99.64 – 116.36 as compared to the 95% confidence limits: 101.82 – 114.18.

27 The t-distributions are more spread out than the normal curve. Thus, the proportion of area beyond a specific value of t is greater than the proportion of area beyond the corresponding value of z. As df increases, the t-distributions more closely resemble the normal curve.

28 a) Married couples: $t = 5.219$, df $= 51$, $p < 0.05$
　　Divorced couples: $t = 0.329$, df $= 22$, not significant

b) Married couples: $t = 1.688$, df $= 51$, not significant
　　Divorced couples: $t = 0.615$, df $= 22$, not significant

Chapter 15

1 $t = 2.795$; reject H_0

2 $t = 1.074$; accept H_0

3 $t = 0.802$; accept H_0

4 a) $t = 3.522$; reject H_0 **b)** $F = 3.00$; accept H_0

5 a) $z = 1.41$; $p = 0.0793$ **b)** $z = -2.12$; $p = 0.9830$

c) $z = -2.12$; $p = 0.0170$ **d)** $z = -5.66$; $p < 0.00003$

6 a) $z = -0.71$; $p = 0.7611$ **b)** $z = -1.06$; $p = 0.8554$

c) $z = -1.41$; $p = 0.9207$ **d)** $z = -1.77$; $p = 0.9616$

7

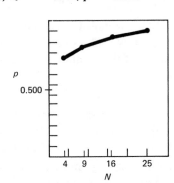

The probability of finding a difference in the correct direction between sample means increases as N increases.

8 $t = 1.275$; df $= 162$

9 $t = 1.63$, df $= 8$; accept H_0

10 $t = 0.77$, df $= 8$, accept H_0

13 a)

$\overline{X}_1 - \overline{X}_2$	f	$p(\overline{X}_1 - \overline{X}_2)$
9	1	0.0083
8	2	0.0165
7	5	0.0413
6	10	0.0826
5	14	0.1157
4	18	0.1488
3	21	0.1736
2	18	0.1488
1	14	0.1157
0	10	0.0826
-1	5	0.0413
-2	2	0.0165
-3	1	0.0083

$N_{\overline{X}_1 - \overline{X}_2} = 121$ $\sum p(\overline{X}_1 - \overline{X}_2) = 1.0000$

b) The mean of the differences $= 3$. The mean of the sampling distribution of differences between means is equal to the difference between the means of the population from which these samples were drawn.

14 a) H_0: $\mu_1 = \mu_2$; that is, the two populations of means from which we are drawing samples are the same.

b) The sampling distribution of differences between means under the null hypothesis.

c)

$\overline{X}_1 - \overline{X}_2$	f	$p(\overline{X}_1 - \overline{X}_2)$
6	1	0.0083
5	2	0.0165
4	5	0.0413
3	10	0.0826
2	14	0.1157
1	18	0.1488
0	21	0.1736
−1	18	0.1488
−2	14	0.1157
−3	10	0.0826
−4	5	0.0413
−5	2	0.0165
−6	1	0.0083

15 a) 0.0496   **b)** 0.0083   **c)** 0.1487   **d)** 0.0166   **e)** 0.0083

16 a) 0.2644   **b)** 0.1487   **c)** 0.0083   **d)** 0.1487   **e)** 0.0000

Chapter 16

1 $F = 47.02$; reject H_0.

2 All comparisons are significant at 0.01 level. It is clear that death rates are lowest in summer, next to lowest in spring, next to highest in fall, and highest in winter.

3 When we use the 0.05 level, one out of every 20 comparisons will be significant *by chance*. Thus, a significant difference at the 0.05 level in one of 10 studies can be attributed to chance.

4 $F = 0.30$, df $= 3/12$

5 $F = 10.43$, df $= 2/12$, HSD $= 7.16$

6 If the F-ratio is less than 1.00, we conclude that the groups probably come from the same population.

7

SOURCE OF VARIATION	SUM OF SQUARES	DEGREES OF FREEDOM	VARIANCE ESTIMATE	F
Between groups	1.083	2	0.5415	0.6228
Within groups	13.042	15	0.8695	
Total	14.125	17		

8

SOURCE OF VARIATION	SUM OF SQUARES	DEGREES OF FREEDOM	VARIANCE ESTIMATE	F
Between groups	919.793	3	306.598	6.43
Within groups	954.167	20	47.708	
Total	1873.96	23		

9

SOURCE OF VARIATION	SUM OF SQUARES	DEGREES OF FREEDOM	VARIANCE ESTIMATE	F
Between groups	26.80	2	13.4	9.348
Within groups	17.20	12	1.43	
Total	44.00	14		

10 All comparisons are significant at 0.01 level. It is clear that group C is superior, next group B, followed by group A.

Chapter 17

1 $N_a = 30$

2 Step 1: The value of $\sigma_{\overline{X}}$ is 0.5.

Step 2: The critical value of $\overline{X}$ is 71.165.

Step 3: The critical value of $\overline{X}$ has a z score in the distribution under H_1 of -1.67.

Thus, the power of the test is 0.9525.

3 0.2877

4 a) (1) 0;   (2) $\beta = 0.1685$;   (3) Power $= 0.8315$

    b) (1) α;   (2) $\beta = 0$;   (3) The concept of power does not apply.

5 a) 0.0526   **b)** 0.1020   **c)** 0.1788   **d)** 0.2877

6 It takes time, effort, and money to conduct research. When the power to reject a false null hypothesis is low, we are in effect sending good money after bad. A small increase in time, effort, and funding may sometimes be sufficient to raise the power of a test to levels that justify the risk.

7 Power is the probability of rejecting a false null hypothesis. If the null hypothesis is true, there is no *false* null hypothesis to reject.

8 When the null hypothesis is false, a rejection of the null hypothesis is the *correct* statistical decision. The concept of type I error simply does not apply.

9 a) 80   **b)** 69   **c)** 47   **d)** 50   **e)** 40

11 Factors affecting power:

Sample size: As N increases, power increases.

α-level: The higher the α-level we set, the greater the power of the test.

Nature of the alternative hypothesis: One-tailed test is more powerful than two-tailed alternative (unless the parameter lies in a direction opposite to the one predicted).

Nature of the statistical test: Parametric tests are more powerful than their nonparametric counterparts (assuming assumptions underlying the use of the parametric test are valid).

Correlated measures: When subjects have been successfully matched on a variable correlated with the criterion, a statistical test that takes this correlation into account is more powerful than one that does not.

Power may be increased by

Increasing sample size.

Setting α high (for example, 0.05 as opposed to 0.01).

Using a one-tailed alternative hypothesis.

Using parametric tests rather than nonparametric.

Using correlated measures, if appropriate.

Chapter 18

1 $\chi^2 = 2.29$; df $= 4$

2 Race and religious affiliation are nominal-level variables, and hence it is not possible to specify the direction of the relationship between the variables.

4 $\chi^2 = 1.94$; df $= 1$

7 $\chi^2 = 6.00$; df $= 3$

8 $\chi^2 = 8.08$; df $= 1$

9 $\chi^2 = 8.87$; df $= 1$

10 The research merely demonstrated that the two variables are related. Since the ownership of dogs was not randomly assigned to the coronary victims, we do not know whether the dog owners differ in other systematic ways from those who do not own dogs. For example, dog owners may differ from nonowners in personality characteristics, dietary habits, family ties, etc. They may also exercise more (walking their dogs).

Index

Abscissa, 58
  defined, 71
Absolute reduction in error, 165
Absolute value of a number, 117
  defined, 126
Addition rule, 276
  defined, 292
  with mutually exclusive events, 278
Adverbs, 21
Alpha α-level, 314
  See also Type I error
  defined, 322
  effect of on power, 421
Alternative hypothesis, 315, 338, 359,
    375, 380, 399, 419, 424, 439
  defined, 323
  effect of on power, 422
Analysis of variance (ANOVA), 392
  assumptions underlying, 399
  between-group sum of squares, 395
  defined, 405
  fundamental concepts of, 398
  HSD, 403
  one-way, 392, 405
  total sum of squares, 393
  within-group sum of squares, 393
ANOVA. See Analysis of variance
A posteriori comparisons, 403
  defined, 405
A priori comparisons, 402
  defined, 405
Arithmetic mean
  calculation of, 95

  defined, 95
  properties of, 97
Arrays, 98, 107
Association
  asymmetric measures of, 160, 181
  measures of for contingency tables,
    160
  nominal measures of, 161, 166, 442
  ordinal measures of, 172
  symmetric measures of, 160, 183
Asymmetric measure of association,
    160
  defined, 181
Average, 6, 93
Average deviation. See Mean deviation

Bar graphs, 58, 59, 67
  defined, 71
Before-after design, 378
  defined, 383
Bell-shaped distribution, 65, 334
Bell-shaped normal curve, 65
Best-fitting straight line, 217, 235
Beta error. See Type II error
Between-group sum of squares, 395,
    396
  degrees of freedom of, 397
Between-group variance
  defined, 405
  estimate of, 398
Bias, 306
  defined, 292
Biased estimates, 343, 373

Biased sampling, 272, 273
Bimodal distribution, 104
Binomial distribution, 305, 312
  defined, 323
Binomial expansion, 309
Binomial sampling distribution, 311
  and binomial expansion, 309
  and enumeration, 306
Bivariate analysis, 151
Bivariate contingency table, 153

Cardinal numbers, 25, 26
  defined, 40
Categorical data, 27
Causal relationship, 232, 233
Causation and correlation, 232
Cell frequencies and percentages, 155
Centile, 89
  defined, 89
Central limit theorem, 335
  defined, 363
Central tendency, measure of, 93, 107
Central value, 94
Chance
  in experiment, 12
  games of, 274
Chi square (χ^2)
  and independence of variables, 438
  limitations in use of, 441
  nominal measures of association
    based on, 442
  one-variable case, 434
  table of, Table B, 476

Classical approach to probability, 274
Class intervals
  grouping into, 53
  true limits of, 55
Coefficients
  contingency, 444, 445, 446
  of correlation, 189, 198, 205, 206, 252, 253
  of determination, 231, 235
  of multiple correlation, 259, 262
  of multiple determination, 260, 262
  of multiple regression, 257, 262
  of nondetermination, 260
  of partial correlation, 249
  phi, 442, 446
  population correlation, 358
Comparisons
  a posteriori, 403, 405
  a priori, 402, 405
  multigroup, 391
  planned, 402, 405
Concordant pair, 172
  calculation of in contingency tables, 175
  defined, 182
Conditional distribution, 154
  defined, 182
Conditional probability, 283
  defined, 292
Conditional tables, 246
  defined, 262
  order of, 262
Confidence intervals, 351
  defined, 363
  for large samples, 355
  for percentages, 356
Confidence limits, 351
  defined, 363
  for large samples, 355
  for percentages, 356
Contingency coefficient, 445
  defined, 446
  disadvantage of, 444
Contingency tables, 155
  bivariate, 153
  calculation of concordant and discordant pairs in, 175
  defined, 182
  measures of association for, 160
  multivariate, 244
  percentaging of, 155
Continuous scales, 30, 31
  defined, 40
Continuous variables, 32, 285
Control
  physical, 250
  statistical, 250
Control group, 12
Control variable, 246
  See also Test variable
  defined, 262
Correlated samples, 378
Correlation
  and causation, 232
  concept of, 189
  multiple, 262
  partial, 248, 249, 253, 262

perfect positive, 195
  zero-order, 253
Correlation coefficient, 189, 262
  defined, 205
  interpretation of, 198
  multiple, 259, 262
  Pearson, 190, 191
  population, 358
  product-moment, 190, 206
  sign of, 191
  test of significance for, 358
  zero-order, 252
Correlation matrix, 197
  defined, 205
Counting units, 31
Covariation, 193
  defined, 206
Cramer's *V*, 444, 446, 447
  defined, 446
Criterion variable, 152
  See also Dependent variable
Critical region, 339, 359, 375, 381, 419, 424, 439
  defined, 363
Critical rejection regions, 348
Critical values, 348
  defined, 363
Cumulative frequency, 56
  defined, 72
Cumulative frequency curve, 63
  defined, 72
Cumulative frequency distribution, 56, 82
  defined, 72
Cumulative frequency ogive. *See*
  Cumulative frequency curve
Cumulative percentage, defined, 72
Cumulative percentage distribution, 56, 82
  defined, 77
Cumulative percentage graph, 82
Cumulative percentiles, 82
Cumulative proportion, 72
Cumulative proportion distribution, 56
  defined, 72

Data, 7, 9, 10, 14
  categorical, 27
  frequency, 27
  grouping of, 51
  nominal, 27
Decile, 89
  defined, 90
Degrees of freedom (df), 346
  of the between-group sum of squares, 397
  defined, 363
Dependence, 292
Dependency ratio, 36
Dependent events, 283
Dependent variable, 151, 152, 244
  defined, 182
Descriptive statistics, 7, 10, 12, 14
Determination
  coefficient of, 231, 235
  multiple, 260, 262

Deviation
  defined, 97, 107
  mean, 116, 119, 126, 192, 220
  standard, 118, 119, 120, 121, 123, 126, 134, 144, 334
df. *See* Degrees of freedom
Diagonal
  negative, 177
  positive, 175
Dichotomous variables, 279
Difference
  direct, 379
  mean, 379, 383
Direct difference method, 379
Direction of a relationship, 157
Directional hypothesis, 315, 316
  defined, 323
Direct relationship. *See* Positive
  relationship
Discontinuous scales. *See* Discrete
  scales
Discordant pair, 172, 177
  calculation of in contingency tables, 175
  defined, 182
Discrete scales, 30, 31
  defined, 40
Dispersion, 94, 113
  defined, 126
Distribution
  bell-shaped, 65, 334
  bimodal, 104
  binomial, 305, 309, 311, 312, 323
  conditional, 154, 182
  cumulative frequency, 82
  cumulative percentage, 82
  F-, 376, 377
  frequency, 52, 54, 55, 56, 72, 94, 153, 306
  grouped frequency, 97, 101
  leptokurtic, 65, 72
  mesokurtic, 65, 73
  multimodal, 104
  negatively skewed, 66, 73
  normal, 123, 311, 312, 334
  percentage, 56
  platykurtic, 65, 73
  positively skewed, 66, 73
  proportion, 56
  rectangular, 66
  sampling, 302, 323, 330, 333, 339, 359, 371, 372, 375, 381, 419, 424, 439
  skewed, 66, 73, 105
  standard normal, 134, 145
  symmetrical, 65
  t-, 345, 347, 364
  U-, 66
  ungrouped frequency, 96, 121
Dual effect, 377

Efficiency of power, 413, 428, 429
Elaboration, 246, 263
Elements, 14
  defined, 8, 9
Empirical approach to probability, 274

Enumeration in construction of binomial sampling distributions, 306
Epsilon (ε). *See* Percentage difference
Errors
  absolute reduction in, 165
  alpha (α), 318, 323, 392, 413, 414, 426
  beta (β), 319, 324, 413, 414, 426
  of estimate, 224, 227, 236
  of measurement, 32
  proportional reduction in, 161, 165, 183
  residual size and direction of, 226
  standard, 224, 227, 236, 330, 332, 334, 357, 363, 364, 372, 379, 383
  in testing statistical hypotheses, 317
  type I (α), 318, 323, 392, 413, 414, 426
  type II (β), 319, 324, 413, 414, 426
Estimates
  biased, 343, 373
  standard error of, 224, 227, 236
  unbiased, 341, 343, 364, 373
  of variance, 396, 398, 405
Estimation
  interval, 350, 363
  of parameters, 9, 10, 340, 350
  point, 340, 350, 363
  of sample variance, 343
  of z statistic, 373
Exhaustive, defined, 72, 292
Exhaustive classes, 53
Exhaustive events, 279
Expansion, binomial, 309
Expected results, 305
Experimental group, 12
Experimental stimulus, 392
Explained variation, 229, 230
  defined, 235
Exponents, 21
Extreme scores, 203

F, interpretation of, 402, 404
Fallacy, *post hoc,* 233, 234, 235
F-distribution, 376
  table for, Table D, 479
  tridimensional nature of, 377
Finite population, 7, 14
Focal cell, 176
F-ratio, 399, 402
  defined, 383, 405
  degrees of freedom in, 397
  homogeneity of variance, 396
  table of critical values, Table D, 479
Freedom, degrees of, 346, 363, 397
Frequency counts, 27
Frequency curves, 62
  cumulative, 63, 72
  defined, 72
  forms of, 65
  skewed, 66
Frequency data, 27
Frequency distribution, 52, 153, 306
  cumulative, 56, 72, 82
  defined, 72

description of, 94
features of, 94
grouped, 54, 55, 72, 97, 101
ungrouped, 96, 121
Frequency polygon. *See* Frequency curve

Games of chance, 274
Gamma. *See* Goodman's and Kruskal's gamma
Goodman's and Kruskal's gamma, 173, 174, 361
  defined, 182
  interpretation of, 178
  limitations of, 178
  test of significance for, 361
Goodman's and Kruskal's tau, 166, 183
  defined, 183
  interpretation of, 171
  limitations of, 171
Goodness-of-fit technique, 446
Grammar of mathematical notation, 20
Graphing techniques, 57
Grouped frequency distribution, 54, 55
  defined, 72
  median of, 101
  raw score method of obtaining mean from, 97
Grouping
  into class intervals, 53
  of data, 51
  of scores, 52

H_0. *See* Null hypothesis
H_1. *See* Alternative hypothesis
Histogram, 60
  defined, 72
Homogeneity of variance, 376, 399
  defined, 383, 405
Homogeneous categories, 26, 27
Homoscedasticity, 228
  defined, 235
HSD test, 403
Hypotheses
  alternative, 315, 323, 338, 359, 375, 380, 399, 419, 424, 439
  directional, 315, 316, 323
  nondirectional, 315, 316, 323
  null, 315, 317, 323, 338, 359, 360, 374, 380, 399, 419, 424, 439
  one-tailed, 316
  two-tailed, 316

Independence
  chi square test of, 438
  defined, 292
  of sample, 371
Independent random sampling, 273
Independent variable, 151, 152, 244
  defined, 182
  percentaging on, 155
  single, 392
Indirect proof, 316
Indirect relationship. *See* Negative

relationship
Inductive statistics, 7, 12, 14
Inferential statistics, 7, 10, 11, 12, 14, 329, 378
Infinite population, 7, 14
Inflated *N,* 446
Insensitivity of median to extreme scores, 103
Interest, variable of, 9
Interquartile range, 115
  defined, 126
Intervals
  class, 53, 55
  confidence, 351, 355, 356, 363
  estimation of, 350, 363
Interval scale, 29
  defined, 40
Interval-scaled variables, 60, 189
Interval variables, 330
Intervening variable, 248
  defined, 262
Intuitive use of probability, 272
Inverse relationship. *See* Negative relationship

Joint occurrence, 280
  defined, 292

Lambda (λ), 161
  defined, 182
  interpretation of, 164
  limitations of, 165
  symmetrical, 161
Law of large numbers, 363
Leptokurtic distribution, 65
  defined, 72
Level of significance, 313
  alpha (α)-level and, 318
  choice of, 318
  and power, 413
  relation to Type I error, 318
Limits
  apparent, 55
  central, 335, 363
  confidence, 351, 363
  confidence interval, 355
  for large samples, 355
  for percentages, 356
  true, 32, 33, 41, 55
Linearity of relationship, 200
  assumption of, 200
Linear regression, 214
  mean deviation computational procedures for, 220
  raw score computational procedures for, 220
Linear relationship equation, 216

Marginals, 155
  defined, 182
Matched-group design, 379
  defined, 383
Mathematical adjectives, 20
Mathematical adverbs, 21

Mathematical notation grammar, 20
Mathematical nouns, 20
Mathematical verbs, 20
Matrix of correlation, 197, 205
Mean, 93, 104, 105
  arithmetic, 95
  calculation of, 95
  comparison to median, mode,
    104–105
  defined, 107
  effect of extreme scores, 98
  population, 134
  properties of, 97–99
  sampling distribution of, 330, 333
  sampling distribution of differences
    in, 371, 372
  and skewed distribution, 105–106
  stability of, 104–105
  standard error of, 330, 332, 334, 363
  standard error of differences in, 372,
    383
  weighted, 99, 107
Mean deviation, 116
  defined, 116, 126
  for linear regression, 220
  for Pearson's r calculation, 192
  for standard deviation calculation,
    119
  for variance calculation, 119
Mean difference, standard error of,
  379, 383
Measurement, 25
  of association for contingency tables,
    160
  of central tendency, 93, 107
  defined, 40
  errors of, 32
  nominal scale of, 27
Median, 93, 104, 105
  calculation
    from a grouped frequency distri-
      bution, 101
    from an array of scores, 100
    from an array with tied scores, 102
  characteristics of, 103
  compared to mean, mode, 104–105
  defined, 100, 107
  effect of a skewed distribution,
    105–106
  of grouped frequency distributions,
    101
  and insensitivity to extreme scores,
    103
Mesokurtic distribution, 65
  defined, 73
Midpoints of class intervals, 97
Mode, 93, 103, 104, 105
  compared to mean, median, 104–105
  defined, 107
  effect of a skewed distribution,
    105–106
Models
  idealized, 305
  normal curve, 285
Multigroup comparisons, 391
Multimodal distributions, 104
Multiple correlation coefficient, 259

defined, 262
Multiple coefficient of determination,
  260, 262
Multiple regression analysis, 254, 256
  visual presentation of, 255
Multiple regression coefficient, 257
  defined, 262
Multiplication rule, 280
  defined, 292
Multivariate contingency table, 244
Multivariate statistical technique, 243
  defined, 262
Mutually exclusive classes, 26, 27, 53
Mutually exclusive events, 279
  addition rule with, 278
  defined, 73, 293

N, 20
  inflated, 446
Negatively skewed distribution, 66
  defined, 73
Negative relationship, 159, 192
  defined, 182, 206
Nominal data, 27
Nominally scaled variables, 59
Nominal measures of association,
  161, 166
  based on chi square, 442
Nominal numbers, 25, 26
  defined, 40
Nominal scale, 27
  defined, 40
Nonchance factors, 313
Nondirectional hypothesis, 315, 316
  defined, 323
Nonexistent relationship, 159
Nonparametric test, 423, 434
Normal curve, 134, 305
  bell-shaped, 65
  defined, 73
  and probability, 285
  table of proportions of area under,
    Table A, 473
Normal distribution, 311, 312, 334
  standard, 134, 145
  and standard deviation, 123
  table of proportions of area under,
    Table A, 473
Normality, assumption of, 134
Notation, 20
Nouns, 20
Null hypothesis, 315, 338, 359, 374,
    380, 399, 419, 424, 439
  concerning ρ, 360
  defined, 323
  rejection of, 317
  testing of, 360
Numbers
  cardinal, 25, 40
  nominal, 25, 40
  ordinal, 25, 40
  true limits of, 32, 41
  types of, 24
  use of, 25
  whole, 31

Occurrence
  joint, 280, 292
  successive, 280
Ogive, defined, 73
One-sample case, parameters known,
  336
One-sample case, parameters unknown
    Student's t, 344
One-tailed hypothesis, 316
One-tailed probability values, 288, 293
    323
One-way analysis of variance, 392
  defined, 405
Operators, 21, 22
Order
  of conditional tables, 246
  of partial correlation coefficients,
    262
  of partial tables, 262
  rank, 28
Ordinally scaled variables, 60
Ordinal measures of association, 172
Ordinal numbers, 25, 26
  defined, 40
Ordinal scale, 27
  defined, 40
Ordinate, 57
  defined, 73

Pair-by-pair comparison
  defined, 182
  logic of, 172
Pairs
  concordant, 172, 175, 182
  discordant, 172, 175, 177, 182
  tied, 172, 183
Panel design, 379
  defined, 383
Parameters, 8, 9, 14
  defined, 14
  estimation of, 9, 340, 350, 364
  interval estimation, 363
  point estimation, 340, 350
  of populations, 10, 301
  unbiased estimate of, 364
Parametric techniques, 433
Parametric test of significance, 423
Partial correlation, 248
Partial correlation coefficient, 249
  defined, 253, 262
  logic of, 249
  order of, 262
Partial table, 246
  defined, 262
  order of, 262
Pearson correlation coefficient, 191
Pearson product-moment correlation
    coefficient, 190
Pearson's r, 190, 192, 194, 195, 206
  assumption of linearity, 199
  calculation of by mean deviation
    method, 194
  defined, 206
  and explained variation, 230
  table of functions of, Table G, 488
  test of significance of, 358

and total variation, 229
and unexplained variation, 230
and *z* scores, 195
Percentage, 37
  confidence intervals for, 356
  cumulative, 72, 82
  defined, 40
  as expression of probability, 275
  limits for, 356
Percentage change, 37
Percentage difference, 159
  defined, 182
Percentage distribution, cumulative, 56, 72, 82
Percentaging
  on the independent variable, 155
  on totals, 157
Percentaging across, 157
Percentaging contingency tables, 155
Percentaging down, 155
Percentile rank, 81, 82, 87
  defined, 90
Percentiles, 83
  cumulative, 82
  defined, 90
Perfect positive correlation, 195
Perfect relationship, 159
  defined, 182
Phi coefficient, 442
  defined, 446
Physical control, 250, 246
Pictograph, 67
Pie chart, 67
Planned comparisons, 402
  defined, 405
Platykurtic distribution, 65
  defined, 73
Point estimation, 340, 350
  defined, 363
Population, 9, 14, 302
  defined, 301, 323
  finite, 7, 14
  infinite, 7, 14
  parameter of, 8, 301
Population correlation coefficient, 358
Population mean, 134
Population parameter estimate, 10
Population standard deviation, 134
Population variance estimate, 341
Positive correlation, 195
Positively skewed distribution, 66
  defined, 73
Positive relationship, 159, 191
  defined, 183, 206
Post hoc fallacy, 233, 234
  defined, 235
Power, 422, 423, 426
  and α-level, 421
  calculation of, 419, 424
  concept of, 413
  and correlated measures, 425
  and sample size, 420
Power efficiency, 413
  defined, 429
  of statistical tests, 428
Power of a test, 414, 429
PRE. *See* Proportional reduction in

error
Prediction, 213
Prediction line, 222
  See also Regression line
Probability, 271, 273, 285, 293
  approaches to, 273, 274
  classical approach to, 274
  conditional, 283, 292
  empirical approach to, 274
  formal properties of, 275
  intuitive use of, 272
  and normal-curve model, 285
  as a percentage, 275
  as a proportion, 275
Probability value, 323, 414
  one-tailed, 288, 293, 323
  two-tailed, 288, 293, 314
Product-moment correlation coefficient, 206
Proof, indirect, 316
Proportion, 36
  cumulative, 72
  defined, 40
  as expression of probability, 275
  standard error of, 357, 364
Proportional reduction in error (PRE), 161, 165
  defined, 183
Proportion distribution, cumulative, 56, 72

Quartile, 89
  defined, 90

r. See Pearson's *r*
Random, defined, 73
Random digit table, 273
  See Table M, 505-508
Randomness concept, 272
Random sample, simple, 272
Random sampling, 8, 9, 10, 14, 51, 272, 273, 293
  defined, 293
  independent, 273
Range, 6, 114
  defined, 126
  interquartile, 115, 126
  restricted, 201,206
  truncated, 201
Rank ordering, 28
Rate, 38
  defined, 39, 41
Ratios, 35
  dependency, 36
  F, 383, 399, 405
  sex, 35
  Student's *t. See* Student's *t*
  t. See Student's *t*
Ratio scale, 29, 60, 189, 330
  defined, 41
Raw score
  and linear regression, 220
  and mean from grouped frequency distribution, 97
  and Pearson's *r*, 194

and standard deviation, 120, 121
  transformation of to *z* score, 136
Rectangular distribution, 66
Reduction
  absolute, 165
  proportional, 161, 165, 183
Reference group, 87
Region, critical, 339, 348, 359, 363, 375, 381, 419, 424, 439
Regression
  linear, 214, 220
  multiple, 254, 255, 256, 257, 262
Regression line, 217, 256
  See also Prediction line
  construction of, 222
  defined, 235
Regression plane, 256
Regression sum of squares, 230
  defined, 235
Rejection regions, 392
  critical, 348
Relationships, 28
  causal, 232, 233
  direction, 157
  existence of, 157
  linear, 200, 216
  negative, 159, 182, 192, 206
  nonexistent, 159
  perfect, 159, 182
  positive, 159, 183, 191, 206
  spurious, 248, 263
  strength of, 157
Residual, defined, 235
Residual size and direction of error, 226
Residual sum of squares, 230
  defined, 236
Residual variance, 224, 227
  defined, 236
Restriction of range, 201
  See also Truncated range
  defined, 206
Rho (*ρ*), Spearman's, 203, 206, 358
  critical values of, Table F, 487
  test of significance for, 360
Robust test, 347
  defined, 363
Rounding, 33

Sample, 8, 14, 302
  biased, 273
  correlated, 378
  defined, 323
  random, 8, 10, 14
  selection of, 272, 373
  simple random, 272, 293
  size of, 420
Sample independence, assumption of, 371
Sample mean, 338
Sample size and power, 420
Sampling
  biased, 272
  independent random, 273
  with replacement, 280
  without replacement, 282
  simple random, 272, 273, 293

Sampling distribution, 302, 339, 359, 375, 381, 419, 424, 439
  binomial, 306, 309, 311
  defined, 323
  of difference between means, 371, 372
  of mean, 330, 333
Scales
  continuous, 30, 31, 40
  discontinuous, 31
  discrete, 30, 31, 40
  interval, 29, 40, 60, 189
  measurement, 26
  nominal, 27, 40, 59
  ordinal, 27, 40, 60
  ratio, 29, 41, 60, 189
  types of, 24
Scatter diagram, 190, 191
  defined, 206
Scattergram, 190
Scatterplot, 190
Scores
  extreme, 203
  grouping of, 52
  standard, 131
Sex ratio, 35
Sigma, 20
Significance
  level of, 312, 313, 314, 323, 338, 359, 375, 381, 413, 419, 424, 439
  nonparametric test of, 423
  parametric test of, 423
  tests of, 358, 360, 361, 422
Simple random sample, 272
Single independent variable, 392
Skew, 105
  negative, 66, 73
  positive, 66, 73
Skewed distribution, 66, 105
  defined, 73
Skewed frequency curves, 66
Slope, 216
Somer's d, 173, 179
  defined, 183
Spearman's rho, 203, 206, 358
  critical values of, Table F, 487
  defined, 206
  testing of other null hypotheses concerning, 360
  tied ranks and, 204
Spurious relationship, 248
  defined, 263
Square root, 21
Standard deviation, 118, 334
  defined, 119, 126
  deviation method of calculating, 119
  estimation of, 123
  interpretation of, 123, 144
  and normal distribution, 123
  population, 134
  raw score method of calculating, 120, 121
Standard error of the difference between means, 372
  defined, 383
Standard error of estimate, 224, 227

defined, 236
Standard error of the mean, 330, 332, 334
  defined, 363
Standard error of the mean difference, 379
  defined, 383
Standard error of the proportion, 357, 364
Standard normal distribution, 134
  areas under, 135
  characteristics of, 134
  defined, 145
  illustrative problems of, 135
Standard score
  concept of, 131
  defined, 146
Statistic, defined, 8, 9, 14
Statistical analysis, functions of, 9
Statistical control, 250
Statistical elaboration, 246
  defined, 263
Statistical hypothesis testing, 312, 315, 317, 336, 344, 373
Statistical inference, 7, 10, 11, 12, 14, 329
  with correlated samples, 378
Statistical testing, 338, 359, 375, 381, 419, 424, 439
  power efficiency of, 428
Statistical variance, estimation of from sample data, 373
Statistics
  defined, 6, 7, 14
  descriptive, 10, 12, 14
  inductive, 7, 12, 14
  inferential, 7, 10, 11, 12, 14, 329
Stimulus, experimental, 392
Strength of a relationship, 157
Student's t, 344, 345, 347, 373, 379
  characteristics of, 347
  critical values of, 363
  defined, 364
  degrees of freedom, 346
  distribution of, 345, 347, 364
  and homogeneity of variance, 376
Subscripts, 20
Successive occurrence, 280
Summation sign, 20, 21
  rules for use of, 22-24
Sum of squares, 99, 120
  between-group, 395, 397
  concept of, 392
  defined, 107, 126, 405
  regression, 230, 235
  residual, 230, 236
  total, 229, 236, 393
  within-group, 393
Symmetrical distribution, 65
Symmetric measure of association, 160
  defined, 183

t. See Student's t
Tau. See Goodman's and Kruskal's tau
Testing

See also specific test
  of hypotheses about the sample mean, 338
  of independence of variables, 438
  nonparametric, 423, 434
  of null hypotheses concerning [rho], 360
  one-variable, 446
  parametric, 423
  power of, 414, 429
  robust, 347, 363
  of significance, 358, 360, 361
  statistical, 338, 359, 375, 381, 419, 424, 428, 439
  of statistical hypotheses, 312, 315, 317, 336, 344, 373
  with unknown parameters, 344
Test variable, 246
  See also Control variable
  defined, 263
Tied pair, 172
  defined, 183
Total sum of squares, 229, 393
  defined, 236
Total variation, 229
  defined, 236
Treatment variable, 392
Trend chart, 67
Tridimensional nature of F-distribution, 377
True limits, 33
  of class intervals, 55
  of numbers, 32, 41
True value, 305
Truncated range, 201
Two-by-four (2×4) table, 155
Two-category variables, 279
Two-sample case, 399, 424
Two-tailed hypothesis, 316
Two-tailed probability value, 288, 293, 314
  defined, 323
Type I (α) error, 318, 392, 413, 414, 426
  defined, 323
Type II (β) error, 319, 413, 414, 426
  defined, 324

U-distribution, 66
Unbiased coin, 306
Unbiased estimate, 341, 343
  defined, 364
Uncertainty, 271
Unexplained variation, 229, 230
  defined, 236
Ungrouped frequency distribution, 96
  in standard deviation, 121
Unordered variables, 27

Value
  absolute, 117, 126
  central, 94
  critical, 348, 363
  probability. See Probability value

true, 305
  of variable, 26
Variability, 94
Variables, 7, 14, 20, 26
  See also Criterion variable
  chi square test of independence of,
    438
  continuous, 32, 285
  control, 246, 262
  criterion, 152
  defined, 41, 14
  dependent, 151, 152, 182, 244
  dichotomous, 279
  independent, 151, 152, 155, 182, 244
  of interest, 9
  interval, 330
  interval-scaled, 60, 189
  intervening, 248, 262
  nominally scaled, 59
  ordinally scaled, 60
  probability, 285
  ratio-scaled, 60, 189, 330
  single independent, 392
  statistical relationship between, 151
  test, 246, 263
  treatment, 392
  two-category, 279
  unordered, 27

  value of, 26
Variance,118
  See also Variation
  analysis of (ANOVA), 392, 398, 399,
    405
  between-group, 405
  defined, 119, 126
  estimates of. *See* Variance estimate
  homogeneity of, 376, 383, 399, 405
  mean deviation method of
    calculating, 119
  one-way analysis of, 392, 405
  population, 341
  residual, 224, 227, 236
  t-ratio and homogeneity of, 376
  within-group, 405
Variance estimate, 396
  between-group, 398
  defined, 405
  obtaining of, 396
  within-group, 398
Variation
  See also Variance
  explained, 229, 230, 235
  total, 229, 236
  unexplained, 229, 230, 236
V, Cramer's, 444, 446, 447
Verbs, mathematical, 20

Weighted mean, 99
  defined, 107
Whole numbers, 31
Within-group sum of squares, 393
Within-group variance
  defined, 405
  estimate of, 398

X, 20
X-axis, 58
  See also Abscissa
  defined, 71

Y, 20
Y-axis, 57
  See also Ordinate
  defined, 73
Y-intercept, 223
Yule's *Q,* 174

Zero-order correlation, 252, 253
Z score
  defined, 146
  estimation of from sample data, 343
  and Pearson's *r,* 195
  transforming raw scores to, 136
Z statistic. *See Z* score

List of Equations (continued)

NUMBER	EQUATION	PAGE
14.9	$t = \dfrac{\overline{X} - \mu_0}{s_{\overline{X}}}$	345
14.10	upper limit $\mu_0 = \overline{X} + t_{0.05}(s_{\overline{X}})$	353
14.11	lower limit $\mu_0 = \overline{X} - t_{0.05}(s_{\overline{X}})$	354
14.14	$\sigma_P = \sqrt{\dfrac{P(1 - P)}{N}}$	357
15.1	$s_{\overline{x}_1 - \overline{x}_2} = \sqrt{\left(\dfrac{SS_1 + SS_2}{N_1 + N_2 - 2}\right)\left(\dfrac{1}{N_1} + \dfrac{1}{N_2}\right)}$ when $N_1 \neq N_2$	373
15.2	$s_{\overline{x}_1 - \overline{x}_2} = \sqrt{\dfrac{SS_1 + SS_2}{N(N - 1)}}$	373
15.3	$t = \dfrac{(\overline{X}_1 - \overline{X}_2) - (\mu_1 - \mu_2)}{s_{\overline{x}_1 - \overline{x}_2}}$	374
15.7	$\sum d^2 = \sum D^2 - \dfrac{(\sum D)^2}{N}$	380
16.2	$SS_{tot} = \sum X_{tot}^2 - \dfrac{(\sum X_{tot})^2}{N}$	393
16.5	$SS_{bet} = \sum \dfrac{(\sum X_i)^2}{N_i} - \dfrac{(\sum X_{tot})^2}{N}$	395
16.6	$SS_{tot} = SS_w + SS_{bet}.$	396
16.7	$df_{bet} = k - 1.$	397
16.9	$df_w = N - k.$	397
16.10	$\hat{s}_w^2 = \dfrac{SS_w}{df_w}$	397
16.11	$F = \dfrac{\hat{s}_{bet}^2}{\hat{s}_w^2}$	398
16.12	$df_{tot} = N - 1.$	402
16.13	$HSD = q_{tx} \sqrt{\dfrac{\hat{s}_w^2}{N}}$	403